Inclusion Compounds
Volume 2

Inclusion Compounds

Volume 2
Structural Aspects of Inclusion Compounds formed by Organic Host Lattices

Edited by

J. L. Atwood
University of Alabama, USA

J. E. D. Davies
University of Lancaster, UK

D. D. MacNicol
University of Glasgow, UK

1984

ACADEMIC PRESS

(*Harcourt Brace Jovanovich, Publishers*)

London   Orlando   San Diego   San Francisco   New York
Toronto   Montreal   Sydney   Tokyo   São Paulo

ACADEMIC PRESS INC. (LONDON) LTD.
24–28 Oval Road,
London NW1

United States Edition Published by
ACADEMIC PRESS INC.
(Harcourt Brace Jovanovich, Inc.)
Orlando, Florida 32887

British Library Cataloguing in Publication Data

Inclusion compounds.
  Vol. 2
  1. Molecular structure
  2. Chemistry, Organic
  I. Atwood, J. L.   II. Davies, J. E. D.
  III. MacNicol, D. D.
  547.1'22   QD461

ISBN 0-12-067102-6

LCCCN 83-72135

Printed in Great Britain by J. W. Arrowsmith Ltd.
Bristol BS3 2NT

Contributors to Volume 2

COLLET, A., *Collège de France, Chimie des Interactions Moléculaires, 11 Place Marcelin Berthelot, 75005 Paris, France*

DAVIES, J. E. D., *Department of Chemistry, University of Lancaster, Lancaster, Lancs LA1 4YA, UK*

DIETRICH, B., *Institut de Chimie, Université Louis Pasteur de Strasbourg, 1 Rue Blaise Pascal, 67008 Strasbourg Cedex, France*

FARINA, M., *Istituto di Chimica Industriale, Università di Milano, Via Venezian 21, 20133 Milan, Italy*

FINOCCHIARO, P., *Department of Chemistry, Università di Catania, Viale A. Doria 6, 95125 Catania, Italy*

GIGLIO, E., *Istituto di Chimica Fisica, Università di Roma, P. le delle Scienze 5, 00185 Rome, Italy*

GOLDBERG, I., *Institute of Chemistry, Tel-Aviv University, Ramat-Aviv, Tel-Aviv 69978, Israel*

HERBSTEIN, F. H., *Department of Chemistry, Technion-Israel Institute of Technology, Haifa 32000, Israel*

MACNICOL, D. D., *Department of Chemistry, University of Glasgow, Glasgow, G12 8QQ, UK*

OLLIS, W. D., FRS, *Department of Chemistry, University of Sheffield, Sheffield, S3 7HF, UK*

SAENGER, W., *Institut für Kristallographie, Freie Universität, Taku Str. 6, D-1000 West Berlin 33, FRG*

SONODA, N., *Faculty of Engineering, Osaka University, Yamada-kami, Suita, Osaka, Japan*

STODDART, J. F., *Department of Chemistry, University of Sheffield, Sheffield, S3 7HF, UK*

TAKEMOTO, K., *Faculty of Engineering, Osaka University, Yamada-kami, Suita, Osaka, Japan*

PREFACE

In September 1980 the Institute of Physical Chemistry of the Polish Academy of Sciences hosted the First International Symposium on 'Clathrate Compounds and Molecular Inclusion Phenomena' at Jachranka, near Warsaw. At this timely meeting, the first devoted entirely to all types of inclusion behaviour, the unanimous opinion of the participants was that every effort should be made to draw together in print the various threads from which the rich tapestry of Inclusion Chemistry is currently being woven.

As a first step in this direction, the proceedings of the conference were published in special issues of the *Journal of Molecular Structure* (Volume 75, Number 1, 1981) and the *Polish Journal of Chemistry* (Volume 56, Number 2, 1982). However, to obtain a more global modern picture of Inclusion Chemistry it was apparent that an up-to-date Comprehensive Treatise would be necessary. In view of the rapid advances being made at present, it was clear that such a work could only be produced on an acceptable timescale, and with a sufficient depth of treatment of recent work, by inviting recognised international authorities to write on their own particular fields of interest. Accordingly, this was the plan chosen for the present work.

Earlier useful books, in English, have appeared on inclusion compounds over the years, each reflecting the state of knowledge at the time of publication, three being *Clathrate Inclusion Compounds*, Reinhold, 1962, by M. Hagen; *Non-Stoichiometric Compounds*, Academic Press, 1964, edited by L. Mandelcorn; and *Clathrate Compounds*, Chemical Publishing Company, 1970 by V. M. Bhatnagar. The most comprehensive of these is undoubtedly the book edited by L. Mandelcorn (1964) and in some ways the present treatise may be regarded as complementary to that work.

The editors note, with pleasure, the greatly increasing interest in inclusion phenomena, as evidenced by recent relevant publications on *specific* aspects of Inclusion Chemistry: *Cyclodextrin Chemistry*, by M. L. Bender and M. Komiyama, Springer-Verlag, 1977; *Host–Guest Complex Chemistry I and II*, edited by F. Vögtle, Springer-Verlag, 1981; *Ionophores and their Structures*, by M. Dobler, Wiley, 1981; *Cyclodextrins and their Inclusion Complexes*, by J. Szejtli, Akademiai Kiado, Budapest, 1982; and *Intercalation Chemistry*, edited by M. S. Whittingham and R. J. Jacobson, Academic Press, 1982. Also a new journal devoted to inclusion compounds *The Journal of Inclusion Phenomena* has been launched by Reidel.

We have great pleasure in dedicating these three volumes to Professor
H. M. Powell, FRS, whose pioneering crystallographic work laid firm
foundations for subsequent work in Inclusion Chemistry.

We wish to thank Professor Powell for kindly agreeing to write the
important introductory chapter; and we are indebted also to all our other
contributors for their help and participation in writing this book. We must
also thank the staff of Academic Press for the efficient way in which the
book has been produced.

The present volume is the second of a three volume series designed to
provide comprehensive coverage of all aspects of inclusion compounds.
Volume 1 is principally concerned with structural and design aspects of
inclusion compounds formed by inorganic and organometallic host lattices,
Volume 2 is concerned with similar aspects of inclusion compounds formed
by organic host lattices, while Volume 3 concentrates on the physical
properties and applications of inclusion systems.

January, 1983

                                                        J. L. Atwood
                                                        J. E. D. Davies
                                                        D. D. MacNicol

Contents

Chapter 11. Inclusion compounds formed by other host lattices
J. E. D. Davies, P. Finocchiaro and F. H. Herbstein

Contents of Volume 1

Structural aspects of inclusion compounds formed by inorganic and organometallic hosts

Contents of Volume 3

Physical properties and applications

Dedicated to
H. M. Powell, FRS
who laid the firm foundation on
which this book is based

1 · STRUCTURE AND DESIGN OF INCLUSION COMPOUNDS: THE CLATHRATES OF HYDROQUINONE, PHENOL, DIANIN'S COMPOUND AND RELATED SYSTEMS

D. D. MacNICOL

University of Glasgow, Glasgow, UK.

1. Introduction

Host molecules possessing at least one phenolic hydroxyl group have played a vital role in the development of the chemistry of crystalline multimolecular inclusion compounds. Viewing this fascinating area in historical perspective one may identify as important landmarks the early *chance* discoveries of key host molecules such as hydroquinone (**1**), phenol (**2**), and Dianin's compound (**3**); the subsequent X-ray elucidation of the crystal structures of adducts of these hosts; and, comparatively recently, the successful *design* of new host molecules belonging to the phenolic class. A number of reviews have appeared,[1-26] and the principal aims of the present chapter are to

INCLUSION COMPOUNDS 2
ISBN 0-12-067101-8

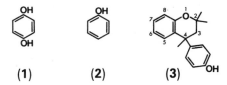

(1)   (2)   (3)

cover recent structural work on phenolic hosts and to describe in some detail the successful synthesis of new hosts by structural modification of a known host, Dianin's compound (3). Other important aspects of the key phenolic class of host molecule are discussed in detail elsewhere, as indicated: infrared and Raman studies (Volume 3, Chapter 2), thermodynamic considerations (Volume 3, Chapter 1), and dielectric and magnetic resonance investigations (Volume 3, Chapter 3). In the present chapter some practical applications are mentioned, and a brief survey of less-studied phenolic hosts is also given. The successful synthesis of new host molecules with no *direct* structural relationship to any known host, based on the recognition of the importance of the (OH···O) hydrogen-bonded hexameric unit found in many phenolic host lattices, represents a significant step forward in host design, and this is considered separately in Volume 2, Chapter 5.

2. Hydroquinone

2.1. β-Hydroquinone clathrates

The history of adducts of hydroquinone, or quinol (1), already reviewed,[1] dates back into the nineteenth century: in 1849 Wöhler[27] found H_2S to be trapped by 1, then, ten years later, Clemm[28] found SO_2 to be similarly retained; and in 1886 Mylius,[29] discovering the inclusion of carbon monoxide by 1, made the very shrewd observation that perhaps the molecules of hydroquinone were somehow able to lock the volatile component into position without chemically combining with it. It was not until the 1940s, however, that the pioneering X-ray studies of Powell and coworkers[30,31a,32–34] firmly established the true cage, or clathrate,[35] nature of these intriguing systems. In fact, hydroquinone can exist in three crystal modifications designated α, β, and γ forms, the α-form being the stable form at room temperature. The monoclinic γ-form is produced by sublimation or by rapid evaporation of a solution of 1 in ether. The β-form is the most versatile, however, and the classical studies of Powell and colleagues[30,31a,32–35] established that three crystallographically distinguishable kinds of β-hydroquinone clathrate host lattice, now termed[36] Types I-III, can exist, all having

Table 1. Selected crystal data for β-hydroquinone clathrates and other forms of **1**

Designation	Space group	Lattice parameters[a]	Guest	Hexamer dimensions (O···O)	Ref.
β-form (Type I)	R$\bar{3}$	a = 16.613(3), c = 5.4746(5) Å, Z = 9	None	2.678 (3) Å	37
β-form (Type I)	R$\bar{3}$	a = 16.616(3), c = 5.489(1) Å, Z = 9 (host)	H$_2$S	2.696 (1) Å	41
β-form (Type II)	R3	a = 16.31(5), c = 5.821(1) Å, Z = 9 (host)	SO$_2$	2.727 (6) Å, 2.733(6) Å	31(b)
β-form (Type II)	R3	a = 16.621(2), c = 5.562(1) Å, Z = 9 (host)	MeOH	2.653(5) Å, 2.779(5) Å	45
β-form (Type II)	R3	a = 16.650(1), c = 5.453(1) Å, Z = 9 (host)	HCl	2.61 (1) Å, 2.77 (1) Å[b]	47
β-form (Type II)	R3	a = 15.946(2), c = 6.348(2), Z = 9 (host)	CH$_3$NC	2.779 (6) Å, 2.800 (6) Å	43
β-form (Type III)	P3	a = 16.003(2), c = 6.245(2), Z = 9 (host)	CH$_3$CN	2.778 Å[c] (mean)	43
α-form	R$\bar{3}$	a = 38.46(2), c = 5.650(3) Å, Z = 54	None	2.677 (3) Å	49
α-form	R$\bar{3}$	a = 38.529, c = 5.66 Å, Z = 54 (host)	SO$_2$	[d]	50
γ-form	P2$_1$/c	a = 8.07, b = 5.20, c = 13.20 Å, β = 107°, Z = 4		[e]	52

[a] For R$\bar{3}$ and R3, the values of a and c given are referred to a hexagonal unit cell (α = β = 90°, γ = 120°).
[b] X-ray values.
[c] Individual values for the three independent [OH]$_6$ rings are 2.792, 2.788, 2.785, 2.782; 2.745, 2.773 Å (e.s.d. 0.006 Å in each case).
[d] Not available.
[e] No hexamers present in structure, see Fig. 6(b).

the same general formula 3 $C_6H_4(OH)_2.xG$, where G represents the encaged guest molecule and x is a site occupancy factor between zero and one. Table 1 gives representative crystal data, mainly selected from recent sources, for β-hydroquinone, as well as for the α- and γ-modifications. (The crystal structures of the α- and γ-forms are discussed below.) As indicated in Table 1, the unsolvated β-form and the corresponding H_2S clathrate correspond to the Type I situation for **1**, and in such cases cavities having $\bar{3}$ (C_{3i}) symmetry are present. Figure 1a shows a stereoview of such a centrosymmetric cage of the unsolvated form. As can be seen the top and bottom of the void are formed by hexagons of hydrogen-bonded oxygen atoms; an ordered arrangement of hydrogen atoms is apparent in the $[OH]_6$ rings and host molecules point alternately above and below the mean plane of the (nearly planar) six oxygen atoms. The hexameric units forming the ceiling and floor of a given cage, as may be seen from Fig. 1b, belong to two identical, but displaced, three-dimensional interlocking networks first defined for the "empty" form by Powell and Riesz.[38] The remarkably low packing coefficient,[39] 0.62 (or 0.59 excluding the hydrogen atoms involved in hydrogen bonding[37]) may be compared with the normal range, 0.65–0.77, for most organic molecular crystals, and demonstrates the realization of an "open" structure with unfilled cavities stabilized by an extended system of

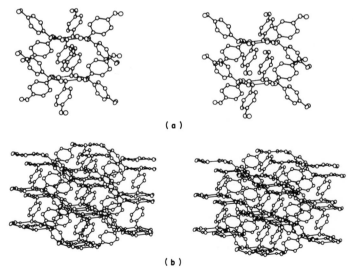

(a)

(b)

Fig. 1. Stereoviews illustrating (a) the construction of a single cage in the unsolvated form of β-hydroquinone and (b) more extended portions of the two identical, but displaced, three-dimensional networks from which cages are constructed. (Both drawn from data of ref. 37.)

hydrogen bonds. In recent X-ray work on the H_2S clathrate, undertaken to define accurately a Type I clathrate, Mak and coworkers[40,41] have found that the H_2S guest molecule, situated in an approximately spherical cavity of mean free diameter c. 4.8 Å, undergoes pronounced thermal motion, particularly in the direction of the centres of the $[OH]_6$ rings, that is, along the c-axis of the crystal. In this centrosymmetric clathrate[42] the results are consistent with rotational disorder of the guest molecule.

In Type II clathrates, such as those formed by 1 with SO_2, MeOH, HCl, or CH_3NC, a lowering of space group symmetry from R$\bar{3}$ to R3 is found, and guest accommodation is provided in cages which are still trigonal, though no longer centrosymmetric. For the relatively long guest molecule methyl isocyanide the cage length, corresponding to the c-spacing (Table 1), is markedly increased compared with the Type I systems already discussed (see below). Figure 2 illustrates the alignment of the CH_3NC along the

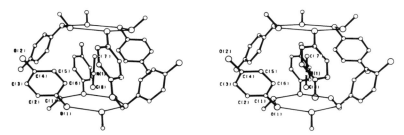

Fig. 2. A stereo-drawing showing a CH_3NC guest molecule trapped inside a cage in the structure of hydroquinone (1). For clarity all hydrogen atoms have been omitted. (Reproduced, by permission, from ref. 43.)

c-axis in its Type II clathrate.[43] Interesting new information has recently become available on the Type II MeOH clathrate, a system previously studied by Palin and Powell[44] using two-dimensional X-ray data. Mak,[45] employing diffractometer data, has found that the encaged MeOH molecule is located in three preferred orientations related by three-fold rotation about the c-axis, one such orientation being shown in Fig. 3. In each orientation, the C–O bond is tilted by 35° from c to facilitate interaction of the hydroxyl group with three phenolic oxygen atoms of the adjacent $[OH]_6$ ring. The inclination of 35° found above is in excellent agreement with the values of 32° below 100 K and 40° at 300 K deduced from recent dielectric studies.[46] In the MeOH clathrate, host–guest interaction is reflected in unequal OH···O hydrogen bonds in the $[OH]_6$ ring (Table 1); and, interestingly, this feature, a marked hydrogen bond length alternation, has also been found by Boeyens and Pretorius[42] in an X-ray and neutron diffraction study of the HCl clathrate

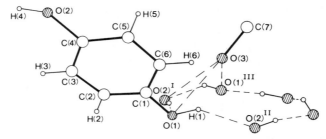

Fig. 3. Host–guest interactions in the methanol clathrate of hydroquinone (**1**). The O(3)–C(7) bond of the CH_3OH guest molecule is inclined at an angle of 35° to the *c*-axis of the crystal. (Reproduced, by permission, from ref. 45.)

of **1**. In this latter study the lowering of symmetry to R3 has been attributed to a large number of weak OH···Cl–H···OH interactions which orient the HCl molecule in its cage, the location of the guest being described as lying preferentially on the surface of a cone, with its generator inclined by 33° to *c* and its apex at the Cl position. The reasonable conclusion has been reached[45] that formation of Type II β-hydroquinone clathrates is favoured by guest molecules of appropriate sizes which can interact appreciably with specific sites in the walls of a clathration cavity. Very recent work[31b] has established that the SO_2 clathrate is also of Type II, and a weak interaction has been observed between the SO_2 molecule, through one of its oxygen atoms, and the $[OH]_6$ ring of the hydroquinone framework. In the acetonitrile clathrate of **1**, the only authenticated Type III system, a further lowering of symmetry from the rhombohedral lattice (R3) of Type II leads to a trigonal lattice, space group P3. There are now three distinct types of trigonal clathrate cavity and all these have the shape of prolate spheroids.[36,43] The three symmetry-independent acetonitrile molecules fit snugly inside these cavities, with, as previously suggested,[33] one guest molecule aligned in the opposite sense to the other two, see Fig. 4. Figure 5 shows electron density sections through the guest molecules; although molecule *c*, in the opposite orientation from molecules *a* and *b*, appears to be displaced from its "idealized" position along the z-axis, the disposition of this molecule with respect to the top $[OH]_6$ ring of its cage is virtually the same as that of the other molecules with respect to their bottom rings.[43] In the markedly unstable CH_3CN and CH_3NC clathrates, which rapidly lose the guest in air,[43] the mean O···O hydrogen bond lengths, 2.778 Å and 2.790 Å respectively, are significantly longer than the corresponding distances of 2.696 Å, 2.69 Å, and 2.716 Å found in the relatively stable[48] H_2S, HCl and MeOH β-hydroquinone clathrates. (There are problems in assigning an e.s.d. to the mean of quantities which are known to be unequal, Table 1.) It is intriguing

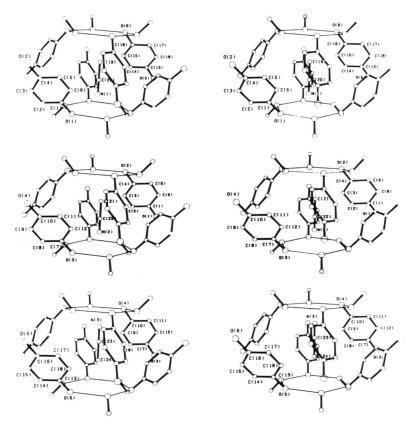

Fig. 4. ORTEP plots showing the guest acetonitrile molecules trapped inside their respective cages in the Type III hydroquinone clathrate. For comparison, all three cages are viewed in equivalent directions, which are approximately parallel to a vector along *a* towards the cell origin. (Reproduced, by permission, from ref. 36.)

that CH_3CN and CH_3NC, both prolate spherical tops and iso-electronic, should yield β-hydroquinone clathrates which are not isomorphous. Microwave spectroscopy has shown that the molecular *skeleton* of CH_3CN is slightly longer than that of CH_3NC, though if one takes into account the appropriate van der Waals radii at the end of each molecule then the "effective molecular length", the crucial factor in host-guest packing here,[43] is expected to be greater for CH_3NC than CH_3CN. In accord with the more sterically demanding nature of CH_3NC, which stretches the β-hydroquinone lattice to its limit in the *c* direction (without producing a Type III system), Table 1, the methyl isocyanide clathrate of **1** is even less stable than its acetonitrile counterpart.

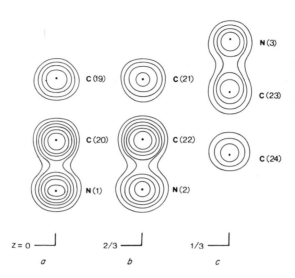

Fig. 5. Electron-density sections through the CH_3CN guest molecules (a) at $(0, 0, z_a)$, (b) at $(\frac{1}{3}, \frac{2}{3}, z_b \simeq \frac{2}{3} + z_a)$, and (c) at $(\frac{2}{3}, \frac{1}{3}, z_c)$ in the hydroquinone (1) clathrate. Contours are drawn at $1e\,Å^{-3}$ intervals starting at $2e\,Å^{-3}$. For each cage, the dotted line represents the mean plane of its top $[OH]_6$ ring. The disposition of guest molecule (c) relative to the top $[OH]_6$ ring of its cage is approximately the same as that of molecules (a) and (b) with respect to their bottom rings. (Reproduced, by permission, from ref. 43.)

2.2. α-Hydroquinone and γ-hydroquinone

The intricate and beautiful structure of α-hydroquinone, the stable form of **1** at room temperature, has recently been fully elucidated in a definitive X-ray study by Wallwork and Powell,[49] and the ability of this form to include small guest species such as carbon dioxide,[35] sulphur dioxide,[50] and argon[34] has now been explained. The space group is R$\bar{3}$ with 54 molecules of **1** in the hexagonal unit cell, and, correspondingly, 3 molecules in each asymmetric unit. As shown in Fig. 6a, of the three crystallographically independent hydroquinone molecules in the asymmetric unit, two are involved in forming two interpenetrating, open, hydrogen-bonded cageworks similar to those found in the structure of β-hydroquinone (cf. Fig. 1) and capable of clathrating small molecules, whereas the third forms double helices consisting of hydrogen-bonded chains of molecules round the three-

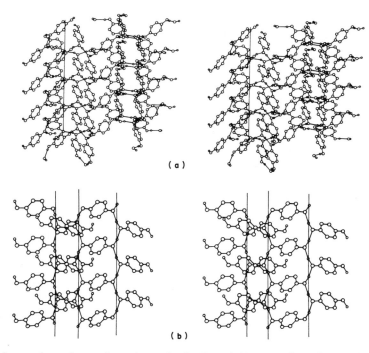

Fig. 6. Stereoviews drawn from data of refs. 49 and 52 respectively illustrating (a) the structure of α-hydroquinone; two crystallographically distinct molecules of **1** form (unoccupied) cages to the right, while a third type forms double helices, spiralling round the three-fold screw axis denoted by the continuous line to the left; and (b) the structure of γ-hydroquinone, built up from sheets of hydrogen-bonded molecules of **1** (the parallel lines represent two-fold screw axes).

fold screw axes, one of which is denoted by a continuous line at the left side of Fig. 6a. The cage works and helices are hydrogen-bonded together in such a way that the interpenetrating cageworks are connected (unlike the β-form) and the two strands of the double helix are connected. The estimated local density, 1.43 g cm^{-3}, for the helical region is high for an organic structure consisting only of light atoms (calculated density[51] for γ-hydroquinone is 1.38 g cm^{-3}) and confirms that this part of the structure cannot include guest molecules. Thus, there being three cages (of $\bar{3}$ symmetry) similar in size to those of the β-form and 54 molecules of **1** per unit cell, the maximum ratio expected for α-hydroquinone clathrates is 18:1, against 3:1 for the β-hydroquinone clathrates. In fact, the hexagonal prism of α-hydroquinone studied[49] was produced by sublimation (which also produces plates of the γ-form) and, for this case, a difference Fourier synthesis indicated that the cages contained a negligible amount of clath-rated air. For comparison, the structure[52] of the γ-form (produced by rapid evaporation of a solution in ether) is shown in Fig. 6. In this case there are no hydrogen-bonded hexameric units and no known inclusion proper-ties; the structure is built up from sheets of hydrogen-bonded hydroquinone molecules, two-fold screw axes, relating centrosymmetric molecules of **1**, being denoted by parallel lines in Fig. 6b. In accord with the instability and pronounced cleavage of the γ-form, these sheets are apparently held together only by weak van der Waals forces.

Returning to the inclusion properties of **1**, the formation of argon, krypton, and xenon clathrates,[34] apparently stable under normal conditions, is of particular interest since almost all[53] condensed phases containing rare gases are stable only under high external pressure of the gases. The formation of the Ar clathrate has been studied[34,54] employing aqueous and nonaqueous solutions of **1**; and solventless methods, involving sublimation[54,55] or direct contact of Ar with the hydroquinone melt,[54] have also been investigated.[56] No stable Ar adducts have been found[34,54] for *p*-aminophenol, *p*-phenyl-enediamine, or 3,6-dihydroxy-pyridazine, which are all isoelectronic with **1** and have related hydrogen-bond forming potential; also two isomers of **1**; catechol and resorcinol, show no evidence of Ar inclusion. This suggests a delicate balance of factors operating in the host packing of **1**. Although the dihydric phenol orcinol (**4**) and the trihydric phenol phloroglucinol (**5**) also fail[54] to include Ar, these compounds do in fact form adducts[57] with HBr (host/guest ratio 2:1 in each case), though the structure of these is not yet known; also of apparently unknown structure are the unstable SO_2 adducts[58] of pyrogallol (**6**) and catechol.

(4)   (5)   (6)

In an early application,[59,60] Powell found that noble gases could be separated by selective clathration in **1**. Another ingenious use, described by Chleck and Ziegler,[61] involves the storage of the almost pure *beta* emitter ^{85}Kr in **1**; this clathrate can be used to monitor air pollution (sensitive to a few parts in 10^9) for contaminants such as ClO_2 and O_3 which oxidize **1** releasing the ^{85}Kr, which is then detected by radiation counters. Other studies of the ^{85}Kr clathrate have been concerned with SO_2 monitoring,[62] fluorine detection,[63] a modified method for ozone monitoring,[64] and storage of ^{85}Kr from nuclear power plant off-gases;[65] while the same clathrate has been used to give surface coatings permanent luminescence,[66] and as a component in a radionuclide battery operating on the photo-voltaic effect.[67] The methyl bromide clathrate of **1** has been employed as a soil fumigant,[68] and other hydroquinone clathrates have been used as useful lubricant and fuel additives.[69]

3. Phenol and simple substituted phenols

The initial observation of inclusion behaviour for phenol itself **2** was made in 1935 by Terres and Vollmer[70] who discovered the H_2S adduct of **2** while investigating the solubility of petroleum and tar constituents in liquid hydrogen sulphide. It is now known that the host **2** forms a series of isomorphous clathrates,[71] space group $R\bar{3}$ with many guest species, for example, hydrogen sulphide,[70–74] sulphur dioxide,[71,72,74,75] carbon dioxide,[71,72,76] carbon disulphide,[71,72] hydrogen chloride,[71,72,74] hydrogen bromide,[71,72,74] methylene chloride,[71,72] vinyl fluoride,[71,72] and xenon.[71,72,74–77] The basic feature of the host structure of **2** is the linking of the OH groups of six phenol molecules by hydrogen bonds such that the oxygen atoms form a hexagon, alternate phenyl groups pointing above and below this hexagon.[71,72] As may be appreciated from Fig. 7, drawn from

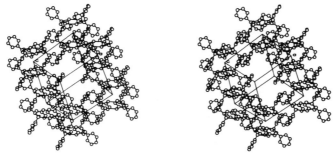

Fig. 7. A stereoview drawn from data of ref. 71 showing the host packing in the clathrates of phenol (**2**); a pair of hexameric units round the rhombohedral lattice point closest to the viewer, and also a pair round the lattice point most remote, have been omitted in order to reveal the long cage, centred at $(\frac{1}{2}, \frac{1}{2}, \frac{1}{2})$, more clearly.

available data,[71] pairs of such sextets are grouped around each corner of the rhombohedral unit cell, which contains twelve phenol molecules; for clarity, the hexameric units round the lattice point closest to the viewer, and also those round the lattice point most remote, have been omitted. As a consequence of this mode of hexamer packing, two crystallographically distinct types of cages are formed (one of each type per unit cell), a small cage with a free diameter of $c.\,4.5\,\text{Å}$ whose centre corresponds to the rhombohedral lattice point $(0, 0, 0)$, and a large cage (running in the direction of the rhomb diagonal) of effective length about $15\,\text{Å}$ and $4\text{--}4.5\,\text{Å}$ in free diameter, centred at $(\tfrac{1}{2}, \tfrac{1}{2}, \tfrac{1}{2})$. Both cages are capable of including suitably sized guest molecules and limiting compositions[18] are:

$12C_6H_5OH\cdot5G_1,$ where G_1 is a molecule of the size of HCl or HBr (four such molecules can be accommodated in the large cage and one in the small cage)

$12C_6H_5OH\cdot4G_2,$ where G_2 is a molecule of the size of SO_2 (one and three guest molecules in the small and large cages respectively)

$12C_6H_5OH\cdot2G_3,$ where G_3 is a molecule of the size of CS_2 (two CS_2 molecules can be located in the large cavity and none in the small cavity)

$12C_6H_5OH\cdot2G_3\cdot G_1,$ for the double clathrate (formed,[71,72] for example, with CS_2 and air as guests).

Recent work has been concerned with the inclusion of noble gases or other volatile species in phenol,[76,78] *p*-fluorophenol,[78,79,54] *m*-fluorophenol,[78,80,54] *o*-fluorophenol,[81] *p*-chlorophenol,[78,82] *p*-cresol,[78,83] *p*-bromo-, ethyl-, t-butyl-, and phenyl-phenols.[78] Barrer and Shanson,[78] in their recent study of phenol and *para*-substituted phenols, have found that facile clathration can be effected by merely agitating the host crystals with small ball bearings, at temperatures even as low as $-196°\,\text{C}$. A number of selective clathration experiments were described in the work,[78] and for **2** at $-78°\,\text{C}$ the separation of Xe and Kr was found to be particularly good, giving almost pure Xe, while use of phenol or *p*-cresol allowed pure Ar to be obtained from neon at $-196°\,\text{C}$.

4. Dianin's compound 3 and related molecules

4.1. General considerations

These molecules are ideally suited for illustrating a principal theme of the present chapter; namely, the first strategy for the design of new crystalline multimolecular host systems. (Further strategies are discussed separately in Volume 2, Chapter 5.) In this initial approach, discussed below, we employed the idea that perhaps judicious modification of a known host might lead to the discovery of new hosts, with properties significantly different from those of the parent. As will be seen presently, a knowledge of the crystal structure of the "parent" system is a prerequisite for the efficient identification of appropriate host modifications.

4.2. Crystal structures of the parent host 3 and its thia-analogue 7

The choice of starting host is important; and the versatile host 4-*p*-hydroxyphenyl-2,2,4-trimethylchroman (3) appeared an attractive parent system. The compound 3, widely known as Dianin's compound, was first prepared by a Russian chemist A. P. Dianin[84] in 1914. (Incidentally, Dianin was a student and biographer of a versatile professor of chemistry who was also a great composer, Alexander Borodin.) Dianin reported that the chroman he obtained by the gaseous HCl catalysed condensation of phenol and mesityl oxide had the remarkable ability to retain certain organic solvents tightly, and in fixed amounts; subsequently 3 has been shown to be capable of including numerous other diverse species, for example, argon,[85,86] sulphur dioxide,[87] iodine,[87] ammonia,[87] decalin,[87] glycerol,[88] sulphur hexafluoride,[89,90] and di-t-butylnitroxide.[91] The molecular structure of Dianin's compound (3) was unambiguously established in the mid-fifties by Baker and coworkers,[87,92] who also prepared over fifty adducts.[87] At this time also, Powell and Wetters in Oxford suggested[93] a probable cage structure for the adducts, and the unsolvated form, on the basis of space group ($R\bar{3}$), unit cell dimensions, and packing considerations. They proposed that (as shown schematically in Fig. 8) hexamer units of host molecules were stacked on

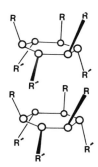

Fig. 8. Schematic representation of the cage structure suggested by Powell and Wetters (ref. 93) for the adducts and unsolvated form of Dianin's compound (3). R and R' correspond to molecules of 3 of opposite configuration and the hydroxyl hydrogen atoms are not shown.

top of each other, leaving cages between the sextet units. Each hexamer unit was made up of six host molecules, linked by a network of hydrogen bonds involving their hydroxyl groups, with three molecules of one configuration (R) pointing upwards, and three of the opposite configuration (R') pointing downwards [Dianin's compound (3) has only one asymmetric centre, C(4)]. Some dozen or so years later in Glasgow, the writer prepared 4-*p*-hydroxyphenyl-2,2,4-trimethylthiachroman (7), a thia-analogue of 3,

(7)

and found[94] that this molecule was a new host sharing the wide-ranging inclusion properties of its parent, Dianin's compound. The sulphur atom in 7 was important since it allowed the heavy-atom method to be used in the crystal structure analysis[95] of the ethanol clathrate of 7. In fact it turns out that the clathrates of 7 are isomorphous and isostructural with the clathrates, and unsolvated form, of Dianin's compound itself; and the empty cage structure of Dianin's compound (3) followed[96] immediately from an electron density map phased on the atomic co-ordinates of the ethanol clathrate of 4-*p*-hydroxyphenyl-2,2,4-trimethylthiachroman (7). A comparison of crystal data for Dianin's compound (3), its thia-analogue (7), and related molecules is given in Table 2. For the space group $R\bar{3}$ the lattice parameters *a* and *c* are referred to a hexagonal unit cell containing 18 host molecules; the *c*-spacing gives the cavity length (and the *a*-spacing reflects the width of the host-columns, *vide infra*). Completely concordant, independent, results were reported from Washington by Flippen *et al.*[97] on the ethanol and chloroform clathrates of 3, and the *n*-heptanol clathrate of 3 was also subsequently studied.[98] Figure 9a illustrates the molecular packing[96] in the unsolvated form of Dianin's compound (3) and, as can be seen, the earlier suggestion[93] about the structure is essentially correct. We do indeed have hexameric units stacked on top of each other leaving cages in between units; in Fig. 9a a molecule of 3 pointing up at the front, and one pointing down at the back have been omitted (apart from their OH oxygen atoms) to reveal the cage region more clearly. At the top and bottom of each cage there are puckered hexagons of six hydroxyl oxygen atoms, linked by a network of hydrogen bonds, and three molecules of one configuration do indeed point upwards, with three of the opposite configuration pointing downwards. Each particular column is infinite in extent and runs parallel

Table 2. A comparison of crystal data for Dianin's compound (3) and related molecules

Compound	Space group	Lattice parameters[a]	Guest	Mole ratio host:guest	Ref.
3	$R\bar{3}$	$a = 26.969\,(3),\ c = 10.990\,(2)$ Å	Ethanol	3:1[b]	97
3	$R\bar{3}$	$a = 27.116\,(3),\ c = 11.023\,(2)$ Å	Chloroform	6:1	97
3	$R\bar{3}$	$a = 27.12\,(3),\ c = 11.02\,(2)$ Å	n-Heptanol	6:1	98
3	$R\bar{3}$	$a = 26.94,\ c = 10.94$ Å	None		96
7	$R\bar{3}$	$a = 27.81,\ c = 10.90$ Å	Ethanol	3:1	95
7	$R\bar{3}$	$a = 27.91,\ c = 10.99$ Å	2,2,5-Trimethylhex-3-yn-2-ol	6:1	100
7	$R\bar{3}$	$a = 28.00,\ c = 11.08$ Å	Di-t-butylacetylene	6:1	101
9	$P2_1/c$	$a = 14.25\,(3),\ b = 6.52\,(1),\ c = 18.67\,(4)$ Å, $\beta = 113.0°,\ Z = 4$	[c]		104
10	$P2_1/n$	$a = 12.91\,(2),\ b = 12.11\,(2),\ c = 9.79\,(1)$ Å, $\beta = 90.3°,\ Z = 4$	[c]		104
11	$R\bar{3}$	$a = 32.392,\ c = 8.423$ Å	[d]		107
12	$R\bar{3}$	$a = 29.22\,(4),\ c = 10.82\,(1)$ Å	Cyclopentane	6:1	104
13	$P2_12_12_1$	$a = 11.777\,(1),\ b = 16.501\,(2),\ c = 8.479\,(1)$ Å, $Z = 4$	[c]		105
14	$R\bar{3}$	$a = 33.629\,(9),\ c = 8.239\,(3)$ Å	Cyclooctane	4.5:1	106,104
16	$R\bar{3}$	$a = 26.936\,(6),\ c = 10.796\,(1)$ Å	Carbon tetrachloride	6:1	108,109
18	$P\bar{4}2_1c$	$a = b = 12.640\,(2),\ c = 17.254\,(4)$ Å, $Z = 8$	[c]		109
19[e]	$P2_12_12_1$	$a = 10.60\,(1),\ b = 13.30\,(2),\ c = 10.08\,(1)$ Å, $Z = 4$	[c]		117
22	$P2_12_12_1$	$a = 10.42\,(2),\ b = 13.69\,(1),\ c = 10.37\,(2)$ Å, $Z = 4$	[c]		117
23	$P2_12_12_1$	$a = 10.66\,(2),\ b = 13.55\,(3),\ c = 10.50\,(1)$ Å, $Z = 4$	[c]		117
23	$R\bar{3}$	$a = 27.063,\ c = 12.074$ Å	Carbon tetrachloride	3:1	117[h]
19, 21[e,f]	$R\bar{3}$	$a = 26.94\,(2),\ c = 11.19\,(2)$ Å	Carbon tetrachloride	6:1	114
20, 21[e,f]	$R3$[g]	$a = 26.64\,(2),\ c = 11.24\,(1)$ Å	Carbon tetrachloride	6:1	114

[a] For $R\bar{3}$ and $R3$, the values of a and c given are referred to a hexagonal unit cell containing 18 host molecules ($\alpha = \beta = 90°$, $\gamma = 120°$); for other space groups unspecified angles are 90°. [b] Ratio from ref. 87. [c] No inclusion behaviour found to date. [d] Crystals obtained by recrystallization from diethyl ether/light petrol. [e] Crystals kindly provided by Professor J. Jacques. [f] Quasiracemate. [g] Probable space group. [h] Initial report: A. D. U. Hardy, D. D. MacNicol, J. J. McKendrick, and D. R. Wilson, *J. Chem. Soc., Chem. Commun.*, 1977, 292.

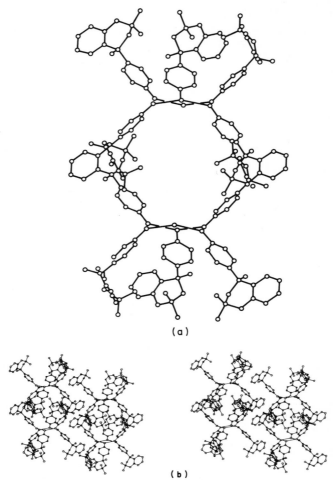

Fig. 9. (a) The structure of unsolvated Dianin's compound (**3**), viewed normal to the *ac*-plane. Two molecules of **3** which lie above and below the cavity as viewed in this direction have been excluded (apart from their hydroxyl oxygen atoms) to show the cage more clearly. (b) A stereoview illustrating the dove-tailing involved in the lateral packing of columns, infinitely extended along *c*, in unsolvated **3**.

to the *c*-axis; it is surrounded efficiently by six other identical columns related by three-fold screw axes (parallel to *c*), such that no significant spaces are left between columns. The dove-tailing involved in the lateral packing is illustrated in Figure 9b, the projections in one column fitting neatly into the indentations of its neighbour. As will be readily appreciated, knowledge of the clathrate structure of the "parent" systems **3** and **7**, throws new light on various regions of structural modification. Important features of **3** and **7** for further consideration, in addition to the hydrogen bonded hexameric unit already mentioned, are the existence of a waist half-way up the cavity formed by six *inward-pointing* methyl groups [four of which may be seen near the cage centre in Fig. 9a], and the fact that the fused benzene ring of the chroman **3**, or the thiachroman **7**, is situated on the *outside* of the columns, where it abuts with neighbouring columns. With this information in mind, the effects of controlled structural modification can now be considered in detail, and with much more insight. In the following section some notable properties of the guest molecules will also be considered.

4.3. Structural modification of Dianin's compound (3)

The first deliberate attempt to modify Dianin's compound was reported[87] in 1956 by Baker and coworkers who successfully prepared the phenolic crystalline homologue (**8**) which has an additional methyl group *ortho* to the hydroxy function. No inclusion properties were found for compound **8** however. As will be seen below, subsequent systematic studies have not only led to new hosts, but also to systems with markedly altered cage geometry.

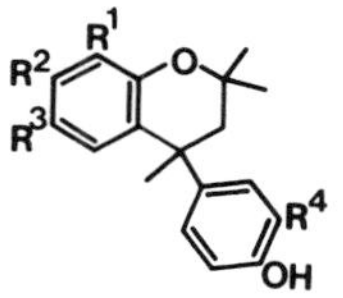

$$(8) \quad R^1 = R^2 = R^3 = H; \ R^4 = Me$$
$$(9) \quad R^1 = R^2 = R^4 = H; \ R^3 = Me$$
$$(10) \quad R^1 = R^3 = R^4 = H; \ R^2 = Me$$
$$(11) \quad R^2 = R^3 = R^4 = H; \ R^1 = Me$$

4.3.1. Replacement of the heteroatom of (3) and early studies of guest conformation

The synthesis of 4-*p*-hydroxyphenyl-2,2,4-trimethylthiachroman (**7**), already mentioned, appears to represent the earliest example of the deliberate preparation[95] of a versatile organic clathrate host of established closed-cage type.[99] The clathrates formed by **7** are, however, almost identical in structure to those of **3**, and in each case guest accommodation is provided in extremely similar hour-glass shaped cavities [see, for example, Fig. 15a]. At this time, we were interested in the possibility of determining the structure, conformation, and orientation of a guest molecule actually present within a clathrate cavity. The central constriction in the cavity of **3** and **7** suggested that the acetylenic alcohol $(CH_3)_3CC \equiv CC(CH_3)_2OH$ might be a suitable guest candidate. An X-ray study[100] revealed that in the clathrate of this acetylene with host **7** all the guest molecules adopt a staggered conformation (with a statistical disorder of the OH and Me groups to conform with the imposed $\bar{3}$ (C_{3i}) symmetry of the cavity. As shown in Fig. 10a, the acetylenic unit of the guest molecule is collinear with the c-axis, the triple bond fitting neatly into the waist of the cavity, leaving a tetrahedral unit in the upper and lower halves of the cavity. Careful choice of the guest gives rise to a "lock and key" situation here, the conformation of the guest being closely controlled by the host: the match between host and guest is illustrated in Fig. 10b, which shows

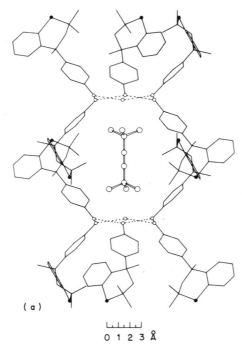

(a)

```
└┴┴┴┘
0 1 2 3 Å
```

Fig. 10. (a) The structure of the clathrate of 4-*p*-hydroxyphenyl-2,2,4-trimethyl-thiachroman (**7**) with 2,5,5-trimethylhex-3-yn-2-ol as guest. Two host molecules have been excluded (apart from their hydroxyl oxygen atoms) to reveal the guest, which is accurately aligned along the c-axis; (b) The van der Waals contacts, as viewed along the c-axis, for a section at $z = 0.26$; the broken lines represent the van der Waals volumes of the atoms comprising the cage and the full lines the approximate van der Waals volume of the guest. (Reproduced, by permission, from ref. 100.)

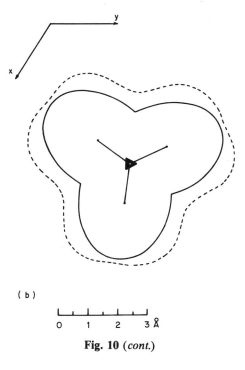

(b)

```
L__L__L__L__L__L__J
0    1    2    3 Å
```

Fig. 10 (*cont.*)

van der Waals contacts for a section about a quarter way up the cavity (at $z = 0.26$). Similar results have also been obtained[101] for the more symmetrical guest, di-*t*-butylacetylene.

Also of conformational interest, the independent X-ray study[98] of the *n*-heptanol clathrate of Dianin's compound **3** revealed the (disordered) guest threading the cavity's central constriction, and the *n*-heptanol has been assigned a *gauche* conformation, consistent with cavity length considerations.

It is relevant to inquire whether the bulk of the guest component has any effect on the cavity shape. Results are summarized in Table 3: on passing from the ethanol clathrate of **7**, which has two sterically undemanding EtOH molecules per cavity, to the increasingly bulky acetylenes indicated, only slight increases in the *a* and *c* axial dimensions are observed. However, significant increases in the O···O distance are found with increasing guest size, and this corresponds to a weakening of the $[OH]_6$ unit's hydrogen bonds, an effect nicely paralleled by the change in hydroxyl stretching frequency. Thus, for bulky guests, a modest expansion of at least the ends of the cage does occur.

Table 3. Variation of lattice parameters, $\nu(OH)$, and $O \cdots O$ distance with guest size for clathrates of **7**

Guest molecule	Host : guest ratio	Lattice parameters	$\nu(OH)$[a]	$O \cdots O$ distance	Ref.
Ethanol	3 : 1	$a = 27.81$, $c = 10.90$ Å	3345 cm^{-1}	2.96 (1) Å	95, 117
2,5,5-Trimethylhex-3-yn-2-ol	6 : 1	$a = 27.91$, $c = 10.99$ Å	3400 cm^{-1}	3.03 (1) Å	100, 117
Di-t-butylacetylene	6 : 1[b]	$a = 28.00$, $c = 11.08$ Å	3435 cm^{-1}	3.07 (1) Å	117

[a] Approximate band maxima.
[b] In subsequent recrystallizations lower incorporations of di-t-butylacetylene were found. The i.r. value quoted was measured (microdisc) using the actual crystal employed for X-ray data collection.[117]

In our early studies, host properties were also found for 4-*p*-hydroxyphenyl-2,2,4-trimethylselenachroman, which corresponds to replacement of the ring oxygen of **3** by a selenium atom,[102] though the crystal structure of these inclusion compounds has not yet been solved. On the other hand, no inclusion properties were found for the sulphone 4-*p*-hydroxyphenyl-2,2,4-trimethylthiachroman 1,1-dioxide, produced[102] by oxidation of **7**. It is noted that the mass spectra of Dianin's compound and its thia, selena, and sulphonyl analogues have been reported and discussed.[102]

4.3.2. Substitution of the ring skeleton of compounds 3 and 7

As the reader will recall, the columns comprising the host structure in **3** are infinite in extent with any given column surrounded by six identical columns, a situation exactly paralleled for **7**. Since the carbon atoms C(5), C(6), C(7), and C(8) of the aromatic ring of the chroman or thiachroman are situated on the outside of the columns (see, for example, Fig. 9), modification at these positions may be expected to affect intercolumn packing. Not unexpectedly, fusion of an additional bulky benzene ring on to **7**, to give **15**, leads to severe column disruption with elimination of inclusion properties.[103] However, fine control between the columns is possible by the introduction of methyl groups at the 6-, 7-, or 8-position of the thiachroman (**7**) to give **12**, **13**, and **14** respectively, and an interesting spectrum of behaviour is

(**12**) $R^1 = R^2 = H$; $R^3 = Me$
(**13**) $R^1 = R^3 = H$; $R^2 = Me$
(**14**) $R^2 = R^3 = H$; $R^1 = Me$

(**15**)

produced.[103,104] Of these only **13** exhibits no inclusion behaviour, and in this case instead of forming discrete, centrosymmetric, [OH]$_6$ units crystallization occurs with *spontaneous resolution* in the orthorhombic space group P2$_1$2$_1$2$_1$ and, as shown in Fig. 11, there are now infinite chains of molecules linked head-to-tail by OH···S hydrogen bonds (length 3.34 Å), the interchain packing leaving no voids for solvent inclusion.[105] The new hosts **12** and **14** are isomorphous with parent **7** [space group R3̄], and comparison of relevant *c*-spacings in Table 2 indicates that a marked decrease in cavity

D. D. MacNicol

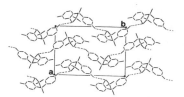

Fig. 11. A view normal to the *ab* plane showing the molecular packing of 4-*p*-hydroxyphenyl-2,2,4,7-tetramethylthiachroman (**13**) in the crystal. Intermolecular hydrogen bonds are denoted by broken lines. (Reproduced, by permission, from ref. 105.)

length to 8.24 Å has occurred for **14**, implying that, for the first time, a fundamental change in cavity geometry has been achieved. As can be seen from the stereoviews in Fig. 12, the "legs" of the hexameric unit of **14** have "splayed out", compared with their disposition in **7**, thus explaining the decrease in the *c*-axial length and the corresponding increase in the *a*-dimension, which reflects an increase in column width. Just how dramatic the change in cavity shape is, may be appreciated from Fig. 13, which shows that the initial hour-glass shaped cavity of **7** has been transformed[104,106]

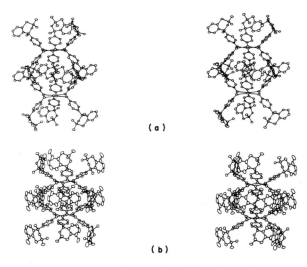

(a)

(b)

Fig. 12. Comparative stereoviews for (a) the 2,5,5-trimethylhex-3-yn-2-ol clathrate of 4-*p*-hydroxyphenyl-2,2,4-trimethylthiachroman (**7**) and (b) the cyclooctane clathrate of 4-*p*-hydroxyphenyl-2,2,4,8-tetramethylthiachroman (**14**). The guest molecules are not shown.

into the "chinese-lantern" contour of **14**. This change in cavity geometry is reflected in selective clathration properties.[103,104] On recrystallization from an equimolar mixture of cyclopentane, cyclohexane, and cycloheptane, **7** greatly favours cyclopentane [respective percentages included: 85%, 10%, 5%], whereas **14**, which has a rounder cavity, favours the larger cyclo-paraffins [20%, 50%, 30%], consistent with the view that van der Waals host-guest attractions are optimized during the crystallization process.

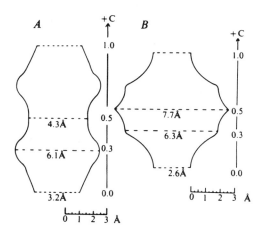

Fig. 13. Section through the van der Waals surface of the cavity for (a) 4-*p*-hydroxyphenyl-2,2,4-trimethylthiachroman (**7**) and (b) 4-*p*-hydroxyphenyl-2,2,4,8-tetramethylthiachroman (**14**), representing the space available for guest accommodation. (Reprinted by permission from *Nature*, Vol. 256, No. 5515, pp. 343–344. Copyright © 1975 Macmillan Journals Limited.)

Introduction[104] of methyl groups at either C(6) or C(7) of Dianin's compound itself does not, in either case, yield a new host; on recrystallization, the homologues **9** and **10** form monoclinic crystals (Table 2) without inclusion of solvent. Interestingly, the recently synthesized[107] 4-*p*-hydroxyphenyl-2,2,4,8-tetramethylchroman (**11**), the 8-methyl homologue of Dianin's compound, returns to a parallel with the thiachroman series and this molecule, like its direct counterpart (**14**), forms clathrates with a substantially reduced cavity length [Table 2, and Fig. 14c]. A more highly puckered arrangement for the six hydroxyl oxygen atoms is found[107] for **11** (±0.38 Å from the mean plane of the oxygens) than for **3** (±0.21 Å; CHCl₃ clathrate[97] cf. Fig. 14, c and a); a situation paralleled[104] for **14** and **7**, where the corresponding displacements are ±0.35 Å, and ±0.22 Å (Fig. 12).

4.3.3. Modification of the substitution at C(2) and C(4) of Dianin's compound (3), the optical resolution of 3, and related quasi-racemic systems
The hour-glass shaped cavity contour of Dianin's compound owes its central constriction to six inward pointing methyl groups, one from each of six molecules of 3. The methyl group from each of the six molecules involved is one of the *gem*-dimethyl groups, namely that *syn* to the *p*-hydroxyphenyl substituent. We reasoned that specific removal of these methyl groups might provide a rather direct method of altering the cavity shape, provided, of course, that these methyls did not play a crucial role in preventing "inward collapse" to a close-packed structure. The synthesis of compound 16, lacking the 2-methyl group *syn* to the *p*-hydroxyphenyl substituent, was approached, paralleling Dianin's original reaction conditions,[84] by the action of anhydrous hydrogen chloride on a mixture of phenol and pent-3-en-2-one and, following silicic acid and gel permeation chromatography a pure material was obtained[108,109] whose spectroscopic properties were consistent with those expected for the required epimer 16. Even more encouraging, this new material was a new host, isomorphous with Dianin's compound (3) and having similar values for the lattice parameters *a* and *c*; given in Table 2.

(and enantiomer)

(16) $R^2 = H$; $R^1 = R^3 = Me$
(17) $R^1 = H$; $R^2 = R^3 = Me$
(18) $R^3 = H$; $R^1 = R^2 = Me$

That the waist methyl groups have indeed been eliminated[108,109] is shown in Fig. 14a and b, which gives comparative stereoviews for 3 and its 2-nor analogue 16. The marked change in cavity shape brought about by removal of the six inward-pointing methyl groups of 3 may be readily appreciated from Fig. 15. As can be seen from Fig. 15b, the waist of the cavity has been completely eliminated and indeed the widest cross-section is now encountered halfway up the cavity. Of particular interest, the new cavity contour for 16 is remarkably close to that predicted by the theoretical removal of the appropriate waist methyl groups of 3. [The formal replacement of methyl by appropriately-positioned hydrogen is denoted by the curved broken lines in Fig. 15a]. Collet and Jacques[110] have prepared the epimer 17 which corresponds to removal of the methyl group on C(2) *anti* to the *p*-hydroxyphenyl substituent, and this compound is also a new host. In contrast

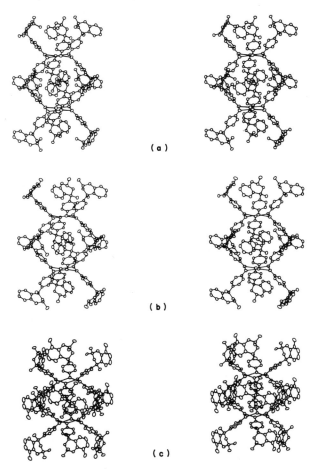

Fig. 14. Comparative stereoviews of the host packing for (a) Dianin's compound (**3**) as chloroform clathrate, drawn from data of ref. 97 and (b) 4-*p*-hydroxyphenyl-*cis*-2,4-dimethylchroman (**16**) as its carbon tetrachloride clathrate[109]; and (c) a clathrate of 4-*p*-hydroxyphenyl-2,2,4,8-tetramethylchroman (**11**), Table 2. The guest molecules are not shown.

compound **18**, prepared[109] to ascertain the effect of removal of the 4-methyl group of **3**, crystallizes, unsolvated, in the tetragonal space group P$\bar{4}$2₁c (Table 2) and there are now infinite chains of glide-related molecules linked head-to-tail by (ether) O···HO hydrogen bonds of length 2.82(1) Å.

In Dianin's compound the centrosymmetric cage is made up of three molecules of one configuration and three of the opposite configuration. Inquiring into the outcome[111] of having only *one* enantiomer present, in

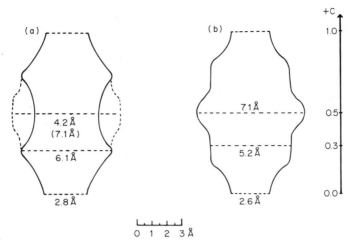

Fig. 15. Section through the van der Waals surface for (a) Dianin's compound (**3**) as chloroform clathrate, replotted from data of ref. 97 (curved broken lines represent the effect of formal replacement of the "waist" methyl groups, see text); and (b) 2-normethyl analogue (**16**) as its CCl$_4$ clathrate. (Reproduced, by permission, refs. 108 and 109.)

which case any cage formed would necessarily be chiral, Brienne and Jacques[112] have optically resolved (**3**) (by crystallization of the diastereoisomeric ω-camphanates, followed by hydrolysis) thereby obtaining S-(−)-Dianin's compound (**19**), which has the absolute configuration shown.[113] Highlighting the importance of the packing of both enantiomers to form cages in **3**, no inclusion compound formation has been found for the resolved material **19**, either for chiral or achiral potential guest components. However, Collet and Jacques[113] have succeeded in preparing chiral clathrate analogues of Dianin's compound, in which chiral cages are present. This was cleverly achieved by constructing suitable quasi-racemic systems related to **3**. Thus an equimolar mixture of **19** and the resolved 2-nor analogue **21** gives rise to a new chiral clathrate host, and a related quasi-racemic chiral host system is also formed by recrystallization of an equimolar mixture of the diastereoisomeric optically active compounds **20** and **21**. It will be noted that for each of these quasi-racemates, the components have inverse configurations at C(4). Preliminary X-ray measurements,[114] Table 2, indicate that the packing[115] in the CCl$_4$ clathrates of these systems is almost certainly directly analogous to that found in **3**, the chiral space group R3 now being encountered, instead of the centrosymmetric space group R$\bar{3}$. Synthetic chiral hosts are of great interest as potential resolving agents since, in principle at least, it should be possible to utilize diastereoisomeric spacial interactions between host and guest to obviate the need for the functionality normally required for conventional resolution methods. An attempt to resolve ethyl α-bromopropionate in (19, 21) was, however, unsuccessful, recrystallization giving an ester of optical purity lower than 0.5%, and possible adverse factors suggested[110,115] to account for this outcome were the quasi-symmetrical form of the cavity, built from three molecules of 19 related to three of 21 by a pseudo-centre of symmetry, and guest disorder in the cavity of high (C_3) symmetry.

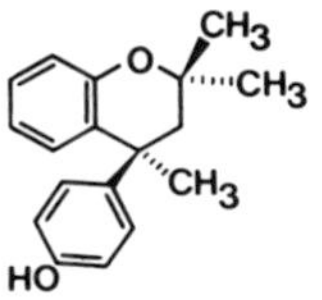

(19) *S*(−)-Dianin's compound

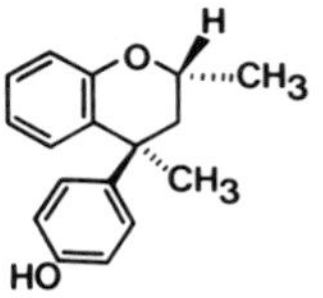

(+)2*R*, 4*S* **(20)** 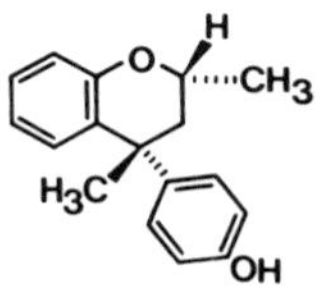   (+)2*R*, 4*R* **(21)**

4.3.4. Changes in the hydrogen-bonding functionality of 3

In view of the key role of the hydrogen-bonded hexameric units which form the floor and roof of each cavity in 3, and other clathrates already discussed, it is of great interest to determine whether another hydrogen-bond forming group might be capable of replacing the OH group without eliminating the clathrate forming ability. In order to investigate this point, we have synthesized[116,117] the amine (22) and the thiol (23) in which the hydroxyl group of 3 has been replaced by the amino or the thiol function. The amine (22) does not form inclusion compounds, but is interesting in that it undergoes *spontaneous* resolution on crystallization, the orthorhombic crystals formed,

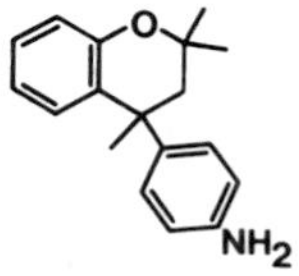

(22) 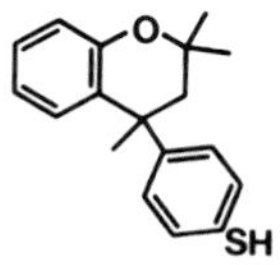   **(23)**

space group P2₁2₁2₁, being isomorphous with *resolved* Dianin's compound (19) and having similar unit cell dimensions (see Table 2). The corresponding thiol (23), prepared[116,117] from 3 by the general method of Newman and Karnes,[118] shows a fascinating duality of behaviour: when (23) is crystallized

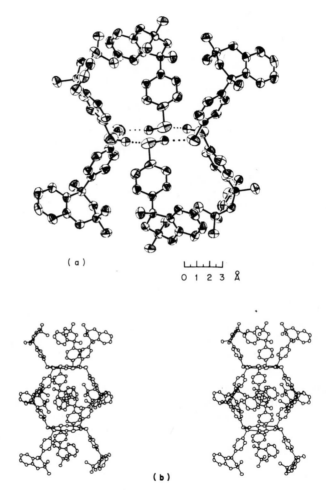

(a)

```
└─┴─┴─┴─┘
0 1 2 3 Å
```

(b)

Fig. 16. (a) A general view of the hydrogen-bonded hexameric unit of 4-*p*-mercap-tophenyl-2,2,4-trimethylchroman (23) in the CCl₄ clathrate (reproduced, by per-mission, from ref. 117); and (b) an illustration (drawn from data of ref. 117) of the hexamer stacking in 23 which is analogous to that found in Dianin's compound 3 itself.

from cyclohexane *spontaneously resolved*, unsolvated, crystals are produced, the enantiomorphous space group $P2_12_12_1$ again being encountered; however, in contrast, crystallization of (23) from carbon tetrachloride yields crystals of the CCl_4 inclusion compound which has the centrosymmetric space group $R\bar{3}$ (and a host–guest ratio of 3 : 1). The relative values of lattice parameters, see Table 2, for the CCl_4 compound of (23) and clathrates of 3 were consistent with analogous packing modes for the two hosts, the increased *c*-spacing for 23 being in keeping with the longer C–S bond in 23, compared with the corresponding C–O bond in 3. Also encouraging, and consistent with the first [SH]$_6$ unit, the i.r. spectrum of the adduct displayed[116] a broad ν(S–H) band, centred at 2506 cm^{-1} ($\Delta\nu_{1/2}^a$ *c.* 70 cm^{-1}), a position compatible[119] with unusually short SH$\cdots$S hydrogen bonding; this may arise from a "supported hydrogen bond" effect.[120] A detailed X-ray study of the CCl_4 clathrate of 23 has confirmed the presence of discrete hydrogen-bonded hexameric units of six thiol molecules comprising the host structure in 23, and one such sextet unit is shown in Fig. 16a. As can be seen, the six molecules in the basic packing unit are linked by a network of SH$\cdots$S hydrogen bonds such that the sulphur atoms form a near planar hexagon; the SH$\cdots$S hydrogen bond is characterized by S$\cdots$S and S$\cdots$H distances[117] of 3.76(1) Å and 2.67(9) Å respectively, and by the S–H$\cdots$S angle of 164(5)°. Interestingly, this represents the first example of a regular, uniform, array of six hydrogen-bonded atoms of the same kind in a crystal, other than for oxygen. Figure 16b illustrates the packing of two hexameric units of 23, which, like 3, are stacked along the *c*-axis such that their bulkier parts interlock to form a cage; in the present case, however, the roof and floor of each cage are made up of hydrogen-bonded hexamers of sulphur atoms, one *c*-spacing (12.07 Å) apart. (The two disordered carbon tetrachloride guest molecules, present in each cage, are not shown.)

The compounds with the OH function of the thiachroman (7) replaced by an amino or a thiol group, 4-*p*-aminophenyl-2,2,4-trimethylthiachroman and 4-*p*-mercaptophenyl-2,2,4-trimethylthiachroman, have also been prepared,[117] but these compounds crystallize without inclusion of solvent.

4.3.5. 2-Phenyl-3-p-(2,2,4-trimethylchroman-4-yl)phenylquinazolin-4(3H)-one (24) and its sulphur analogue (25)

The remarkable inclusion properties of the quinazolinone (24) were discovered[121,122] when it was characterized as a synthetic intermediate in the conversion of Dianin's compound into the already discussed amine 22. Compound 24 has been found to be a particularly versatile host forming stable inclusion compounds with a wide range of solvents and important classes of guest[122] are cycloalkanes, cyclic ethers and ketones, alcohols, and

(24) Z=O
(25) Z=S

aromatic molecules. For the cycloalkanes, for example, we found[122] that molecules ranging in size from cyclopropane to cyclodecane could be included in **24**. An X-ray study, which employed quartet relationships in the direct-method analysis,[122] established that the methylcyclohexane compound of **24** has a true clathrate structure. The cell is triclinic, space group P$\bar{1}$, and there are two host and one methylcyclohexane guest molecule in the asymmetric unit. The packing of **24** is such that large closed cavities are formed, and as can be seen from Fig. 17, the floor of one such cage is made up of molecules of **24** of one crystallographic type (denoted by fainter lines) while the walls are formed by molecules of the second type (darker lines). The roof is then made up of another layer of host molecules (not shown) of the first type, a unit cell translation away along *a*. In each clathrate cage there are two methylcyclohexane guest molecules and these are approximately centred at the enantiomerically related points G and G'. The cavity's shape and size may be appreciated from the contours in Fig. 18, which represent only half the large centrosymmetric cage.[122] Preliminary X-ray work[122] has established that $BrCF_2CF_2Br$ and cyclohexane also form triclinic compounds with **24**, and, as for methylcyclohexane above, these have a 2:1 host–guest ratio; though a 1:1 compound, which has the monoclinic space group P2_1/c, has been obtained on recrystallization of **24** from *t*-butanol.

The related compound **25** derived from thiachroman **7** also exhibits significant inclusion properties.[122]

4.4. Applications of Dianin's compound (3) and related system 7

Dianin's compound (**3**) exhibits useful selective clathration properties[123] allowing efficient separation of certain hydrocarbon mixtures. For example, 2-methylhexane (b.p. 90.05° C) could be recovered from an equimolar mixture with 2,3-dimethylpentane (b.p. 89.78° C) in a predicted[123] purity of 99% by a single clathration with **3**. The SF_6 clathrate of **3** is of interest[89,90] as a convenient means of storage and controlled release of sulphur hexafluoride, a gas of considerable use in the electrical industry. Other notable applications involve the use[124] of amine complexes of **3** as polymerizing agents in

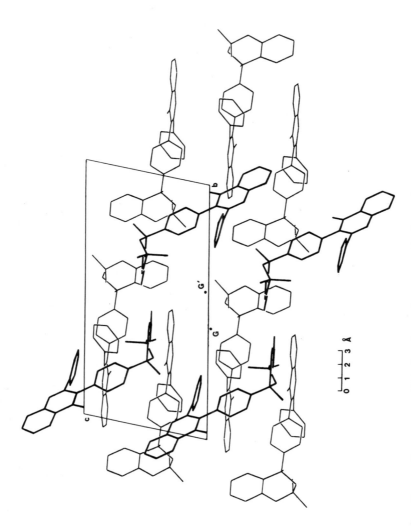

Fig. 17. A packing diagram for compound **24**, showing a view perpendicular to the *bc* plane. The points G and G' represent the approximate centres of the two methylcyclohexane guest molecules in the cavity. The cage is completed by several host molecules (which are closer to the viewer); these have been omitted for clarity. (Reproduced, by permission, from ref. 122.)

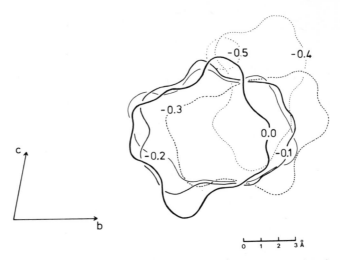

Fig. 18. The space available in the cage in cross-section at various fractional levels parallel to the *bc* plane, in **24**. The contours shown represent only half the large centrosymmetric cage. (Reproduced, by permission, from ref. 122.)

the preparation of epoxy and urethane resins; and the use[125] of the $(CF_3SO_2)_2CH_2$ clathrate of **3** as a latent curing catalyst in cationic polymerization. The diethylamine clathrate of the same host can be used as developer for the production of heat-sensitive copying sheets.[126] In addition, the highly toxic compound dimethylmercury can be handled with comparative safety in the form of its clathrate[127] with the thiachroman host (**7**). A known weight of this clathrate (host–guest ratio 6 : 1) can be melted or dissolved to release a known quantity of the organomercurial guest when required.

5. Other complex phenolic host molecules

A recent X-ray study[128] of the acetone adduct of the naturally-occurring compound guayacanin (**26**) has revealed the presence of hydrogen-bonded

(26)

hexameric units analogous to those found in clathrates already discussed. In this inclusion compound, which has a host–guest ratio of 3:1, the space group R$\bar{3}$ is again encountered. Figure 19 shows a view of the crystal structure looking down the (hexagonal) *c*-axis, and each host cluster, such as the one shown, is centred on a point of $\bar{3}$ symmetry. As is apparent in this sextet unit there is a very pronounced displacement (±0.92 Å) of oxygen atoms alternately above and below the mean plane of the six oxygens, the OH···O hydrogen bond length being 2.82(1) Å. The acetone guest molecules are disordered and two such guest species are situated between adjacent hexameric host clusters along *c*.

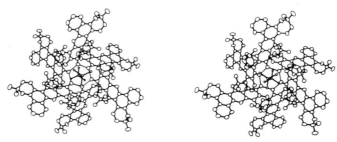

Fig. 19. Stereoscopic view of the molecular packing in guayacanin (**26**), looking down *c* with *a* horizontal. The thermal ellipsoids are drawn at the 50% level except for the (disordered) acetone guest molecules; for these atoms an arbitrary temperature factor of 1.0 Å^2 was used. (Reproduced, by permission, from ref. 128.)

For over thirty years 2'-hydroxy-2,4,4,7,4'-pentamethylflavan (**27**) has been known to be capable of forming crystalline complexes with many ethers, ketones, and amines.[129] A recent detailed ESR study of the nitroxide spin label 2,2,6,6-tetramethyl-4-piperidinol-1-oxyl as guest in **27** has been described,[130] and the free radical guest species was found to be well oriented at room temperature in the host lattice. Only very recently, however, has the crystal structure of a complex of host **27** been elucidated.[131] The stoichiometry[129] of the complex studied was 2:1:2, for host **27** to 1,4-dioxan and water guest species respectively. This inclusion compound is triclinic, space group P$\bar{1}$, and the 1,4-dioxan guest molecule is located at a crystallographic centre of symmetry, while the host (**27**) and water molecules are

(27)

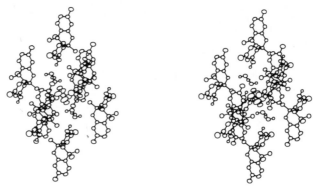

Fig. 20. A stereoview of the host–guest packing in the 2:1:2 adduct of compound **27** with 1,4-dioxan and water. All hydrogen atoms have been omitted except those involved in hydrogen-bonding (reproduced, by permission, from ref. 131).

situated in general positions, see Fig. 20. Contrasting with the discrete $[OH]_6$ units of for example **3**, the OH proton of **27** is directly involved in a hydrogen bond to a water molecule which in turn uses its protons to form hydrogen bonds to the dioxan guest and the ether oxygen of another molecule of **27**. Thus there is no OH···O hydrogen bonding between molecules of **27**. The host molecule **27** has a half-chair conformation for its oxygen-containing ring with C(2) and C(3) displaced by 0.34 Å and 0.29 Å above and below the mean plane of the fused benzene ring. Although 2′-hydroxy-2,2,4,8,3′-pentamethylflavan (**28a**), isomeric with **27**, does not exhibit inclusion behaviour, crystalline complexes are formed[132] by both the 2′-hydroxy-flavans **28b** and **28c**; with guests 1,4-dioxan and morpholine, **28b** forms adducts with a host–guest ratio of 2:1, while **28c** forms 1:1 adducts with

(**28a**) $R^1 = R^2 = R^3 = R^6 = H; \; R^3 = R^4 = Me$
(**28b**) $R^1 - R^6 = H$
(**28c**) $R^2 - R^5 = H; \; R^1 = R^6 = Me$
(**28d**) $R^2 = R^3 = R^5 = H; \; R^1 = R^6 = Me; \; R^4 = NO_2$
(**28e**) $R^2 = R^3 = R^5 = H; \; R^1 = R^6 = Me; \; R^4 = NH_2$
(**28f**) $R^2 - R^5 = H; \; R^1 = Me; \; R^6 = Cl$
(**28g**) $R^2 - R^5 = H; \; R^1 = Me; \; R^6 = Br$
(**28h**) $R^1 = R^3 = R^4 = R^6 = H; \; R^2 = R^5 = Et$
(**28i**) $R^3 = R^4 = H; \; R^1 = R^2 = R^5 = R^6 = Me$

the same guests, a situation inviting further crystallographic investigation of the 2'-hydroxyflavan series. Other hosts of this type are the nitro compound **28d** and corresponding amine **28e**,[132] the halogen-containing flavans **28f** and **28g**,[133] and also **28h, 28i, 29**[126,134] and related molecules.[135] On the practical side, an interesting and useful method for the isolation and purification of pyrimidines, employing complexation by hosts **27** and **28c**, has been described.[136] Lawton[126] has employed amine addition complexes of **27, 28c, 28h, 28i**, and **29** as developers for the production of heat-sensitive copying sheets. Similarly, the morpholine adduct of **28b** can be employed as an image forming material.[137]

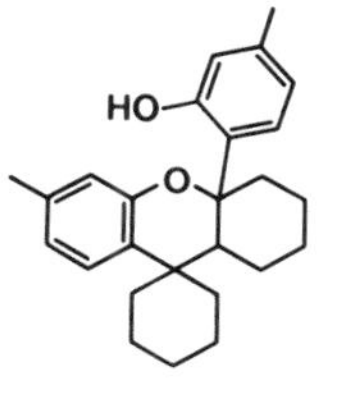

(29)

An X-ray study of the inclusion compound formed by the dibromocannabicyclol **30** on recrystallization from light petroleum has been described.[138] In this triclinic adduct, which has space group P$\bar{1}$ and host–guest ratio *c.* 2:1, the molecular architecture of host **30** has been established, though the guest component, probably a mixture of pentane and hexane, is highly disordered.

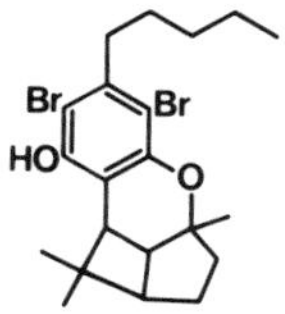

(30) 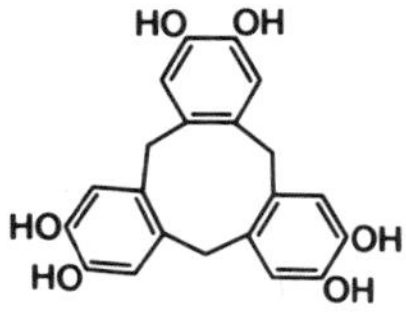   **(31)**

The marked inclusion properties of cyclotricatechylene (**31**) are considered separately in Chapter 4, Volume 2, however in Table 4 it is of interest to bring together some other hosts of widely diverse structure, all of which have polyhydric phenolic character. The inclusion properties of wightianone (**32**),[139] hosts of the 2-(2-arylindan-1,3-dion-2-yl)-1,4-naphthohydroquinone type represented by **34a** and **34b**,[140] and the novel hosts **35a–35c**[141] are here noted, and the other systems in Table 4 will be discussed briefly below.

Table 4. Other, polyhydric, phenolic host molecules forming crystalline adducts

Host	Typical guest molecules (host–guest ratio)[a]	Ref.
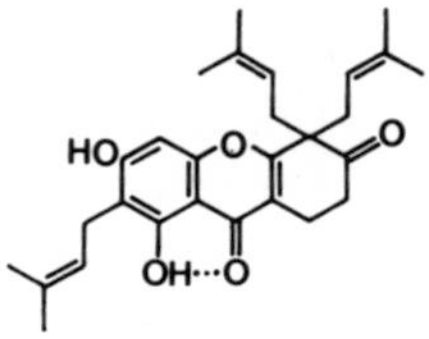 (32)	palmitic acid (4:1)	139
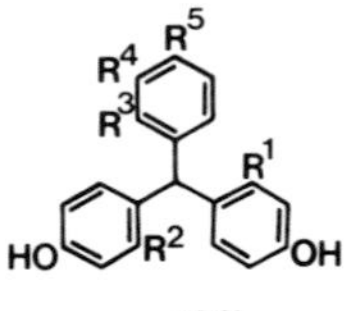 (33)		
(33a) $R^1 = R^2 = R^4 = R^5 = H$; $R^3 = Cl$	benzene, heptane, octane, decane, oct-1-ene	142
(33b) $R^1 = R^2 = R^3 = R^5 = H$; $R^4 = NO_2$	benzene (1:2), toluene (1:1)	143
(33c) $R^1 – R^4 = H$; $R^5 = Br$	benzene (2:3), heptane, octane, decane, oct-1-ene	144, 142
(33d) $R^3 = R^5 = H$; $R^1 = R^2 = Me$; $R^4 = Br$	benzene (1:1), p-xylene (1:1), chlorobenzene (1:1), bromobenzene (1:1), heptane, octane, decane, oct-1-ene	144, 142
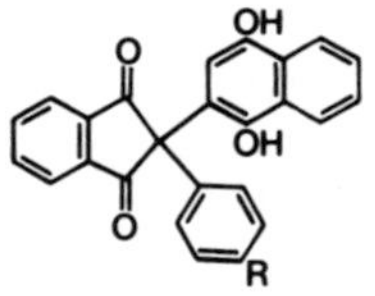 (34)		
(34a) $R = H$	methanol (1:1), ethanol (1:2), acetone (1:1), benzene (1:1), o-xylene (2:1)	140
(34b) $R = Cl$	methanol (1:1), ethanol (1:1), benzene (1:2), o-xylene (1:1)	

Table 4—*continued*

Host	Typical guest molecules (host–guest ratio)[a]	Ref.

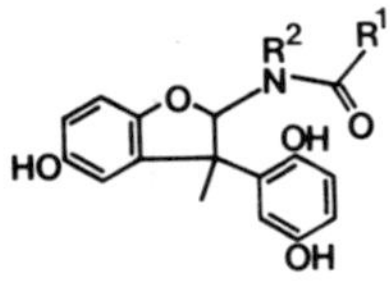

(35)

(35a) $R^1 = R^2 = Et$	1,4-dioxan-d_8	141
(35b) $R^1 = Et; R^2 = Me$	diethyl ether, acetone-d_6 carbon tetrachloride	
(35c) $R^1 = i$-Pr; $R^2 = Me$	1,4-dioxan-d_8	

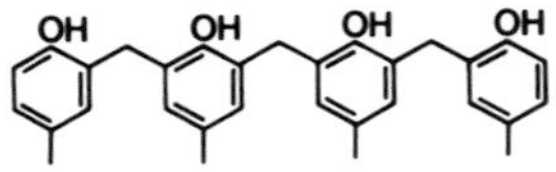

(36) 1,2-dichloroethane (1:1) 145

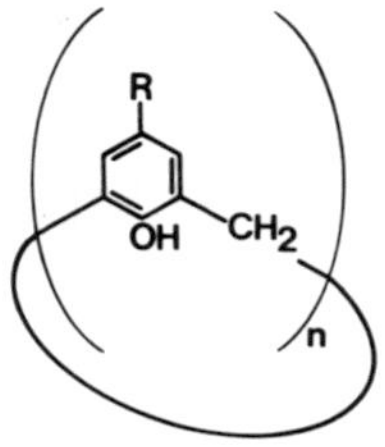

(37)

(37a) $n = 4$, R = t-Bu	toluene (1:1), benzene, p-xylene, anisole, chloroform	148, 149, 146
(37b) $n = 4$, R = 1,1,3,3-tetramethylbutyl	toluene (1:1)	150
(37c) $n = 5$, R = H	acetone (1:2)	147
(37d) $n = 6$, R = t-Bu	methanol/chloroform[b]	146
(37e) $n = 8$, R = t-Bu	c	146

[a] Host–guest ratio, where available.
[b] Complex contains as guest two molecules of methanol and one of chloroform.
[c] Only very labile complexes formed.

Compounds **33a–33d** represent just four of many derivatives of 4,4′-dihydroxytriphenylmethane exhibiting host properties,[142–144] the *para*-hydroxyl groups in two of the rings being an important structural feature here. Although more work is required to elucidate the structures of the different types of complex formed by members of this series, preliminary work[142] indicates that channel type complexes may be formed with *n*-alkanes, *n*-alkenes, and the branched molecules 2,2,4-trimethylpentane, diisobutene, and squalane. Polymerization of isoprene, induced by γ-rays, was accomplished[142] in the complexes with **33a**, **33c**, and **33d**; the ratio of *cis*-1,4- to *trans*-1,4-addition was however, little different from that in emulsion polymerization, though, a sign of steric control, 1,2- and 3,4-addition was reduced almost to zero. (For a detailed consideration of the fascinating topic of inclusion polymerization see Chapter 10, Volume 3.)

Although the structure of the 1 : 1 adduct[145] of the linear novolak tetramer (**36**) remains unknown, much recent interest has centred round cyclic analogues known as the calixarenes (Gk. *calix*, cup; arene, aromatic rings), both with respect to a systematic structural investigation of factors responsible for inclusion compound formation in this series; and also to their use, suitably elaborated, as potential *enzyme models*. Particularly significant, very recent, contributions have been made in calixarene chemistry by Andreetti *et al.* in Parma[146] and by Gutsche and coworkers in Wisconsin.[147] Here attention will be focussed on derivatives **37a–37e**, in which the phenolic groups have not been substituted and are therefore free to form hydrogen bonds. Calix[4]arene **37a**, a cyclic tetraphenol derived from the base-catalysed condensation of formaldehyde and *para*-t-butylphenol, forms clath-

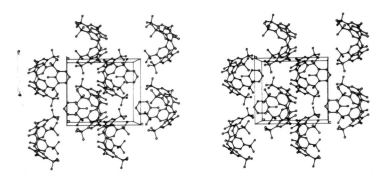

Fig. 21. A stereoview illustrating the host–guest packing in the clathrate of the calix[4]arene (**37a**) with toluene. A second, symmetry-related guest orientation, with toluene rotated by 90° about its C—CH$_3$ bond direction, is not shown (drawn from data of ref. 148).

rates with aromatic molecules such as toluene,[148] benzene, *p*-xylene, and anisole.[149] An X-ray analysis of the toluene adduct[148] of **37a**, which has the tetragonal space group P4/n, has established that the host molecule **37a**, lying on a four-fold proper rotation axis, does indeed have the form of a chalice and that the oxygen atoms are linked by four intramolecular hydrogen bonds [O···O, 2.670(9) Å]. The host and guest molecules are illustrated in Fig. 21, for which an orientation of the guest rotated by 90° (not shown) is equally probable. In contrast to the clathrates formed by **37a**, the calix[4]arene **37b**, derived from *p*-(1,1,3,3-tetramethylbutyl)phenol and formaldehyde, forms mainly channel-type inclusion compounds with aromatic guest molecules.[150] Very recently, the structure of an adduct of a calixarene with an odd number of phenolic units has been determined.[147] The inclusion complex (1:2) for calix[5]arene **37c** with acetone is orthorhombic, space group Pna2₁, and there are four host and eight acetone guest molecules in the unit cell. The macrocyclic host **37c** again has a chalice-like shape, the conformation being determined mainly by intramolecular (OH···O) hydrogen bonding.[147] There are two non-equivalent acetone guest molecules which interact with host **37c** in different ways, as shown in Fig. 22; one guest, at the top, interacts with the [OH]₅ crown unit while the second acetone molecule lies on the inside of the chalice with its methyl groups making

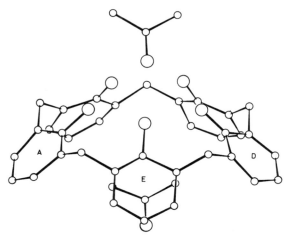

Fig. 22. An illustration of the host–guest interactions in the inclusion compound of the calix[5]arene (**37c**) with acetone. The host molecule has symmetry close to C_s; and two crystallographically non-equivalent acetone guest molecules can be seen above and below the host, interacting with it in different ways. (Reproduced, by permission, from ref. 147.)

contacts with the two (unlabelled) aromatic rings furthest from the viewer. In contrast to the formation of a very stable complex of the calix[6]arene **37d**, only labile complexes have been found for the octamer **37e**.[146] Doubtless, the calixarenes constitute an important class of host molecule which merits further detailed study in the future.

6. Concluding remarks

In the present chapter, in tracing the study of phenolic host molecules, a class of host representing inclusion chemistry in microcosm, one observes an evolution from early chance discoveries to rational host design. The beautiful host structure of β-hydroquinone clathrates, and other phenolic host lattices, have been described in some detail. In contrast to the situation for hydroquinone where structural modification gives no isostructural host lattices, for Dianin's compound, which almost certainly corresponds to a relatively deep well on the crystal structure-free energy surface, judicious host modification leads to new isomorphous clathrates with significantly altered cage geometry in certain cases. Currently under consideration in Glasgow is the possibility of using the increasingly powerful computer programs now available for the calculation of potential energy minima in molecular crystals to identify favourable parent hosts, and indeed, for a given host, to suggest suitable points of structural modification to produce a desired change in void geometry. In the first strategy for the synthesis of new hosts attention has been focussed on structural modification of a known host molecule. In Chapter 5, Volume 2, emphasis is placed on the significance of a *group* of molecules, the hydrogen-bonded hexameric building block commonly encountered in the clathrates already discussed, and in the second strategy, analogy with this molecular grouping has led to the synthesis of an important new class of host molecule known as the hexahosts. In this chapter, the fascinating question of molecular symmetry, corresponding to a third strategic principle in host design, will also be considered.

Acknowledgment

The author is indebted to Dr. P. R. Mallinson for preparing a number of stereoviews used in this chapter.

References

1. H. M. Powell, in *Non-stoichiometric Compounds*, (ed. L. Mandelcorn) Academic Press, New York, 1964, ch. 7.
2. M. Hagan, *Clathrate Inclusion Compounds*, Reinhold, New York, 1962.
3. V. M. Bhatnagar, *Clathrate Compounds*, Chemical Publishing Co., New York, 1970.
4. F. Cramer, *Einschlussverbindungen*, Springer-Verlag, Berlin, 1954.
5. K. Pollmer, *Z. Chem.*, 1979, **19**, 81.
6. E. J. Fuller, *Encycl. Chem. Process. Des.*, 1979, **8**, 333.
7. E. C. Makin, *Kirk-Othmer Encycl. Chem. Technol.*, 3rd ed., 1979, **6**, 179.
8. D. D. MacNicol, J. J. McKendrick and D. R. Wilson, *Chem. Soc. Rev.*, 1978, 7, 65.
9. J. E. D. Davies, in *Molecular Spectroscopy*, (eds J. A. Sheridan, D. A. Long, and R. F. Barrow) (Specialist Periodical Reports), The Chemical Society, London, 1978, Vol. 5, ch. 2.
10. S. G. Frank, *J. Pharm. Sci.*, 1975, **64**, 1585.
11. C. Solacolu and I. Solacolu, *Stud. Cert. Chem.*, 1973, **21**, 1307.
12. D. C. McKean, in *Vibrational Spectroscopy of Trapped Species*, (ed. H. E. Hallam) Wiley, London, 1973, ch. 8.
13. C. A. Fyfe, in *Molecular Complexes, Vol. 1*, (ed. R. Foster) Elek Science, London, 1973, ch. 5.
14. L. C. Fetterly, in *Non-Stoichiometric Compounds*, (ed. L. Mandelcorn) Academic Press, New York, 1964, p. 491.
15. W. C. Child Jr., *Quart. Rev.*, 1964, **18**, 321.
16. C. Asselineau and J. Asselineau, *Ann. Chim. (Paris)*, 1964, **9**, 461.
17. J. F. Brown, *Sci. Amer.*, 1962, **207**, 82.
18. L. Mandelcorn, *Chem. Rev.*, 1959, **59**, 827.
19. F. Cramer, *Angew. Chem.*, 1956, **68**, 115.
20. H. M. Powell, *Recl. Trav. Chim. Pays-Bas*, 1956, **75**, 885.
21. F. D. Cramer, *Rev. Pure Appl. Chem.*, 1955, **5**, 143.
22. G. Montel, *Bull. Soc. Chim. Fr.*, 1955, 1013.
23. H. M. Powell, *J. Chem. Soc.*, 1954, 2658.
24. F. Cramer, *Angew. Chem.*, 1952, **64**, 437.
25. W. Schlenk, *Fortschr. Chem. Forsch.*, 1951, **2**, 92.
26. G. Zilberstein, *Bull. Soc. Chim. Fr.*, 1951, **18**, D.33.
27. F. Wöhler, *Justus Liebigs Ann. Chem.*, 1849, **69**, 297.
28. A. Clemm, *Justus Liebigs Ann. Chem.*, 1859, **110**, 357.
29. F. Mylius, *Chem. Ber.*, 1886, **19**, 999.
30. D. E. Palin and H. M. Powell, *Nature (London)*, 1945, **156**, 334.
31. (a) D. E. Palin and H. M. Powell, *J. Chem. Soc.*, 1947, 208. (b) T. M. Polyanskaya, V. I. Alekseev, V. V. Bakakin and G. N. Chekhova, *J. Struct. Chem. (Engl. Transl.)*, 1982, **23**, 101.
32. D. E. Palin and H. M. Powell, *J. Chem. Soc.*, 1948, 815.
33. S. C. Wallwork and H. M. Powell, *J. Chem. Soc.*, 1956, 4855.
34. H. M. Powell, *J. Chem. Soc.*, 1950, 298, 300, 468.
35. H. M. Powell, *J. Chem. Soc.*, 1948, 61.
36. T. C. W. Mak and K.-S. Lee, *Acta Crystallogr.*, 1978, **B34**, 3631.
37. S. V. Lindeman, V. E. Shklover and Yu. T. Struchkov, *Cryst. Struct. Comm.*, 1981, **10**, 1173.

38. H. M. Powell and P. Riesz, *Nature (London)*, 1948, **161**, 52.
39. A. I. Kitaigorodsky, *Molecular Crystals and Molecules*, Academic Press, New York, 1973, p. 18.
40. T. C. W. Mak, J. S. Tse, C.-S. Tse, K.-S. Lee and Y.-H. Chong, *J. Chem. Soc., Perkin Trans. 2*, 1976, 1169.
41. W.-C. Ho and T. C. W. Mak, *Z. Kristallogr.*, 1982, **161**, 87.
42. J. C. A. Boeyens and J. A. Pretorius, *Acta Crystallogr.* 1977, **B33**, 2123.
43. T.-L. Chan and T. C. W. Mak, *J. Chem. Soc., Perkin Trans. 2*, 1983, 777.
44. D. E. Palin and H. M. Powell, *J. Chem. Soc.*, 1948, 571.
45. T. C. W. Mak, *J. Chem. Soc., Perkin Trans. 2*, 1982, 1435.
46. J. A. Ripmeester, R. E. Hawkins and D. W. Davidson, *J. Chem. Phys.*, 1979, **71**, 1889.
47. (a) J. C. A. Boeyens and J. A. Pretorius, *Acta Crystallogr.*, 1977, **B33**, 2120. (b) cf. R. G. Copperthwaite, *J. Chem. Soc., Chem. Commun.*, 1976, 707 (X-ray photoelectron study).
48. See, for example, H. G. McAdie, *Can. J. Chem.*, 1966, **44**, 1373 (thermal analysis).
49. S. C. Wallwork and H. M. Powell, *J. Chem. Soc., Perkin Trans. 2*, 1980, 641.
50. G. N. Chekhova, T. M. Polyanskaya, Yu. A. Dyadin and V. I. Alekseev, *J. Struct. Chem. (Engl. Transl.)*, 1975, **16**, 966.
51. W. A. Caspari, *J. Chem. Soc.*, 1926, 2944.
52. K. Maartmaan-Moe, *Acta Crystallogr.*, 1966, **21**, 979.
53. However, see H. Gies, F. Liebau and H. Gerke, *Angew. Chem., Int. Ed. Engl.*, 1982, **21**, 206 ("Dodecasils"; stable inclusion of Ar, Kr, and Xe).
54. J. E. Mock, J. E. Myers and E. A. Trabant, *Ind. Eng. Chem.*, 1961, **53**, 1007.
55. Cf. R. W. Coutant, *J. Org. Chem.*, 1974, **39**, 1593 [methane, oxygen, and nitrogen clathrates of **1**].
56. (a) M. F. Pushlenkov and V. A. Ignatov, *J. Gen. Chem. U.S.S.R.*, 1974, **44**, 1. (b) Yu. N. Kazankin, A. A. Palladiev and A. M. Trofimov, *J. Gen. Chem. U.S.S.R.*, 1972, **42**, 2600, 2604 (Kr and Xe inclusion). (c) Yu. N. Kazankin, A. A. Palladiev and A. M. Trofimov, *Radiokhimiya*, 1972, **14**, 847, 900 (Ar, Kr, Xe inclusion); *Chem. Abs.*, 1973, **78**, 76317, 76317. (d) Yu. N. Kazankin, A. A. Palladiev and A. M. Trofimov, *U.S.S.R.*, 239270, 1974; *Chem. Abs.*, 1974, **81**, 123784. (e) A. A. Palladiev, Yu. N. Kazankin and K. A. Kurrinen, *U.S.S.R.*, 816955, 1981; *Chem. Abs.*, 1981, **95**, 100002.
57. M. Gomberg and L. H. Cone, *Justus Liebigs Ann. Chem.*, 1910, **376**, 236.
58. L. I. Kashtanov and N. V. Kazanskaya, *Trudy Moskov Inzhenerno-Ekonom. Inst.*, 1954, No. 1, 139; *Chem. Abs.*, 1956, **50**, 11270g.
59. H. M. Powell, British Patent 678312-3, 1952; *Chem. Abs.*, 1953, **47**, 3532a.
60. See also, e.g. F. Schumann, *Rep.-Staatl. Amt Atomsicherheit Strahlenschutz DDR 1977, SAAS-220, Wiss. Tech. Konf. RGW Bearbeitung Beseit. Radioakt. Abfalle Dekontam. Oberflachen, 4th*, 74–82; *Chem. Abs.*, 1978, **88**, 142360.
61. D. J. Chleck and C. A. Ziegler, *Nucleonics*, 1959, **17**(9), 130 and *Int. J. Appl. Radiat. Isot.*, 1959, **7**, 141.
62. J. Pruzinec and J. Tolgyessy, *Zb. Pr. Chemickotechnol. Fak. SVST 1973–1974 (Pub. 1977)*, 123–7; *Chem. Abs.*, 1979, **90**, 43291.
63. C. O. Hommel, F. J. Brousaides and R. L. Bersin, *Anal. Chem.*, 1962, **34**, 1608.
64. J. Pruzinec, J. Tolgyessy and M. M. Naoum, *Radiochem. Radioanal. Lett.*, 1974, **20**, 73 and *Chem. Abs.*, 1975, **82**, 148981.
65. (a) F. Schumann, *Zentralinst. Kernforsch., Rossendorf. Dresden, [Ber.]* 1977, *ZfK-340, Jahresbericht*, 25–7 and *Chem. Abs*, 1978, **89**, 203198. (b) H. Shimojima,

Y. Nakayama, K. Matsumoto and H. Hyodo, *Proc. Intern. Conf. Peaceful Uses At. Energy, (3rd), Geneva*, 1964, **14**, 314 and *Chem. Abs.*, 1966, **65**, 5209.

66. D. J. Chleck and C. A. Ziegler, U.S. Patent 3,084,062, 1963 and *Chem. Abs.*, 1963, **59**, 4659*h.*
67. J. W. Leonhardt, R. Goeldner and H. Schlegel, East German Patent 137,631 and *Chem. Abs.*, 1980, **92**, 101244.
68. B. E. Bryant, U.S. Patent 3,076,742, 1963 and *Chem. Abs.*, 1963, **59**, 5079*f.*
69. R. E. Cover, U.S. Patent 3,314,884, 1967 and *Chem. Abs.*, 1967, **67**, 13598.
70. (a) E. Terres and W. Vollmer, *Petroleum Z.*, 1935, **31**, 1 and *Chem. Abs.*, 1935, **29**, 8302. (b) See also, E. Terres and K. Thewalt, *Brennstoff-Chem.*, 1957, **38**, 257.
71. M. V. Stackelberg, A. Hoverath and Ch. Scheringer, *Z. Elektrochem.*, 1958, **62**, 123.
72. M. V. Stackelberg, *Recl. Trav. Chim. Pays-Bas*, 1956, **75**, 902.
73. B. A. Nikitin, M. P. Koval'skaya and M. F. Pushlenkov, *Izvest. Akad. Nauk. S.S.S.R., Odtel Khim. Nauk.*, 1951, 661 and *Chem. Abs.*, 1952, **46**, 5944*h.*
74. B. A. Nikitin and M. P. Koval'skaya, *Izvest. Akad. Nauk. S.S.S.R., Odtel. Khim. Nauk.*, 1952, 24 and *Chem. Abs.*, 1952, **46**, 6919*b.*
75. L. I. Kashtanov and L. N. Sokolova, *J. Gen. Chem. USSR*, 1951, **21**, 1619.
76. S. A. Allison and R. M. Barrer, *Trans. Faraday Soc.*, 1968, **64**, 549, 557.
77. P. H. Lahr and H. L. Williams, *J. Phys. Chem.*, 1959, **63**, 1432.
78. R. M. Barrer and V. H. Shanson, *J. Chem. Soc., Faraday Trans. 1*, 1976, **72**, 2348.
79. (a) Yu. N. Kazankin, F. I. Kazankina, A. A. Palladiev and A. M. Trofimov, *Dokl. Akad. Nauk. SSSR*, 1972, **205**, 1128 and *Chem. Abs.*, 1972, **77**, 172102. (b) Yu. N. Kazankin, F. I. Kazankina, A. A. Palladiev and A. M. Trofimov, *J. Gen. Chem. USSR*, 1973, **43**, 2650. (c) M. F. Pushlenkov and V. A. Ignatov, *J. Gen. Chem. USSR*, 1974, **44**, 2347. (d) Yu. N. Kazankin, F. I. Kazankina, A. A. Palladiev and A. M. Trofimov, U.S.S.R. Patent 411,062, 1974 and *Chem. Abs.*, 1974, **80**, 119701.
80. Yu. N. Kazankin, A. A. Palladiev and A. M. Trofimov, *J. Gen. Chem. USSR*, 1973, **43**, 2648.
81. Yu. N. Kazankin, A. A. Palladiev and A. M. Trofimov, *J. Gen. Chem. USSR*, 1972, **42**, 2363.
82. B. A. Nikitin and E. M. Ioffe, *Dokl. Akad. Nauk SSSR*, 1952, **85**, 809 and *Chem. Abs.*, 1953, **47**, 394.
83. (a) A. M. Trofimov and Yu. N. Kazankin, *Radiokhimiya*, 1965, **7**, 288 and *Chem. Abs.*, 1966, **64**, 2999. (b) A. M. Trofinmov and Yu. N. Kazankin, *Radiokhimiya*, 1966, **8**, 720 and *Chem. Abs.*, 1967, **66**, 61399. (c) A. M. Trofimov and Yu. N. Kazankin, *Radiokhimiya*, 1968, **10**, 445 and *Chem. Abs.*, 1968, **69**, 92527. (d) For studies on dimethyl and trimethyl phenols see also, E. Terres and K. Thewalt, *Brennstoff-Chem.*, 1957, **38**, 257 and *Chem. Abs.*, 1958, **52**, 1948.
84. (a) A. P. Dianin, *J. Russe. Phys. Chem. Soc.*, 1914, **46**, 1310. (b) For later syntheses see G. G. Kondrateva, *Metody Polnch. Khim. Reaktivov Prep.*, 1969, **20**, 199 and *Chem. Abs.*, 1972, **76**, 113017. (c) D. B. G. Jaquiss, Ger. Patent 2,335,854, 1974 and Chem. Abs., 1974, **81**, 26162.
85. W. Baker and J. F. W. McOmie, *Chem. Ind. (London)*, 1955, 256.
86. R. M. Barrer and V. H. Shanson, *J. Chem. Soc., Chem. Commun.*, 1976, 333.
87. W. Baker, A. J. Floyd, J. F. W. McOmie, G. Pope, A. S. Weaving and J. H. Wild, *J. Chem. Soc.*, 1956, 2010.
88. H. L. Hoffman, G. R. Breeden and R. W. Liggett, *J. Org. Chem.*, 1964, **29**, 3440.
89. L. Mandelcorn, N. N. Goldberg and R. E. Hoff, *J. Am. Chem. Soc.*, 1960, **82**, 3297.

90. L. Mandelcorn, R. W. Auxier and C. W. Lewis, U.S. Patent 2,949,424, 1960 and *Chem. Abs.*, 1961, **55**, 11364.
91. A. A. McConnell, D. D. MacNicol and A. L. Porte, *J. Chem. Soc. (A)*, 1971, 3516.
92. W. Baker, J. F. W. McOmie and A. S. Weaving, *J. Chem. Soc.*, 1956, 2018.
93. H. M. Powell and B. D. P. Wetters, *Chem. Ind. (London)*, 1955, 256.
94. D. D. MacNicol, *Chem. Comm.*, 1969, 836.
95. D. D. MacNicol, H. H. Mills and F. B. Wilson, *Chem. Comm.*, 1969, 1332.
96. H. H. Mills, D. D. MacNicol and F. B. Wilson, unpublished results.
97. J. L. Flippen, J. Karle and I. L. Karle, *J. Am. Chem. Soc.*, 1970, **92**, 3749.
98. J. L. Flippen and J. Karle, *J. Phys. Chem.*, 1971, **75**, 3566.
99. Previously (see ref. 124) an analogue of **3** with an additional OH group *meta* to the phenolic hydroxyl of **3** had been prepared and found to form adducts: the detailed nature of these complexes is, however, unknown.
100. D. D. MacNicol and F. B. Wilson, *Chem. Comm.*, 1971, 786.
101. A. D. U. Hardy and D. D. MacNicol, unpublished results, cited in ref. 117.
102. B. S. Middleditch and D. D. MacNicol, *Org. Mass Spectrom.*, 1976, **11**, 212.
103. A. D. U. Hardy, J. J. McKendrick and D. D. MacNicol, *J. Chem. Soc., Chem. Commun.*, 1974, 972.
104. A. D. U. Hardy, J. J. McKendrick and D. D. MacNicol, *J. Chem. Soc., Perkin Trans. 2*, 1979, 1072.
105. A. D. U. Hardy, J. J. McKendrick and D. D. MacNicol, *J. Chem. Soc., Perkin Trans. 2*, 1977, 1145.
106. D. D. MacNicol, A. D. U. Hardy and J. J. McKendrick, *Nature (London)*, 1975, **256**, 343.
107. A. A. Freer, D. D. MacNicol and A. Murphy, paper in preparation.
108. A. D. U. Hardy, J. J. McKendrick and D. D. MacNicol, *J. Chem. Soc., Chem. Commun.*, 1976, 355.
109. J. H. Gall, A. D. U. Hardy, J. J. McKendrick and D. D. MacNicol, *J. Chem. Soc., Perkin Trans. 2*, 1979, 376.
110. A. Collet and J. Jacques, *J. Chem. Soc., Chem. Commun.*, 1976, 708.
111. S. H. Wilen, *Topics in Stereochem.*, 1971, **6**, 128.
112. B. J. Brienne and J. Jacques, *Tetrahedron Lett.*, 1975, 2349.
113. A. Collet and J. Jacques, *Isr. J. Chem.*, 1976/77, **15**, 82.
114. A. D. U. Hardy and D. D. MacNicol, unpublished results.
115. See also, J. Jacques, A. Collet and S. H. Wilen, *Enantiomers, Racemates, and Resolutions*, Wiley, New York, 1981, p. 280.
116. A. D. U. Hardy, D. D. MacNicol, J. J. McKendrick and D. R. Wilson, *Tetrahedron Lett.*, 1975, 4711.
117. A. D. U. Hardy, J. J. McKendrick, D. D. MacNicol and D. R. Wilson, *J. Chem. Soc., Perkin Trans. 2*, 1979, 729.
118. M. S. Newman and H. A. Karnes, *J. Org. Chem.*, 1966, **31**, 3980.
119. M. R. Crampton, in *The Chemistry of the Thiol Group*, Part 1, (ed. S. Patai) Wiley, 1974, ch. 8.
120. L. C. Fetterly, in *Non-stoichiometric Compounds*, (ed. L. Mandelcorn) Academic Press, 1964, p. 503.
121. A. D. U. Hardy, D. D. MacNicol and D. R. Wilson, *J. Chem. Soc., Chem. Commun.*, 1974, 783.
122. C. J. Gilmore, A. D. U. Hardy, D. D. MacNicol and D. R. Wilson, *J. Chem. Soc., Perkin Trans. 2*, 1977, 1427.

123. (a) A. Goldup and G. W. Smith, *Sep. Sci. Technol.*, 1971, **6**, 791. (b) D. H. Desty, A. Goldup and D. G. Barnard-Smith, Brit. Patent 973 306, 1964 and *Chem. Abs.*, 1965, **62**, 2655. (c) cf. K. J. Harrington and C. P. Garland, *Sep. Sci. Technol.*, 1982, **17**, 1339 [sorption by 3].
124. C. K. Johnson, Fr. Patent 1,530,511, 1968 and *Chem. Abs.*, 1969, **71**, 13 717.
125. J. E. Kropp, M. G. Allen and G. W. B. Warren, Ger. Patent 2,012,103, 1970; *Chem. Abs.*, 1971, **74**, 43074.
126. W. R. Lawton, Belg. Patent 632,833, 1963 and *Chem. Abs.*, 1964, **61**, 3851*f.*
127. R. J. Cross, J. J. McKendrick and D. D. MacNicol, *Nature (London)*, 1973, **245**, 146.
128. R. Y. Wong, K. J. Palmer, G. D. Manners and L. Jurd, *Acta Crystallogr.*, 1976, **B32**, 2396.
129. W. Baker, R. F. Curtis and M. G. Edwards, *J. Chem. Soc.*, 1951, 83.
130. (a) W. Smith and L. D. Kispert, *J. Chem. Soc., Faraday Trans. 2*, 1977, 152. (b) See also F. Ohzeki, L. D. Kispert, C. Arroyo and M. Steffan, *J. Phys. Chem.*, 1982, **86**, 4011 (related ENDOR study).
131. D. D. MacNicol and P. R. Mallinson, *J. Incl. Phenom.*, 1983, **1**, 169.
132. W. Baker, R. F. Curtis and J. F. W. McOmie, *J. Chem. Soc.*, 1952, 1774.
133. W. Baker, D. F. Downing, A. E. Hewitt-Symonds and J. F. W. McOmie, *J. Chem. Soc.*, 1952, 3796.
134. W. Baker, J. F. W. McOmie and J. H. Wild, *J. Chem. Soc.*, 1957, 3060.
135. K. Yamada and N. Sugiyama, *Bull. Chem. Soc. Jpn.*, 1965, **38**, 2057; 2061.
136. M. P. V. Boarland, J. F. W. McOmie and R. N. Timms, *J. Chem. Soc.*, 1952, 4691.
137. T. Ohta and S. Togano, Japan Patent, 75,131,533, 1975 and *Chem. Abs.*, 1976, **84**, 114208.
138. W. M. Bandaranayake, M. J. Begley, B. O. Brown, D. G. Clarke, L. Crombie and D. A. Whiting, *J. Chem. Soc., Perkin Trans. 1*, 1974, 998.
139. F. M. Dean, H. Khan, N. Minhaj, S. Prakash and A. Zaman, *J. Chem. Soc., Chem. Commun.*, 1980, 283.
140. (a) L. P. Zalukaev and L. G. Barsukova, *J. Gen. Chem. USSR*, 1972, **42**, 606. (b) L. P. Zalukaev and L. G. Barsukova, *Vysokomol. Soedineniya, Ser A*, 1973, **15**, 2185 and *Chem. Abs.*, 1974, **81**, 14490.
141. L. Kozerski, *J. Mol. Struct.*, 1981, **75**, 95.
142. G. B. Barlow and A. C. Clamp, *J. Chem. Soc.*, 1961, 393.
143. J. E. Driver and S. F. Mok, *J. Chem. Soc.*, 1955, 3914.
144. J. E. Driver and T. F. Lai, *J. Chem. Soc.*, 1958, 3219.
145. R. F. Hunter, R. A. Morton and A. T. Carpenter, *J. Chem. Soc.*, 1950, 441.
146. C. D. Gutsche, B. Dhawan, K. H. No and R. Muthukrishnan, *J. Am. Chem. Soc.*, 1981, **103**, 3782; and references therein.
147. M. Coruzzi, G. D. Andreetti, V. Bocchi, A. Pochini and R. Ungaro, *J. Chem. Soc., Perkin Trans. 2*, 1982, 1133; and references therein.
148. G. D. Andreetti, R. Ungaro and A. Pochini, *J. Chem. Soc., Chem. Commun.*, 1979, 1005.
149. Unpublished results, cited in ref. 147.
150. (a) G. D. Andreetti, A. Pochini and R. Ungaro, *J. Chem. Soc., Perkin Trans. 2*, 1983, 1773. (b) G. D. Andreetti, A. Mangia, A. Pochini and R. Ungaro, 2nd International Symposium on Clathrate Compounds and Molecular Inclusion Phenomena, Parma, 1982, Abstract Book, p. 42.

2 · INCLUSION COMPOUNDS OF UREA, THIOUREA AND SELENOUREA

K. TAKEMOTO and N. SONODA
Osaka University, Osaka, Japan

Since the accidental discovery of the urea adducts about forty years ago, the chemistry of urea and thiourea inclusion compounds has received much attention and is still a subject of continuing interest.[1-3] This chapter is concerned with the formation, structure and properties of a variety of inclusion compounds of urea and thiourea. In addition to these inclusion compounds both urea and thiourea form a number of binary hydrogen bonded complexes, but these will not be discussed in this chapter. Inclusion compounds of selenourea have not been thoroughly studied as yet, and all the published data are reviewed in this chapter.

1. Inclusion compounds of urea

In 1940, Bengen found by chance that urea can form a crystalline adduct with octyl alcohol. In successive publications he showed that a variety of

INCLUSION COMPOUNDS 2
ISBN 0-12-067101-8

linear organic compounds with six or more carbon atoms form such adducts readily with about six molecules of urea.[4,5] A small amount of solvent, such as water or alcohols was assumed to act as a catalyst for the adduct formation. It was also pointed out that the adducts thus formed decomposed into their constituents either by heating or by redissolving in a solvent. Although the nature and the structure of the adducts were not known at that time, particular attention seemed to have been paid to the peculiar feature that the adducts could only be formed with *n*-alkanes and their linear derivatives, but not with branched alkanes. Comprehensive studies were then carried out by Schlenk[6] and by Zimmerschied *et al.*[7], who showed that the adducts belonged to the newly discovered class of host-guest or inclusion compounds.

1.1. Formation of urea inclusion compounds

The crystalline inclusion compounds of urea can be prepared in a very simple way: for example, if a saturated methanol solution of urea is treated with alkanes such as octane and decane, hexagonal crystals are immediately formed. The alternative preparative method consists of dissolving urea in the organic compound to be included by heating, and then cooling to give crystals of the inclusion compounds. In some cases, the inclusion compounds can be formed by pouring the organic compound onto urea in the solid state.

It has been found that urea will form inclusion compounds with an extremely diverse range of organic compounds. Some examples are given in the following paragraphs.

Inclusion compounds are readily formed with *n*-alkanes and *n*-alkenes provided that the hydrocarbon has *six* or more carbon atoms. This lower limit is demonstrated by the fact that whilst an inclusion compound is formed with *n*-hexane, no such compound is formed with *n*-pentane.[7] This observation defines the lower limit of guest molecule size for the formation of stable inclusion compounds.

Some degree of substitution of the carbon chain is allowed: e.g. 1-bromohexane forms an inclusion compound whilst 2-bromooctane does not. On the other hand 2,2-difluorooctane does form an inclusion compound, probably because of the smaller size of the fluorine atoms. In the case of oxygen derivatives of *n*-alkanes, neither the hydroxyl group in 2-octanol, nor the carbonyl group in 2-heptanone prevents inclusion compound formation.

The effect of substitution is clearly seen with a series of C_{14} compounds where 1-cyclopentylnonane and 2-, 3-, and 4-methyltridecane form inclusion compounds, but 2,4-dimethyldodecane, 3-ethyldodecane, and 1-cyclohexyloctane do not form inclusion compounds.[8] Competitive inclusion has also been studied between tetradecane and its isomers in binary mixtures.[9] The more highly branched compounds form an insignificant amount (0.5–2%) of inclusion compound even when present at $\geqslant 75\%$ content in the binary mixture. The less highly branched hydrocarbons give about 1% inclusion compound at $\leqslant 25\%$ content and 8.5 to 10% inclusion compound at $\leqslant 90\%$ content in the binary mixture.

Among the long chain esters of sufficient length, derivatives of lightly branched alkanes can form inclusion compounds e.g. esters in which either the acid or the alcohol portion of the molecule bears a methyl group. However, if both the acid and the alcohol portions bear methyl substituents, no inclusion compound formation can be observed. Larger groups such as ethyl and phenyl groups also prevent inclusion compound formation.

Urea forms inclusion compounds not only with hydrocarbons and their alcohol, ester and ether derivatives, but also with aldehydes, ketones, carboxylic acids, dicarboxylic acids, amines, nitriles, thioalcohols and thioethers, provided that their main chain consists of six or more carbon atoms. Thus, inclusion compounds of urea with methyl octadecenoates, octadecadienoates and octadecatrienoates, with methyl stearate, stearic acid, and with the three types of ethylenic C_{18} acids can be prepared. The degree of unsaturation or the isomerism of the double bonds does not seem to affect the ability of the guest molecule to form inclusion compounds.[10]

Urea also forms inclusion compounds with a number of butene and butadiene derivatives, which on γ-irradiation can form stereo-regular polymers,[11] (see Chapter 10, Volume 3).

Aromatic compounds would be expected from their size and shape to be capable of forming inclusion compounds provided that they are not highly substituted, but benzene and its smaller homologues never form such inclusion compounds. If however the benzene carries a long chain substituent, e.g. octadecylbenzene, then an inclusion compound will be formed.

Some organic compounds such as sulphur compounds and peroxides are known to inhibit inclusion compound formation. For example, *n*-hexadecane containing 0.25% added inhibitor does not form an inclusion compound until 0.4% methanol is also added to the system.[7]

Although only a little work seems to have been done on the decomposition of the inclusion compounds, it is evident that the decomposition processes depend on the nature of the guest molecule. The inclusion compounds of volatile compounds tend to decompose slowly, even at room temperature, whereas that of cetane, for example, is stable at room temperature and only begins to decompose just near its melting point.[6]

1.2. Composition of urea inclusion compounds

The composition of urea inclusion compounds, i.e. the number of urea molecules per guest can be determined using several techniques, including a calorimetric method.[7] In general, however, there seems to be no simple molecular ratio of urea to the number of carbon atoms of the guest compound to be included. This fact indicates quite clearly that no classical coordination theory can be applied and points to their nature as inclusion compounds. The relationship of the number of carbon atoms of guest molecule with the moles of urea required to form inclusion compounds is shown in Fig. 1 for paraffins, and in Fig. 2 for carboxylic acids. Both figures show that the number of moles of urea required for the formation of the

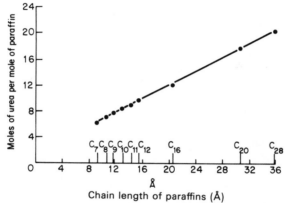

Fig. 1. Urea–paraffin inclusion compounds: relationship of composition with chain length. Reproduced with permission from *Justus Liebigs Ann. Chem.*, 1949, **565**, 204.

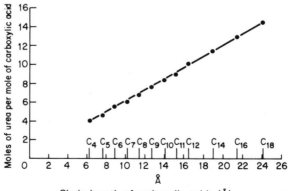

Fig. 2. Urea–carboxylic acid inclusion compounds: relationship of composition with chain length. Reproduced with permission from *Justus Liebigs Ann. Chem.*, 1949, **565**, 204.

inclusion compound is directly proportional to the chain length of the guest molecule.

A detailed study has been made of the chemical composition and packing coefficient of the urea inclusion compounds with *n*-pentane and *n*-hexane.[12] The urea-paraffin ratio and the temperature, heat and entropy of decomposition of the inclusion compounds are found to increase with the paraffin chain length for C_5 to C_{28} paraffins.[13]

When urea-octane inclusion compounds containing different amounts of octane are mixed, the octane molecule is reported to be redistributed from the inclusion compound of higher concentration to the one of lower concentration.[14] It is not however clear from the abstract whether or not this is a solid state reaction.

1.3. Structure and properties of urea inclusion compounds

The urea inclusion compounds generally crystallize in long, hexagonal prisms or occasionally as hexagonal plates (for example, with cetyl alcohol). The crystals of the urea inclusion compounds can easily be distinguished from the tetragonal prisms of urea (Fig. 3).

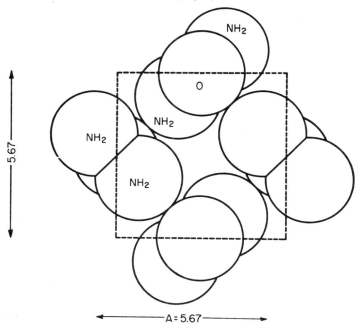

Fig. 3. The tetragonal urea structure. Reproduced with permission from *Acta Crystallogr.*, 1952, **5**, 224.

The crystalline structure of the urea-linear hydrocarbon inclusion compounds has been determined by X-ray diffraction.[6,15] As Fig. 4 schematically illustrates, the unit cell is hexagonal and the structure has six urea molecules per unit cell. The general features of the structure and the nitrogen positions of the urea are obtained directly from an implication diagram or Patterson–Harker sections. It becomes clear from the analysis that the urea molecules form a hollow channel structure in which the n-hydrocarbon molecules are enclosed, and that the hydrocarbons are in an extended planar zigzag configuration with their long axis parallel to the c axis.[15]

Fig. 4. Cross section of the urea–normal hydrocarbon inclusion compound (modified picture by M. Hagan[2] from the original: A. E. Smith, *Acta Crystallogr.*, 1952, **5**, 224).

The above structure accounts for the selectivities mentioned in Section 1.1. The channels have a diameter of about 5.25 Å and this is compared in Fig. 5 with the cross sections of n-octane, benzene, 3-methylheptane and 2,2,4-trimethylpentane. It can be seen that whilst 3-methylheptane can be accommodated in the channel, 2,2,4-trimethylpentane cannot be accommo-

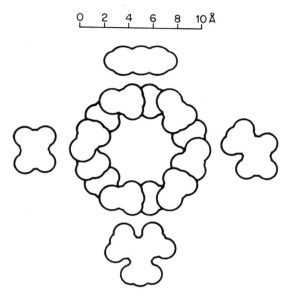

Fig. 5. Cross section of the cavity in the urea channel compared with the size of *n*-octane (left), benzene (top), 3-methylheptane (right) and 2,2,4-trimethylpentane (bottom). Reproduced with permission from *Justus Liebigs Ann. Chem.*, 1949, **565**, 204.

dated. The determining factor as to whether or not a molecule can act as a guest molecule is thus its ability to fit into the space available in the urea channel.

An X-ray study has also been made on the tetragonal crystalline structure of urea, and the interatomic distances have been determined by Fourier and least-squares analysis.[16]

X-ray powder patterns and lattice constants have been reported for the inclusion compounds with alkenoic acids and their esters,[17] and with the guest hydrocarbons 2-methyldodecane, 2-methylpentadecane, 2-methylheptadecane, and 2-methyleicosane.[18] Inclusion compounds with *n*-alkanones have been prepared and studied by X-ray diffraction to investigate the possibility of separating and isolating ketones on the basis of their abilities to form inclusion compounds.[19]

It has been proposed that the determination of the length of the included molecule can be used to differentiate between *cis* and *trans* isomers of ethylenic inclusion compounds.[20] This is possible since a *cis* double bond shortens a molecule 0.9 Å, while a *trans* double bond shortens it 0.15 Å compared to the length of the saturated molecule.

The preparation of several uncharacterized inclusion compounds of liquid paraffin mixtures and hexadecane have been reported.[21] The crystals are said to be hexagonal or tetragonal depending on the preparative conditions.

It has also been reported that the hexagonal crystals of several urea/n-alkane inclusion compounds transform into orthorhombic crystals at low temperature.[22]

The urea-vinyl chloride inclusion compound crystallizes as needles from an isopropanol solution of the components at 263 K. The crystal structure has been determined, and it is found that the channel structure of the urea molecules is rhombohedral. The crystal lattice disintegrates during the polymerization of the guest, and is transformed into polycrystalline tetragonal urea and polyvinyl chloride.[23]

Since n-alkanes are present in the urea host lattice in an extended *planar zig-zag arrangement* of the carbon skeleton, these inclusion compounds present an opportunity of studying the alkane guest in this configuration. Infrared[24] and Raman[25] studies have been reported and assignments made for the vibrations of the host lattice and of the guest molecules.

A recent study[73] reported the preparation of microcrystals of n-alkane inclusion compounds when finely powdered urea was ultrasonically agitated in the liquid alkanes. Raman studies of these microcrystals were interpreted on the basis that the host lattice consisted of a hexagonal polymorph capable of including the n-alkane guest in a *non-planar conformation* with alternating *trans* and *gauche* bonds. It should be pointed out that these deductions were made solely on the basis of a Raman study, and since the Raman spectra of these microcrystals bear some resemblance to the Raman spectrum of tetragonal urea, further work needs to be done to preclude the possibility that the products are simply tetragonal urea containing very small amounts of trapped n-alkane.

There are also marked temperature dependent variations in the band width of the guest asymmetric CH_3 stretching mode band and this has been ascribed to rotation of the methyl groups.[26] A study of the temperature dependence of the spectrum of the n-heptylazide guest molecule also suggests a rotation of the azide group with respect to the rest of the guest molecule.[27]

NQR,[28] NMR[29] and differential thermal analysis have been used to study the phase transitions and guest molecule motions in urea inclusion compounds. Inclusion compounds of urea have also been used in gas chromatography.[30] All of these topics are discussed in greater detail in Chapters 1 and 6 in Volume 3.

The formation of trapped electrons in urea–n-octane and urea–1-octene inclusion compounds was first reported by Nowak and Stachowicz.[31] Single crystals of the urea–6-undecanol and urea–1-decanol inclusion compounds were irradiated by X-rays at 77 K, and the free radicals generated examined by using ESR spectroscopy. In this case, the long-lived free radicals are those formed by removal of a hydrogen atom from the carbon attached to the hydroxyl group.[32] Further ESR study reveals the formation of trapped electrons in γ-irradiated crystals of a variety of urea inclusion compounds.[33] Recently, the temperature dependent ESR spectra of peroxy polyethylene radicals in urea–polyethylene complex have been simulated by modified Bloch equations.[34] The properties of the urea–polyethylene inclusion compound have also been determined by electron microscope observations and thermal analysis.[35]

γ-Irradiation of the inclusion compounds of urea with butyl acetate and nonanoate, and with decyl pentanoate, has been studied at 77 K, and a mechanism for the formation of the paramagnetic centres has been proposed.[36] γ-Irradiation of urea inclusion compounds with hydrocarbons generates alkyl radicals and captured electrons. In the case of the urea 1-heptene inclusion compound, cation-radicals are also formed, which are then transformed into alkyl free radicals.[37] Further examples of free radicals which can be trapped in host lattices can be found in Chapter 2, Volume 3.

The hexagonal lattice of urea inclusion compounds occurs in two mirror image forms. This asymmetry can be used for resolving racemates by crystallizing part of a racemic mixture as guest in the urea host lattice.[38] Schlenk has given numerous examples which prove that homologs having the same configuration have the same lattice structure. Correlations between homologous α-methyl, α-chloro-, α-bromo, α-amino, and α-mercaptocarboxylic acids, and between β-methyl- and β-chlorocarboxylic acids were verified and established. It was found that in the series of L-β-chlorocarboxylic acids, the optical rotation is (+) up to L-β-chlorovaleric but (−) from L-β-chlorocaproic acid onwards.[39] In a subsequent study the inclusion lattice configuration of asymmetric guest compounds were examined.[40] Schlenk also studied the lattice configuration of six 1,2- and three 1,3-dimethylalkanes and of various groups of methylalkanones, and verified that the lattice of the optically levorotatory inclusion crystal had a right-handed spiral configuration, i.e. had the right-handed screw axis as a symmetry element.[41]

Further examples of the enantiomeric selectivity of host lattices can be found in Chapter 9, Volume 3.

1.4. Heat of inclusion compound formation

Considerable heat evolution is observed during inclusion compound formation, the magnitude of which provides information about the physiochemical nature of the formation. This heat of formation has been measured by Zimmerschied and coworkers by a calorimetric method for several classes of linear organic compounds in the presence of an excess of urea.[7] As can be seen from Fig. 6, the heats of formation are closely proportional to the length of the carbon main chain of the compounds to be included. Extrapolation of the *n*-alkane curve in this figure affords a zero heat of reaction near two carbon atoms, which suggests that the terminal carbon atoms are not involved in the inclusion compound formation reaction, and that each methylene group is associated in some manner with a definite amount of urea.

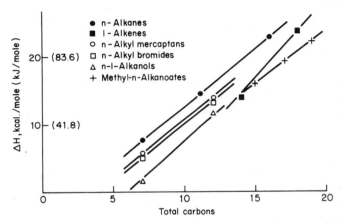

Fig. 6. The heat of formation of urea inclusion compounds. Reproduced with permission from *Ind. Eng. Chem.*, 1950, **42**, 1300.

From the curve of urea–*n*-alkane inclusion compounds in Fig. 6, the heat of inclusion compound formation for each carbon atom was estimated to be 6.7 kJ mol^{-1}. By comparing this value with a variety of heats of physical transformations, the heat of formation per carbon atom is found to be far greater than those of the solid state crystal transitions of hydrocarbons. Urea inclusion compound formation appears to resemble adsorption of hydrocarbons on solid surfaces more closely than any other familiar physical process.

The heat of formation has been measured in detail for a number of urea inclusion compounds (Table 1),[6] from which it is possible to calculate the increment for each additional carbon atom of the guest molecule. The distance between two chain ends of the guest molecules included in a urea channel is found to be constant at about 2.4 Å, and Schlenk was able to explain why smaller paraffin molecules cannot form inclusion compounds.

Table 1. *Heat of formation of urea inclusion compounds*

Organic guest	kJ mol^{-1} urea	kJ mol^{-1} organic guest
n-octane	4.22	29.9
n-decane	4.60	38.1
n-hexadecane	5.18	62.3
methylethylketone	4.48	17.9
diethylketone	4.90	23.0
dipropylketone	5.14	30.8
n-octanol	3.38	22.6
n-butyric acid	5.71	22.7
methyl *n*-butyrate	4.32	23.3

The heats of formation of inclusion compounds of 2-methyl paraffins, ranging from C_{13} to C_{21}, were measured later by using a twin-type conduction calorimeter. The values on nine examples of 2-methyl paraffin adducts are in the range of 5.22 to 7.31 kJ mol^{-1} of urea and 54.7 to 114.1 kJ mol^{-1} of 2-methyl paraffin.[42] Heat capacities and entropies of the inclusion compounds of 1-alkenes was also measured in the temperature range from 12 to 300 K.[43]

2. Inclusion compounds of thiourea

In 1947, Angla reported that thiourea forms crystalline complexes with a variety of organic compounds at room temperature, when they are mixed either with solid thiourea or its saturated aqueous or alcoholic solutions.[44] Schlenk also reported similar results.[45] It was demonstrated that these thiourea complexes were similar to the urea complexes and were in fact inclusion compounds. The thiourea host lattice however has a channel diameter of approximately 6.1 Å compared with the corresponding value of 5.25 Å for the urea host lattice. Thiourea will consequently accommodate larger guest molecules than urea for example, not only highly branched hydrocarbons, ketones, esters and halides, but also 5-, 6- and 8-membered ring compounds and condensed ring systems.

2.1. Formation of thiourea inclusion compounds

The thiourea inclusion compounds are prepared in a similar way to the urea inclusion compounds. They can be prepared by direct mixing of thiourea and the guest molecule, and by dissolving and crystallizing from methanol solution. Some organic compounds act as inhibitors to prevent the formation of the inclusion compounds. For example, ethanol, water and hexadecane are known to lower the yield of the thiourea–cyclohexane inclusion compound.[46]

As mentioned above, thiourea forms channels of larger diameter than those of urea, so that a number of highly branched hydrocarbons, for example, trimethylpentane and larger cyclic compounds such as cyclopentane, cyclohexane and decalin can be included in the thiourea channels. One of the largest compounds found to form a thiourea inclusion compound is 2,6,9,13,16-pentamethylheptadecane which is approximately 22 Å long, while the smallest one is chloroform. On the other hand, linear *n*-paraffins cannot form inclusion compounds with thiourea because of their loose fit in the large thiourea channels. In the case of highly branched linear paraffins, the length of the stretched molecules gives a linear relationship to the moles of thiourea necessary for forming the inclusion compounds[45] (Fig. 7). Thiourea forms inclusion compounds smoothly with a series of ω,ω'-dicyclohexyl paraffins. In this case, the relationship is not linear, but a zigzag line (Fig. 8), which relates exactly to the fitting of the size and shape of the guest molecules in the thiourea channel.

Among aliphatic compounds, those having at least two methyl side groups, or other groups of similar size and shape, (e.g. halogen atoms), form inclusion compounds: 2-bromooctane is an example. The introduction of more side groups on the paraffin chain causes the formation of more stable

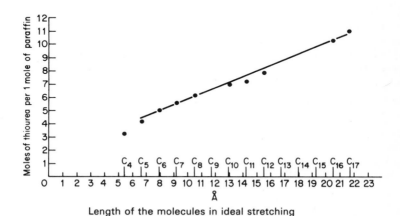

Fig. 7. Thiourea adducts of highly branched paraffins; dependence of molar ratios on chain lengths. Reproduced with permission from *Justus Liebigs Ann. Chem.*, 1951, **573**, 142.

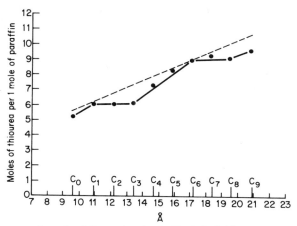

Length of the molecules in ideal stretching

Fig. 8. Thiourea adducts of ω,ω'-dicyclohexyl paraffins; dependence of molar ratios on chain lengths. Reproduced with permission from *Justus Liebigs Ann. Chem.*, 1951, **573**, 142.

thiourea inclusion compounds. Longer side groups, such as ethyl or propyl, seem to prevent the formation of inclusion compounds. The stability of thiourea inclusion compounds generally increases as the length of the guest molecule increases, similar to the situation with urea inclusion compounds. The relative position of the side groups has been found to affect the stability of the inclusion compounds formed: 2,3-, 2,5-, 2,6-, 2,7-, 2,8-, 2,9- and 2,10-dimethyl paraffins are found to form inclusion compounds, whereas the corresponding 2,4-dimethyl compounds do not. If, however, a third methyl group is introduced in any position on the carbon chain, then the trimethyl compounds can act as guest molecules.

Saturated cyclic compounds form inclusion compounds much more readily than unsaturated and aromatic compounds. Thus, cycloparaffins ranging from cyclopentane to cyclooctane form inclusion compounds with thiourea. In general, heterocyclic compounds show no ability to form inclusion compounds with thiourea. It seems that in these cases, the size and shape of the molecule and also the degree of saturation is important. Thus the non-aromatic heterocycles 2,2,4-tetramethyltetrahydrofuran and 4,4-dimethyl-1,3-dioxane form inclusion compounds with thiourea.

Among aromatic compounds, benzene and its simple homologues form no inclusion compounds with thiourea. However, Dyadin and coworkers have reported recently that benzene can be included in thiourea in the presence of acetic acid at 0° C and 20° C, and that the molar ratio of benzene to thiourea is 1:2.4. This inclusion compound is stable at 59.5 to 88 weight % benzene in the equilibrium liquid phase.[47] Introduction of branched alkyl

groups and cycloparaffin residues onto the aromatic ring generally leads to
the formation of inclusion compounds.

Condensed cyclic compounds cannot form inclusion compounds unless
they are saturated. e.g. Naphthalene and anthracene do not form inclusion
compounds with thiourea, but tetrahydronaphthalene does. With a series
of naphthalenes, the introduction of only one methyl group is enough to
give them the ability of acting as guest molecules in thiourea; 1-methyl-,
2-methyl-, 2,3-dimethyl-, and 1,6-dimethylnaphthalenes are known to form
inclusion compounds with thiourea. Recently, the dispiro compound 1, and
the tricyclotridecenones (2; R = H, and Br) have been found to form
inclusion compounds.[48] Alkyl- and alkenyl substituted norbornane and
norbornene derivatives also form thiourea inclusion compounds.[49]

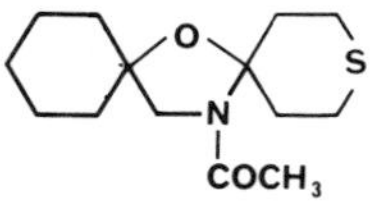

(1) 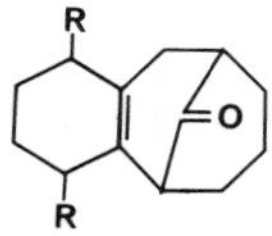   (2)

It is interesting to note that some of the guest molecules which can form
inclusion compounds readily with thiourea, can show a co-inclusion
phenomenon and introduce other smaller molecules, which on their own
cannot form inclusion compounds with thiourea because of their loose fit
in the channels.[45] The strength of the co-inclusion effect of some guest
molecules is as follows:

> diisobutylene > isooctane > triisobutylene > 2,2,3-trimethyl-
> butane > decahydronaphthalene > cyclohexane > dicyclohexyl

And the compounds, which can be introduced by the above guest molecules
into the thiourea channels, are in the following order:

> 1,3-dimethylcyclohexane > dipentene > benzene >
> 3-methylheptane > heptane > decane > 1-octene > toluene,
> m-xylene > o-xylene, p-xylene.

This co-inclusion effect has also been observed with urea inclusion
compounds.

2.2. Structure and properties of thiourea inclusion compounds

Thiourea inclusion compounds are in general rhombohedral crystals. The
arrangement of thiourea molecules in the crystal is similar to that of urea

molecules in the urea inclusion lattice. Figure 9 compares the lattices of urea and thiourea inclusion compounds.[45]

The structure of thiourea inclusion compounds containing chloroform and other chlorine substituted hydrocarbons have been studied in detail. An inclusion compound with the composition of chloroform and thiourea in a 1:2.25 molar ratio dissolves incongruently in acetic acid. A guest molecule packing in the channels according to the principle of maximum close packing was suggested. This model is in agreement with some previous

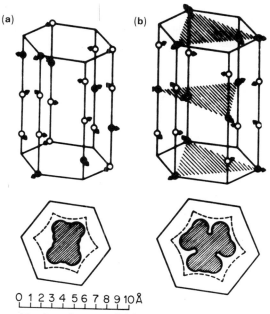

0 1 2 3 4 5 6 7 8 9 10Å

Fig. 9. Fundamental lattice of urea and thiourea inclusion compounds. Reproduced with permission from *Justus Liebigs Ann. Chem.*, 1951, **573**, 142.

studies of thiourea inclusion compounds with chloroform, carbon tetra-chloride, tetrachloroethylene, pentachloroethane and hexachloroethane.[50]

NMR, NQR and differential thermal analysis have been used to study the phase transitions and guest molecule motions in thiourea inclusion compounds.[51] These techniques are discussed in greater detail in Chapter 1, Volume 3.

The infrared[52,53] and Raman[54] spectra of the inclusion compounds with monohalocyclohexanes have been studied in detail since the thiourea host lattice displays an interesting selectivity for the axial conformer. In the gaseous and liquid phases the equatorial conformer is the more abundant, whilst the solid phase consists of the equatorial conformer alone. The thiourea inclusion compounds thus provide the only means of obtaining the monohalocyclohexanes in a configuration whereby the axial conformer is the more abundant.

The configuration of *trans*-2,3-dichloro-1,4-dioxan as a guest molecule also differs from the liquid and crystal phase conformation.[55] The liquid and crystal phases have a structure containing diaxial C–Cl bonds with the ring in a deformed chair conformation, whilst in the guest molecule the C–Cl bonds become antiparallel with the ring in an ideal chair conformation.

An ESR study has been reported on some γ-ray-irradiated alkyl- and alkenyl substituted norbornane and norbornene derivatives as guest molecules in the thiourea host lattice.[49] The irradiation leads to the formation of norbornyl free radicals by the loss of hydrogen atoms at the substituent sites or by partial hydrogenation of the double bonds. A reaction model based upon geometrical control by the molecular packing within the channels has been proposed.

Molecules as large as ferrocene can form inclusion compounds with thiourea. X-ray crystallographic studies at 295 and 100 K have been carried out on ferrocene included in the thiourea lattice. The rhombohedral crystals of the high temperature form consist of thiourea molecules forming a honeycomb of channels by spiralling with a pitch of 120° parallel to the *c* axis. Within these channels, sites of point symmetry 32 are occupied by the ferrocene iron atoms. The cyclopentadienyl rings are disordered and the time-averaged picture shows regions of three-dimensionally delocalized cyclopentadienyl electron density around the iron atoms.[56]

Anisotropic reorientation of ferrocene molecules in the channels of the 3:1 thiourea-ferrocene inclusion lattice produces an unusual temperature-dependent relaxation of the electric field gradient tensor. In this study, theoretical equations appropriate to anisotropic relaxation were developed.[57]

Nickelocene and other metallocenes are found to form a similar type of inclusion compound with thiourea. The pure ferrocene inclusion compound has thiourea and ferrocene in 3:1 molar ratio and shows a well-defined first-order, reversible phase transition at 162 K.[58] The technique of inclusion compound formation can be used to separate ferrocene from 1,1'-diethylferrocene, and also cyclopentadienyltricarbonylmanganese from acetylcyclopentadienyltricarbonylmanganese.[59]

Some reactions of the guest molecule in the thiourea channels have been reported. The hydrogen abstraction of a photostimulated allylic radical from its next neighbour molecules, and the thermal recovery of the allylic radical from photoirradiation in a thiourea inclusion compound have been studied, using saturated and unsaturated hydrocarbons as the guest molecules.[60]

Butadiene and its derivatives are easily included in thiourea channels. Thus the γ-ray induced polymerization of isobutene and the copolymerization of isobutene and vinylidene chloride have been studied.[61] Chatani and coworkers have carried out structural studies of the channel polymerization of 2,3-dichlorobutadiene in the thiourea inclusion compound, and the crystal structure of the polymer obtained has been analysed.[62,63] The structural change during polymerization has also been investigated, and they found that under γ-ray or X-ray irradiation the monomers combine with each other in the thiourea channels without destroying the channel structure and that the polymer chain in the channel has the same conformation as in the polymer crystal. The length of channel occupied by one monomer is 6.25 Å in the monomer-thiourea complex, which is shortened to 4.80 Å in the polymer-thiourea complex. A later paper is concerned with detailed structural evidence for the specificity of the thiourea channel polymerization of two 2,3-disubstituted 1,3-butadienes.[64] Chapter 10, Volume 3 deals fully with the topic of inclusion polymerization.

3. Industrial applications of urea and thiourea inclusion compounds

Since the urea and thiourea host lattices display a selectivity which is essentially governed by the size of the available channels, much work has been carried out to exploit this selectivity in separating the components of mixtures on an industrial scale. Urea has been used to separate *n*-alkanes from branched or cyclic compounds [65] whilst thiourea has been used to separate benzene and cyclohexane from *n*-heptane.[66] These two examples are well documented in the quoted references and no further details will be given in this chapter.

4. Inclusion compounds of selenourea

The preparation of selenourea is most easily accomplished by adding hydrogen selenide to a solution of cyanamide. This

$$H_2N-C\equiv N + H_2Se \;\rightarrow\; H_2N-\underset{\underset{Se}{\parallel}}{C}-NH_2$$

method was first reported by Verneuil in 1884,[67] but has since been modified.[68] In particular Hope reported a high yield method employing a 50% aqueous solution of cyanamide.[69]

Selenourea is a colourless crystalline solid, but the reported melting point varies due to decomposition. Hope reports that selenourea has a melting point around 235° C, and can be kept in a refrigerator with no apparent changes other than a moderate darkening.[69]

van Bekkum and coworkers first showed that selenourea, like urea and thiourea, can form inclusion compounds with a variety of hydrocarbons.[70] The guest molecules used were 4-tert-butyl-1-neo-pentylbenzene, 1,4-di-neo-pentylbenzene, trans-1,4-di-iso-propylcyclohexane, trans-1-tert-butyl-4-iso-propylcyclohexane, trans-1,4-di-tert-butylcyclohexane, cis-1,4-di-tert-butylcyclohexane, trans-1-tert-butyl-4-neo-pentylcyclohexane, cis-1-bromo-4-tert-butylcyclohexane, bicyclo[3.3.1]nonane, adamantane and camphor. All of these compounds can also be included in the thiourea channel. Though the difference in channel diameter between thiourea and selenourea inclusion compounds is small, selenourea seems to be much more selective in its choice of guest molecule. For example, trans-1-tert-butyl-4-neo-pentylcyclohexane forms an inclusion compound with selenourea, whereas the cis-isomer is not included. Thiourea shows no observable preference in this case.

Table 2 shows the densities of urea, thiourea and selenourea and their inclusion compounds. An X-ray crystallographic study was also made of the rhombohedral selenourea inclusion compounds. It is found that there are more striking differences in the unit cell dimensions of the selenourea inclusion compounds than are observed for the thiourea inclusion compounds. Adaptation of the selenourea lattice to the shape and size of the included molecules is apparently relatively easy. A reasonably linear relation also exists between the number of carbon atoms per channel period and the a axis.[70] The hexagonal lattice of selenourea has unit cell parameters[71]

Table 2. The densities of urea, thiourea and selenourea and their inclusion compounds. Data taken from van Bekkum et al.[70]

	Urea derivative $(g\ cm^{-3})$	Inclusion compounds $(g\ cm^{-3})$
Urea	1.30	1.20
Thiourea	1.40	1.10
Selenourea	2.08	1.60 ~ 1.65

of $a = 15.37$ Å, $c = 13.08$ Å and $Z = 27$ whereas in the rhombohedral inclusion compounds[70] a varies between 16.21 and 16.71 Å and c varies between 12.81 and 12.97 Å. Thiourea and selenourea have also been reported to have very similar electron diffraction patterns.[72]

References

1. F. Cramer, *Einschlußverbindungen*, Springer Verlag, Berlin, Göttingen, Heidelberg, 1954.
2. M. Hagan, *Clathrate Inclusion Compounds*, Reinhold Publishing Corp., New York, 1962.
3. K. Takemoto, *Chemistry of Inclusion Compounds* (in Japanese), Tokyo Kagaku Dojin, Tokyo, 1969.
4. M. F. Bengen and W. Schlenk, *Experientia*, 1949, **5**, 200.
5. M. F. Bengen, *Angew. Chem.*, 1951, **63**, 207.
6. W. Schlenk, *Justus Liebigs Ann. Chem.*, 1949, **565**, 204.
7. W. J. Zimmerschied, R. A. Dinerstein, A. W. Weitkamp and R. F. Marschner, *Ind. Eng. Chem.*, 1950, **42**, 1300.
8. S. R. Sergienko, A. A. Ovezov and A. Aidogdyev, *Dokl. Akad. Nauk SSSR*, 1975, **223**, 1150; *Chem. Abst.*, 1975, **83**, 192506.
9. A. A. Ovezov, A. Aidogdyev and S. R. Sergienko, *Izv. Akad. Nauk Turkm. SSR, Ser. Fiz.-Tekh., Khim. Geol. Nauk*, 1977, **6**, 72. (*Chem. Abs.*, 1978, **88**, 79912.)
10. A. Strocchi, G. Bonaga and A. Galletti, *Ann. Chim.*, 1974, **64**, 703. (*Chem. Abs.*, 1976, **84**, 89555.)
11. K. Takemoto and M. Miyata, *J. Macromol. Sci., Rev.*, 1980, **18**, 83.
12. G. N. Chekhova, Yu. A. Dyadin and D. D. Trotsenko, *Izv. Sib. Otd. Akad. Nauk SSSR, Ser. Khim. Nauk*, 1977, **2**, 38. (*Chem. Abs.*, 1977, **87**, 38825.)
13. G. N. Chekhova and Yu. A. Dyadin, Deposited DOC.VINITI, 1976, 1644-76. (*Chem. Abs.*, 1978, **89**, 5887.)
14. K. I. Patrilyak, *Zh. Fiz. Khim.*, 1980, **54**, 2203. (*Chem. Abs.*, 1981, **94**, 29975.)
15. A. E. Smith, *Acta Crystallogr.*, 1952, **5**, 224.
16. P. Vaughan and J. Donohue, *Acta Crystallogr.*, 1952, **5**, 530.
17. J. Radell and B. W. Brodman, *Can. J. Chem.*, 1965, **43**, 304.
18. Y. Oshima, F. Chida and H. Ohnuma, *Nippon Kagaku Kaishi*, 1978, 99.
19. B. W. Brodman and J. Radell, *Separ. Sci.*, 1967, **2**, 139. (*Chem. Abs.*, 1967, **67**, 76992.)
20. N. Nicolaides and F. Laves, *J. Am. Chem. Soc.*, 1958, **80**, 5752.
21. B. N. Ivanov, Sh.Sh. Bashkirov, L. M. Kozlov and A. S. Khramov, *Zh. Prikl. Khim.*, 1977, **50**, 2057. (*Chem. Abs.*, 1977, **87**, 192331.)
22. Y. Chatani, Y. Taki and H. Tadokoro, *Acta Crystallogr.*, 1977, **B33**, 309.
23. Y. Chatani, K. Yoshimori and Y. Tatsuta, *Polymer Prepr., Am. Chem. Soc., Div. Polymer Chem.*, 1978, **19**(2), 132.
24. R. A. Durie and R. J. Harrisson, *Spectrochim. Acta*, 1962, **18**, 1505.
25. V. Fawcett and D. A. Long, *J. Raman Spectrosc.*, 1975, **3**, 263.
26. R. A. MacPhail, R. G. Snyder and H. L. Strauss, *J. Am. Chem. Soc.*, 1980, **102**, 3976.

27. R. K. Gosavi and C. N. R. Rao, *Indian J. Chem.*, 1967, **5**, 162.
28. R. Clément, M. Gourdji and L. Guibé, *J. Magn. Reson.*, 1975, **20**, 345.
29. Y. Chatani, H. Anraku and Y. Taki, *Mol. Cryst. Liq. Cryst.*, 1978, **48**, 219.
30. E. Smolkova, L. Feltl and J. Vsetecka, *Chromatographie*, 1979, **12**, 147.
31. Z. Nowak and W. Stachowicz, *Nukleonika*, 1969, **14**, 1113.
32. G. B. Birrell and O. H. Griffith, *J. Phys. Chem.*, 1971, **75**, 3489.
33. T. Ichikawa, *J. Phys. Chem.*, 1977, **81**, 2132.
34. S. Schlick and L. Kevan, *J. Am. Chem. Soc.*, 1980, **102**, 4622.
35. K. Monobe and F. Yokoyama, *J. Macromol. Sci., Phys.*, 1973, **8**, 295.
36. A. T. Koritskii, A. P. Kuleshov, L. B. Soroka and V. I. Trofimov, *Khim. Vys. Energ.*, 1974, **8**, 437. (*Chem. Abs.*, 1975, **82**, 30664.)
37. A. L. Karasev, *Khim. Vys. Energ.*, 1972, **6**, 444. (*Chem. Abs.*, 1973, **78**, 22483.)
38. W. Schlenk, *Justus Liebigs Ann. Chem.*, 1973, 1145.
39. W. Schlenk, *Justus Liebigs Ann. Chem.*, 1973, 1156.
40. W. Schlenk, *Justus Liebigs Ann. Chem.*, 1973, 1179.
41. W. Schlenk, *Justus Liebigs Ann. Chem.*, 1973, 1195.
42. Y. Oshima, H. Ohnuma, Y. Akai and H. Ohashi, *J. Japan Petrol. Inst.*, 1976, 665.
43. D. J. Gannon and N. G. Parsonage, *J. Chem. Thermodyn.*, 1972, **4**, 745. (*Chem. Abs.*, 1972, **77**, 131563.)
44. B. Angla, *Compt. rend.*, 1947, **224**, 402 and 1166.
45. W. Schlenk, *Justus Liebigs Ann. Chem.*, 1951, **573**, 142.
46. A. A. Krasnov, B. V. Klimenok and V. I. Telle, *Zh. Fiz. Khim.*, 1977, **51**, 294. (*Chem. Abs.*, 1977, **86**, 188931.)
47. Yu. A. Dyadin, G. N. Chekhova and T. Ya. Arapova, *Izv. Sib. Otd. Akad. Nauk SSSR, Ser. Khim. Nauk*, 1977, 45. (*Chem. Abs.*, 1977, **86**, 178209.)
48. A. S. Ciobanu, D. V. Dinu and N. Dinu, *Bul. Inst. Politeh. Bucuresti, Ser. Chim. Metal.*, 1979, **41**, (2), 31. (*Chem. Abs.*, 1980, **93**, 45943.)
49. A. Faucitano, A. Buttafava and F. F. Martinotti, *J. Phys. Chem.*, 1981, **85**, 367.
50. G. N. Chekhova, Yu. A. Dyadin and T. V. Rodionova, *Izv. Sib. Otd. Akad. Nauk SSSR, Ser. Khim. Nauk*, 1979, (5) 78. (*Chem. Abs.*, 1980, **92**, 100212.)
51. R. Clement, C. Mazieres, M. Gourdji and L. Guibe, *J. Chem. Phys.*, 1977, **67**, 5381.
52. (a) K. Fukushima, *J. Mol. Struct.*, 1976, **34**, 67. (b) K. Fukushima and K. Sugiura, *J. Mol. Struct.*, 1977, **41**, 41.
53. J. E. Gustavsen, P. Klaeboe and H. Kvila, *Acta Chem. Scand., Ser. A*, 1978, **32**, 25.
54. A. Allen, V. Fawcett and D. A. Long, *J. Raman Spectrosc.*, 1976, **4**, 285.
55. K. Fukushima and S. Takeda, *J. Mol. Struct.*, 1978, **49**, 259.
56. E. Hough and D. G. Nicholson, *J. Chem. Soc., Dalton Trans.*, 1978, 15.
57. T. C. Gibb, *J. Phys. C*, 1976, **9**, 2627.
58. R. Clement, R. Claude and C. Mazieres, *J. Chem. Soc., Chem. Commun.*, 1974, 654.
59. A. N. Nesmeyanov, G. B. Shulpin and M. I. Pybinskaya, *Dokl. Akad. Nauk SSSR*, 1975, **221**, 624. (*Chem. Abs.*, 1975, **83**, 58989.)
60. T. Ichikawa, *J. Phys. Chem.*, 1979, **83**, 1358.
61. H. P. Bohlmann and C. Schneider, *Proc. Tihany Symp. Radiat. Chem.*, 1976, **4**, 411. (*Chem. Abs.*, 1978, **88**, 105829.)
62. Y. Chatani, *Prog. Polym. Sci. Japan*, 1974, **7**, 149.
63. Y. Chatani, S. Nakatani and H. Tadokoro, *Macromol.*, 1970, **3**, 481.
64. Y. Chatani and S. Nakatani, *Macromol.*, 1972, **5**, 597.

65. L. C. Fetterly, in *Non-Stoichiometric Compounds*, (ed. L. Mandelcorn) Academic Press, New York, 1964, Ch. 8, p. 491.
66. E. J. Fuller, *Enclyp. Chem. Process. Des.*, 1979, **8**, 333. (*Chem. Abs.*, **90**, 167395c.)
67. A. Verneuil, *Bull. Soc. Chim. Fr.*, 1884, **41**, 599.
68. Houben-Weyl, *Methoden der Organischen Chemie*, Georg Thieme Verlag, Stuttgart, 1955, **9**, 1187.
69. H. Hope, *Acta Chem. Scand.*, 1964, **18**, 1800.
70. H. van Bekkum, J. D. Remijnse and B. M. Wepster, *J. Chem. Soc., Chem. Commun.*, 1967, 67.
71. Yu. D. Kondrashev and N. A. Andreeva, *Zh. Strukt. Khim.*, 1963, **4**, 454. (*Chem. Abs.*, 1963, **58**, 7042.)
72. V. F. Dvoryankin and E. D. Ruchkin, *Zh. Strukt. Khim.*, 1962, **3**, 342. (*Chem. Abs.*, 1963, **58**, 7452d.)
73. J. J. Blaha and G. J. Rosasco, *J. Raman Spectrosc.*, 1981, **11**, 75.

3 · INCLUSION COMPOUNDS OF PERHYDROTRIPHENYLENE

M. FARINA

Università di Milano, Milan, Italy

1. Introduction

Perhydrotriphenylene ($C_{18}H_{30}$) was synthesized in 1963[1] during research into low molecular weight models of optically active polymers and into the exceptions to the common nomenclature of organic stereochemistry.[2-4] It

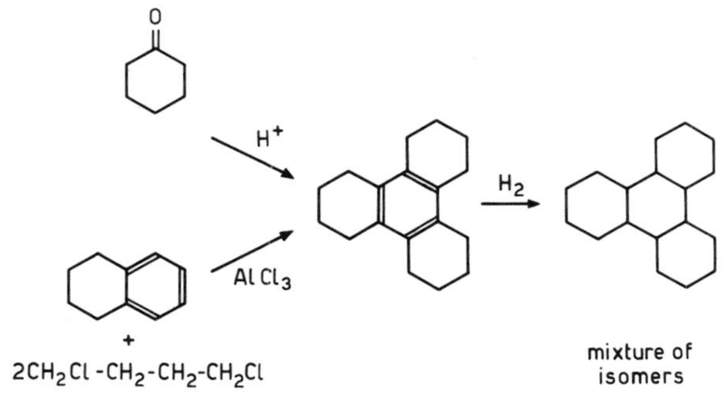

$2CH_2Cl-CH_2-CH_2-CH_2Cl$

mixture of isomers

INCLUSION COMPOUNDS 2
ISBN 0-12-067101-8

was prepared by hydrogenation of dodecahydrotriphenylene, $C_{18}H_{24}$, which in turn had been obtained from cyclohexanone according to Mannich[5] or from tetralin and 1,4-dichlorobutane according to Reppe.[6]

The hydrogenation produces a mixture of stereoisomers (theoretically ten stereoisomers can exist[7,8]) from which the *trans-anti-trans-anti-trans* isomer is easily isolated, it being the most abundant (about 60%) and the least soluble. Only this isomer, whose name is abbreviated to PHTP, will be considered here because of its solid-state properties.

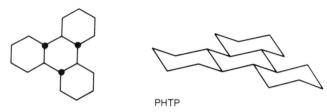

PHTP

X-ray determination of the structure of PHTP met with some initial difficulties precisely because of the formation of inclusion compounds containing crystallization solvents, as will be described later on. Its equatorial conformation and molecular symmetry are clearly indicated by the simplicity of the NMR spectra, and by the value of the spectral parameters. In the ^{13}C-NMR spectrum, for example, only three signals of similar intensity are found at 47.16, 30.17 and 26.60 ppm from TMS (CH, α-CH_2, β-CH_2 respectively). PHTP has D_3 molecular symmetry: i.e., a chiral molecule existing in enantiomeric forms. Its resolution is a highly complex and time-consuming process,[9,10] but nonetheless the availability of one and the same host both in its racemic form and in the optically active one offers interesting possibilities for studies on the structure and the reactivity of its inclusion compounds. The chemical behaviour of PHTP has been studied essentially in connection with its stereochemical properties.[8,10] Amongst the most interesting properties, mention must be made of its sensitivity to acid catalysts ($AlCl_3$, SbF_5, etc.) by the action of which it undergoes isomerization into other tetracyclic hydrocarbons, such as perhydronaphthacene[11] and dimethylperhydropyrene.[12] This shows that PHTP is one of the least stable hydrocarbons of formula $C_{18}H_{30}$, having considerable steric interactions between the lateral rings. Its stability is further diminished because of its symmetry: the entropic contribution $R \ln \sigma$, where σ ($=6$) is the symmetry number, is close to 15 J mol^{-1} K^{-1}. Nonetheless in the absence of catalysts PHTP is perfectly stable even at high temperature and lends itself very well to thermodynamic studies in the solid and liquid phases, as will be discussed later.

2. The preparation and composition of PHTP inclusion compounds

The formation of PHTP inclusion compounds was discovered right at the beginning of the investigations. And, as happened for the majority of the compounds described in this book, the discovery was completely unexpected. After the first clues observed during the structural resolution of PHTP crystallized from *n*-heptane, unequivocal confirmation was obtained by use of other solvents, particularly $CHCl_3$ and CCl_4, the presence of which within the crystals could easily be detected by diffractometry and thermogravimetry.

Once the existence of inclusion compounds had been ascertained, research was undertaken in order to determine their physical and chemical properties, and, on the other hand, the structural requirements for an organic compound to be included as a guest. The results of this preliminary research, to be found in the first publication on the subject,[13] showed the exceptional variety of guest molecules with regard to shape, size and chemical constitution; the stability of the inclusion compounds, as evidenced by their melting or decomposition points, was frequently higher than those of both host and guest, and also noteworthy is the ease with which PHTP forms inclusion compounds with macromolecular substances.

Of the organic compounds capable of acting as guests mention may be made of not only a great number of linear compounds such as aliphatic hydrocarbons, mono- and dicarboxylic aliphatic acids and their esters, linear alcohols, dialkylethers, etc., but also of branched compounds such as 2,2,4-trimethylpentane as well as cyclic compounds such as benzene, toluene, cyclohexane, or dioxan, and spherical or quasi-spherical molecules such as CCl_4, $CHCl_3$, $CHBr_3$, etc.[13–15] As a paradox, it may be stated that it took longer to find a simple compound which would not form an inclusion compound with PHTP: the most convenient and useful was found to be, at least under the usual working conditions, methylethylketone (MEK).

Many inclusion compounds obtained with guests of adequate length and low volatility show congruent melting over 125° C, which is the melting temperature of pure PHTP. This feature, which can easily be determined with a hot-plate microscope or by DSC, represents a simple analytical means of ascertaining the presence of the inclusion compound.

From the outset, however, the most interesting aspect of the behaviour of PHTP appeared to be its high affinity for macromolecular substances. At that time, only one adduct with a polymeric guest was known, namely that between urea and polyoxyethylene[16–19] and this inclusion compound

has low thermal stability. A few years earlier Brown and White had succeeded in polymerizing unsaturated monomers included in urea and
thiourea,[20,21] yet they provided no direct experimental proof of the formation
of inclusion compounds containing polymers. Even after this, the preparation of polymeric adducts was found to be a difficult and extremely slow
process: months of milling are required to obtain the inclusion compound
between urea and polyoxyethylene,[18] whilst for the urea-polyethylene compound use has to be made of an exchange reaction between the inclusion
compound urea-hexadecane and the polymer dissolved in boiling xylene.[22]

In the case of PHTP this reaction proceeds by mixing the two components
in the molten state and is complete in a few seconds.[23] The liquid mixture
solidifies suddenly, even around 150 °C, turning into a high-melting crystalline mass. As examples, the compounds PHTP-polyethylene and PHTP-1,4-
trans-polybutadiene melt at around 180 °C, more than 50 K above the
melting point of pure PHTP. This propensity towards the inclusion of
macromolecular compounds led us to conduct an in-depth study of the
polymerization of various unsaturated monomers included in PHTP.[24–26]
The crystal lattice imposes upon the process a high degree of regularity:
thus crystalline polymers can be obtained which sometimes show optical
activity and an extended-chain morphology. The greater part of PHTP
research is concentrated in this field. A detailed examination of inclusion
polymerization, both in PHTP and in other hosts, is given elsewhere (see
Chapter 10, Volume 3).

The ease with which PHTP inclusion compounds are formed is not limited
to macromolecular guests but can equally well be observed for other classes
of guests and may be considered to be one of the most salient properties
of PHTP as a host,[13,14,23,26,27] linked essentially to its hydrocarbon nature
and its solubility in many compounds.

The preparation of the inclusion compounds has been achieved by many
methods. For example, they can be formed by crystallization of PHTP from
the liquid guest (e.g., *n*-heptane) or by dissolving PHTP and guest together
in a suitable solvent (MEK) and allowing it to crystallize. It is also possible
to first prepare separate solutions of PHTP and of the guest in MEK: when
the inclusion compound is of low solubility (as in the case of long chain
hydrocarbons), mixing the two solutions leads to immediate precipitation
of the inclusion compound. If the guest is a solid, the two components can
be melted together, as described for polymers; the inclusion compound
crystallizes at high temperature or after cooling, according to its melting
point. In most cases, however, it is not necessary to start from a homogeneous
liquid phase or to use extraneous solvents. With solid guests, brief cold
grinding of the two components may be used. For liquid guests, the inclusion
compound may be formed by direct contact with solid PHTP, which does

not go into solution, and it is possible to carry out a solid phase transformation from pure host to inclusion compound. For volatile guests it is enough to expose the PHTP crystals to vapours of the guest in order to obtain the inclusion compound in a very short time (within 5–30 min.). In this last case the difference with respect to urea is very clear: urea requires low temperatures ($-78°$ C), the presence of promoters such as methanol, or reaction times lasting from many hours to days or even weeks.[21] (See Chapter 2, Volume 2)

The stoichiometry of PHTP inclusion compounds is not, in general, simple.[13–15] The arrangement of guest molecules inside the channels follows the criterion of total filling of the available space. Determination of the host/guest ratio may be achieved by crystallographic examination or by means of thermogravimetric analysis as well as by UV, IR and NMR spectroscopy, gas chromatography, chemical analysis, extraction procedures or gas-volumetric techniques. This last-mentioned method consists of the measurement of a volatile guest absorbed by a known quantity of PHTP, once equilibrium has been reached.[28,29] In most cases no appreciable variation of stoichiometric ratio with temperature has been observed (see, however, the case of $CHCl_3$, described below). Some composition data are reported in Table 1.

Table 1. Composition of some PHTP inclusion compounds

Guest molecule	X-ray analysis wt.%	Independent analysis wt.%	Method
n-Heptane	8.33[a]	8.5	A
n-Heptane	8.33	8.5	B
n-Nonane	8.75[a]	8.7	C
Cyclohexane	11.72[b]	11.7	B
Dioxan	12.97[b]	12.9	B
$CHCl_3$ (low temperature form)	19.50[b]	19.4	B
CCl_4	20.54[a]	19.8	B
Butadiene	9.89[b]	10.1	A
cis-Pentadiene	nd	7.5$_5$	A
trans-Pentadiene	nd	7.0$_5$	A
trans-trans-Hexa-2,4-diene	nd	7.4$_5$	A
2,5-Dimethylhexa-2,4-diene	nd	9.9$_5$	A
Caprylic acid (C_8)	10.47[a]	10.48	D
Lauric acid (C_{12})	10.37[a]	10.47	D

[a] Experimental value drawn from the repeat distances along the *c*-axis.
[b] Exact value derived from the stoichiometric ratio in the crystal cell.
Methods: A, gas-volumetric analysis; B, thermogravimetric analysis; C, infrared analysis; D, acidimetric titration.

The detection of inclusion compounds and the determination of their purity can be achieved by examination of the X-ray powder spectrum,[14] making use of the presence (or absence) of specific peaks for one of the two structures. For example the diffraction peak at $d = 7.60$ Å indicates the presence of excess PHTP. DSC curves are also very useful in ascertaining the presence and purity of high melting inclusion compounds.[13,23,24,27] As already observed, they generally show congruent melting with a very sharp single peak analogous to that of a pure compound. The presence of a eutectic point close to 125° C is often indicative of excess PHTP. Another method, demonstrating the presence of inclusion compounds with volatile guests, which is rather complicated but has in some cases given unambiguous results obtainable in no other way, consists in measuring their decomposition pressures.[30,31] The pure inclusion compound, or even a mixture of the inclusion compound with PHTP (i.e., when there is a deficiency of guest with respect to the stoichiometric value) has a lower vapour pressure than that of the pure liquid guest. A more extensive treatment of this topic will be given later in Section 4.

To complete these general observations, I would like to mention an unconventional way of determining the structure of PHTP inclusion compounds containing two different guests. In the particular case where the two guests are unsaturated compounds capable of polymerizing inside the channels, the structure of the polymer obtained after irradiation may be considered to be the transcription in macromolecular terms of the situation existing in the channels before polymerization.[32,33] Experimental findings for about ten monomer pairs have shown in all cases the existence of a copolymer having a Bernoullian (or near) distribution of the monomer units. This means there is no segregation of the two guests in different channels or even in different crystals. Under some very reasonable kinetic hypotheses we may say that inside each channel the two guests succeed each other in random fashion without undergoing any notable influence from the adjacent molecules.

3. Crystal structure of PHTP inclusion compounds

Many aspects of the solid state behaviour of PHTP are linked to its high degree of molecular symmetry, its rigidity and its inability to exercise strong intermolecular interaction because of its hydrocarbon nature: it is worth mentioning its low melting entropy (<0.2 J g^{-1} K^{-1}),[34] and hence its rather high melting point (125.2° C), and its tendency to polymorphism both in the pure state and as an inclusion compound with various guests.[13,14] The information on structure presented here is drawn mainly from the studies of Allegra, Colombo, Immirzi and their fellow workers.[13–15,35–38]

Two crystalline forms of pure PHTP exist, both of which are monoclinic (Table 2); one of these is stable and melts at 125.2° C, the other, which is metastable throughout the temperature range taken into consideration, melts at 117° C.[34] The metastable form is obtained by chance, but can be revealed in a DSC experiment by the amount of undercooling during crystallization. It is converted into the stable form after a few hours at room temperature, or more rapidly at a higher temperature. The thermal behaviour of pure PHTP can be included in that of the monotropic polymorphic systems.[34]

When crystallized from *n*-heptane, PHTP contains 8.3% of solvent, which can be eliminated only by prolonged heating (typical conditions for drying are: 1 mm Hg and 50 °C). Structural investigation has shown that each unit cell contains two PHTP molecules and 0.45 molecules of *n*-heptane (Table 2).[13–15,35] The host structure is composed of stacks of superimposed PHTP molecules arranged in a hexagonal symmetry (Fig. 1). At the vertices of the cell there are channel-type cavities the diameters of which, bounded by the van der Waals surfaces of the host molecules, are about 5 Å. The guest molecules in these channels are not arranged in fixed positions and give rise to continuous streaks in the rotating-crystal films. From the separation of these streaks it is possible to determine a 10.69 Å repeat distance along the channel axis, this being about 0.5 Å shorter than that expected for a planar zig-zag arrangement.

The examination of an isomorphous series of inclusion compounds with *n*-alkanes.[15,35] shows that an increase in the molecular weight of the guest corresponds to a slight reduction in the lateral dimensions of the cell, and that the repeat distance p_n along the channel axis can be expressed, in a first approximation, by the following equation (valid for *n* between 5 and 16):

$$p_n (\text{Å}) = 1.215 (n-1) + 3.44 \qquad (1)$$

where *n* is the number of carbon atoms of the guest hydrocarbon. This behaviour is analogous to that found for urea–*n*-alkane inclusion compounds.[39,40] A better approximation, valid over a wider range of molecular weights and providing a reasonable explanation of the phenomenon at a molecular level, can be obtained if we take into account the simultaneous presence in the channel of planar zig-zag molecules (conformation T, T, T, T, . . .) and of shortened molecules. For these latter the hypothesis of a T, G, T, G', T, G, T, G', . . . conformation has been advanced, the lateral dimensions of which are compatible with the diameter of the channel. (G and G' represent plus and minus *gauche* conformations, +60° and −60°, respectively.) By minizing the free energy of the system we obtain

Table 2. Crystal cell parameters of PHTP and of some of its inclusion compounds

Guest	Space group	Unit cell dimensions (Å)	U (Å³)	Z_{host}	Z_{guest}	Melting point (°C)
none (stable form)	P2₁/a	$a = 19.59$; $b = 15.39$; $c = 5.36$; $\gamma = 109.7°$	1522	4	—	125.2[a,b]
none (metastable form)	C2/c	$a = 16.94$; $b = 10.41$; $c = 9.73$; $\gamma = 113.5°$	1574	4	—	117[a]
n-heptane	P6₃/m	$a = b = 14.40$; $c = 4.78$; $\gamma = 120°$	858	2	0.45	120[b]
chloroform (room temperature form)	P6₃/m	$a = b = 25.08$; $c = 4.78$; $\gamma = 120°$	2600	6	3	58[c]
cyclohexane	R$\bar{3}$	$a = b = 25.55$; $c = 43.02$; $\gamma = 120°$	24 300	54	21	117[b]
dioxan	R$\bar{3}$	$a = b = 25.11$; $c = 28.68$; $\gamma = 120°$	15 660	36	15	119[b]
benzene					0.76	nd
toluene	monoclinic	$a = 15.70$[d]; $b = 14.72$[d]; $c = 4.78$; $\gamma = 121°$	890[d]	2	0.61	nd
bromoform					0.64	nd
butadiene	P2₁/m	$a = 13.35$; $b = 14.72$; $c = 4.78$; $\gamma = 115.3°$	849	2	1	nd

[a] Determined by DSC.
[b] Extrapolated value from vapour pressure measurements.
[c] First-order solid–solid transition.
[d] Average value.

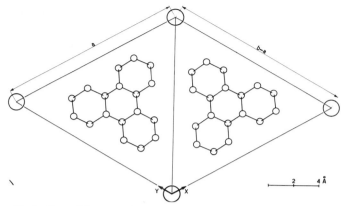

Fig. 1. *x-y* Projection of the crystal cell of the inclusion compound PHTP–*n*-heptane.

Equation 2, which is in excellent agreement with the experimental find-ings[15] (Table 3):

$$p_n = 1.275\,(n-1) + 3.65 - 0.155\,(n-1)(Ky^{(n-3)}/(1 + Ky^{(n-3)})) \qquad (2)$$

where $K = 2.5$ and $y = 0.915$. The parameter y is linked to the different stabilities of the two conformations examined.

For alkanes longer than C_{30} the molecular repetition can no longer be observed: instead of this it is possible to observe a repeat distance of 2.55 Å corresponding to the succession of two CH_2 in TT conformation.

A similar analysis carried out on inclusion compounds containing long chain fatty acids or dialkylethers showed the same shortening of the hydro-carbon moiety.[15] As expected, fatty acids are dimeric in the inclusion state. In the ether series an additional shortening of the repeat distance, localized near the oxygen atom, was observed. This fact was attributed to the presence of twisted conformations around the C–O bonds.

The PHTP–2,4,4–trimethylpentane inclusion compound is isomorphous with the preceding ones and represents an interesting example of the flexibility of the channels.[14,15] In order to accomodate the bulkier guest, the *a* and *b* axes of the crystal cell are about 0.3 Å longer than in the *n*-alkane inclusion compounds. The repeat distance of the guest (8.65 Å) is compar-able with that of *n*-pentane. From these results, and taking into account the geometric constraints due to the channel walls, a conformation has been proposed for the guest molecule, having the methylene hydrogens eclipsed with respect to the isopropyl methyl groups.

As regards the host lattice it should be noted that space group $P6_3/m$ allows for the presence of a crystallographic mirror plane coinciding with

78 M. Farina

Table 3. Repeat distance of linear hydrocarbons C_nH_{2n+2} included in PHTP

	Repeat distance	
n	Experimental	Calculated[a,b]
5	8.32	8.33
6	9.43	9.52
7	10.69	10.71
8	12.00	11.91
9	12.98	13.11
10	14.36	14.33
11	15.67	15.55
12	16.80	16.77
13	18.00	18.01
14	19.24	19.25
16	21.58	21.75
18	24.24	24.28
20	27.13	26.83
22	29.18	29.40
24	31.98	31.98
36	2.55[c]	—
polyethylene	2.55[c]	—

[a] Calculated according to Equation 2. [b] Standard deviation: ±0.12 Å. [c] Repeat distance of a –$(CH_2–CH_2)$– segment.

the mean molecular plane of PHTP.[14,35] Since PHTP does not have mirror symmetry, two enantiomeric molecules must statistically be present in the same crystal site. A complete random substitution does not however seem acceptable for steric reasons, nor would this agree with the melting behaviour of the inclusion compounds, of both racemic PHTP and of the single enantiomers*. Though the whole phase diagram has not yet been determined in detail, the difference of about 17 K between the two melting points[10,34] excludes the possibility of ideal solid solutions. The most acceptable hypothesis is that of the existence of some short-range order:[42] according to this model a certain number of (R) PHTP molecules stacked one upon the other are followed by a series of (S) molecules and so on.

Inclusion compounds isomorphous with those already described have been observed for other long-chain guests (ethers, carboxylic acids, esters, etc.) and for CCl_4 and 2,2,4-trimethylpentane.[14,15,35] Other globular guests, however, give rise to different crystal structures.

* The crystal structure of (−)PHTP inclusion compounds is essentially identical to that of (±)PHTP. The only significant difference is in the space group: $P6_3/m$ for the racemic compound, $P6_3$ for the optically active one, because of the absence of the mirror plane.[10,41]

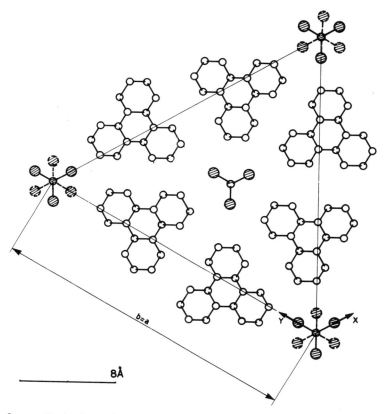

Fig. 2. *x-y* Projection of a part of the crystal cell of the inclusion compound PHTP–chloroform showing the different symmetry of the two channels. Guest molecules are shaded.

At room temperature the inclusion compound with chloroform presents a cell with a volume three times that of the earlier ones (Table 2).[14,36] The host lattice is very similar to that found in the inclusion compounds with *n*-alkanes, whilst considerable differences are observed with regard to the included molecules (Fig. 2). In the first place the absence of continuous streaks in the rotating-crystal photographs suggests that chloroform molecules are arranged along the channels with the same repeat distance as occurs for the PHTP molecules. The host/guest ratio is exactly 6/3, yet the guest molecules do not occupy equivalent sites in the cell: two of them are sited on the three-fold axes and one on the six-fold hexagonal axis existing on the vertices of the cell. Along this axis $CHCl_3$ molecules rotated by 60° follow each other statistically in such a way as to simulate a hexagonal symmetry. In this inclusion compound there are therefore two different types of channel with different symmetries and dimensions.

Through a study of the $P–T$ phase diagrams,[34] a first-order transition has been observed around 58° C, with a slight variation of the stoichiometric ratio. Regarding the structure of the new form we might advance the hypothesis that the features described above disappear and that the form in question contains equivalent $CHCl_3$ molecules oriented at random.

Third and yet more complex kind of structure is found in the inclusion compounds PHTP-cyclohexane and PHTP-dioxan (Table 2).[14,37] The crystal cells of these have a volume respectively 27 and 18 times greater than that of the compounds with *n*-alkanes. The dimensions *a* and *b* are close to those of the inclusion compound with $CHCl_3$, whilst the *c* axis is exactly 9 and 6 times greater. Given the considerable similarity of the two structures, I shall describe only the one with cyclohexane (Fig. 3). The PHTP molecules are not stacked precisely but are slightly displaced on the horizontal plane in such a way that their centre of gravity describes a 9/1 helix with a 0.4 Å radius around the ternary crystallographic axis. In addition to its 54 PHTP molecules, the cell contains 21 cyclohexane molecules in three equivalent channels, properly interlocked with the PHTP molecules and statistically arranged in many orientations. A further crystal structure has been found for inclusion compounds with benzene, toluene and $CHBr_3$, the unit cell being monoclinic with practically constant dimensions (Table 2).[14] The repetition distances of the guest molecules along the channel are: 6.30 Å for benzene, 7.90 Å for toluene and 7.50 Å for bromoform. Unfortunately the crystals examined to date have not been found suitable for an accurate definition of the structure.

Butadiene also forms a monoclinic inclusion compound, with a stoichiometric ratio of $2:1$[43] (Table 2 and Fig. 4). Its structure differs from that of the compound with *n*-heptane by a slight translational displacement of the stacks of PHTP molecules, giving rise to different values for *a* and *b* and to a value of γ less than 120° C. The channel is no longer cylindrical but is now approximately elliptical, with the greater axis forming an angle of about 45° C to the *b* axis. The butadiene molecules included in the channel are strongly inclined with respect to the *c* axis, so that the same repetition distance as in PHTP is found (4.78 Å). As a result of this the continuous streaks between the layer lines in the rotating-crystal photographs taken along *c* are not found. The short repeat distance between the guest molecules is such that the terminal atoms of successive monomers are very close to each other (3.50 Å). This may explain the ease with which butadiene included in PHTP polymerizes under the action of γ or *X*-rays. We have demonstrated, however, that the conditions required for inclusion polymerization to take place are in general less stringent.

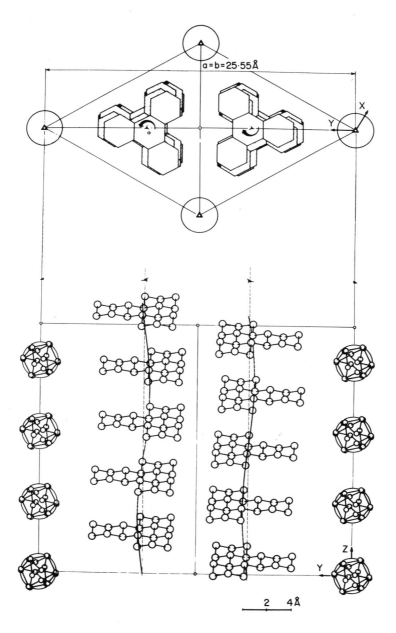

Fig. 3. Top and front view of the crystal cell of the inclusion compound PHTP–cyclohexane.

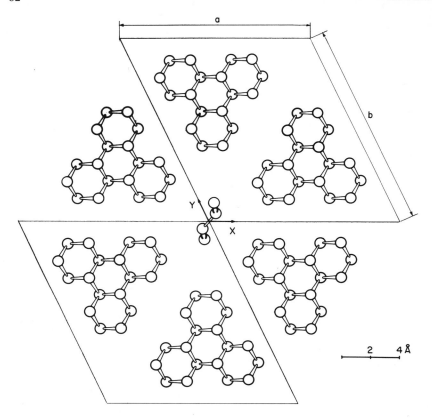

Fig. 4. *x-y* Projection of the crystal cell of the inclusion compound PHTP–butadiene.

Inclusion compounds with macromolecular guests also present various crystal structures.[13,14,24,43] The adducts with polyethylene, with 1,4-*trans*-polybutadiene and with polyoxyethylene are isomorphous with that of *n*-heptane. In the first case, a repeat distance of 2.55 Å is observed for the included component; as stated above, this distance is equal to that observed in crystalline polyethylene in which the molecule is in the .. TTTT .. conformation. In the second case, the adduct has 2:1 stoichiometry as for monomeric butadiene, and the repeat distance coincides with the length of the *c* axis of the crystal cell (4.78 Å). In crystalline 1,4-*trans*-polybutadiene the repeat distance is 4.85 Å[44] for form I, which is stable at room temperature, and 4.68 Å[45] for form II, which is stable at temperatures above 70° C, having a disordered conformation. The experimental finding can reasonably be justified if we admit a slight variation of the internal rotation angles with respect to I, or the presence of a partial disorder analogous to that observed in II.

The repeat distance of polyoxyethylene included in PHTP is 6.87 Å [24] and refers to two monomer units. Its conformation is considerably more elongated than has been observed either in the crystal state (19.3 Å for 7 monomer units),[46] or in the inclusion compound with urea (9.12 Å for 3 monomer units)[19] and is only slightly shorter than that calculated for a *trans*-planar conformation (7.21 Å for two monomer units). For this conformation, a highly-symmetric model has been proposed with the following succession of internal rotation angles: C(1)–C(2) = 180°, C(2)–O = ±141°, O–C(1) = ∓141°.[24]

In their turn, isotactic 1,4-*trans*-polypentadiene and 1,4-*trans*-poly-2,3-dimethylbutadiene form inclusion compounds whose powder spectra show evidence of structures different from the preceding ones. The difficulty of obtaining monocrystals suitable for diffractometric investigation has so far made it impossible to obtain any further information on these and other polymer-containing inclusion compounds of PHTP.[14]

4. Phase diagrams of binary systems forming inclusion compounds

In contrast with the far-reaching structural research carried out on channel inclusion compounds, very little had been done until a few years ago with respect to the study of their stability or their behaviour in phase transitions. I should mention a series of studies by Rheinboldt on deoxycholic acid and on apocholic acid,[47,48] and a few findings by Schlenk Jr.[49] on inclusion compounds with urea. The validity of Clapeyron's law in solid-vapour equilibrium in the case when the inclusion compound and the pure host have different crystal structures had been the subject of a discussion that ended in a positive result.[50]

The structural variety of inclusion compounds formed with PHTP, the possibility of varying their stability progressively (by varying the length of the included molecule) and of passing from incongruent to congruent transitions, the availability of inclusion compounds with highly-volatile guests (butadiene, pentane, $CHCl_3$, etc.) and of other adducts with macromolecular guests rendered this field a highly promising one for systematic research. This has in fact been carried out with the aim of determining the consequences of thermodyanic behaviour on inclusion polymerization, but the findings also have a general value in the understanding of the relationships between properties and structure in this class of binary adducts.

During this research use was made in my laboratory of two complementary techniques: DSC and vapour pressure measurements; the former was aimed at determining the *T-x* diagrams of non-volatile systems, the latter the *P-T* diagrams of systems containing low-molecular-weight guest. Under certain conditions, the two series of data can usefully be compared. For example it is possible to make use of DSC even in the case of volatile components if loss of substance by evaporation is kept to a minimum (sealed sample holders with small free volume), and it is also possible to convert the *P-T* diagrams into *T-x* diagrams by formulating various hypotheses as to the state of the system in the various phases and then making use of the resulting thermodynamic equations. From the point of view of the phase rule, inclusion compounds in PHTP are to be considered as low-stability binary adducts analogous to hydrates, solvates, crystalline racemic compounds, to certain complex salts, to intermetallic compounds and to binary semiconductors.[31] Their stability is limited to the solid state: melting or dissolution are to be considered as true decomposition of the inclusion compound. The thermodynamic properties of binary adducts were studied by Roozeboom a century ago[51] and described in various texts dealing with the phase rule.[52–55]

As was done in the previous section when dealing with the structural data, the PHTP-*n*-heptane inclusion compound is described in detail, as an example to illustrate the entire class.[27,31] Figures 5 and 6 show, respectively, the condensed *T-x* phase diagram and the *P-T* projection of the *P-T-x* diagram. In the hypothesis of non-miscibility between the solid phases

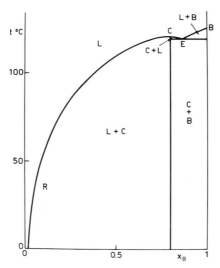

Fig. 5. *T-x* Phase diagram of the binary system PHTP–*n*-heptane.

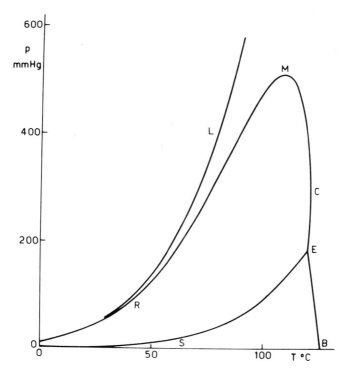

Fig. 6. *P-T* Projection of the solid-liquid-vapour phase diagram of the binary system PHTP–*n*–heptane.

(remembering that pure PHTP and its inclusion compounds crystallize in different space groups), of an ideal behaviour in the liquid phase, of constant composition of the inclusion compound, and neglecting the vapour pressure of component B and the variation of the enthalpic parameters with temperature, the curve B–E of Fig. 5 (solubility curve of PHTP in *n*-heptane) is described by Equation 3:

$$\ln x_B = -(\Delta H_B / R)(1/T - 1/T_B) \tag{3}$$

where ΔH_B and T_B are the melting enthalpy and the melting temperature of host component B (PHTP). The curve R–C–E (solubility curve of the inclusion compound) is described by Equation 4:

$$\ln(x_A / x_{0A}) + n \ln(x_B / x_{0B}) = -(\Delta H_C / R)(1/T - 1/T_C) \tag{4}$$

where T_C and ΔH_C are the melting temperature and the melting enthalpy (this latter being referred to one mole of the guest component) of the crystalline inclusion compound, x_{0A}, x_{0B} and $n(=x_{0B}/x_{0A})$ are the molar fractions of the A (guest) and B (host) components and the stoichiometric host/guest ratio in the inclusion compound.

Similarly, the B–E curve of Fig. 6 (vapour pressure of saturated solution of PHTP in *n*-heptane) is described by Equation 5:

$$\ln(1-(P/P_A)) = -(\Delta H_B/R)(1/T - 1/T_B) \tag{5}$$

where P_A is the vapour pressure of pure A (*n*-heptane). The R–M–C–E curve (vapour pressure of saturated solution of inclusion compound) is represented by Equation 6:

$$\ln((n+1)P/P_A) + n\ln(((n+1)/n)(1-(P/P_A)))$$
$$= -(\Delta H_C/R)(1/T - 1/T_C) \tag{6}$$

To date this curve has received little or no attention, all interest being focussed on the S–E curve, representing the decomposition pressure of the inclusion compound according to the reaction:

Inclusion compound (cryst.) → PHTP (cryst.) + guest (vapour) to which Equation 7 can be applied.

$$\ln P = -\Delta H_v/RT + D \tag{7}$$

It is most important to underline that the intersection of this line with the liquidus curves does not occur at point C (corresponding to the composition and vapour pressure of the pure inclusion compound), but at point *E*, which, because of the coexistence of four phases (vapour, liquid, crystalline inclusion compound and crystalline PHTP) is an invariant point (quadruple point or eutectic point). It follows that the decomposition pressure of the inclusion compound at any temperature is influenced both by ΔH_V and by the *P* and *T* values of the eutectic point, which determine the value of the constant *D* of Equation 7.

Equation 7 can be converted into Equation 8:

$$\ln(P/P_A) = -(\Delta H_{dec}/RT) + D' \tag{8}$$

which refers to decomposition of the crystalline inclusion compound into solid PHTP and liquid guest. The diagrams of $\log P$ or $\log P/P_A$ vs. $1/T$ are very useful for the study of the stability of the inclusion compounds (Figs. 7 and 8).

Agreement between experimental findings and the data calculated according to Equations 4–8 is very good. Of the thermodynamic parameters, T_C, T_E, ΔH_V and ΔH_{dec} can be evaluated with considerable accuracy (Table 4).

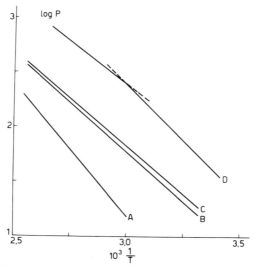

Fig. 7. Decomposition pressure of some inclusion compounds vs. $1/T$ (see equation 7). The curves A, B, C, D refer respectively to the inclusion compounds with n-heptane, dioxan, cyclohexane and chloroform.

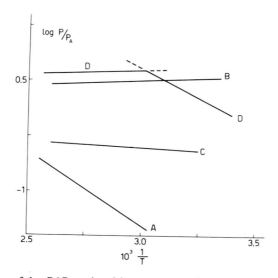

Fig. 8. Variation of the P/P_A ratio with temperature for some inclusion compounds (see Equation 8). The curves A, B, C, D refer respectively the the inclusion compounds with n-heptane, dioxan, cyclohexane and chloroform.

Table 4. Thermodynamic parameters of some PHTP inclusion compounds

Guest	ΔH_v (kJ mol^{-1})	ΔH_{dec} (kJ mol^{-1})	T_E (°C)	P_E (mm Hg)	T_C (°C)
n-heptane	45.2	11.7	119	187	120
cyclohexane	31.8	0.8	113	342	117
dioxan	34.7	−0.8	115	340	119
chloroform[a]	38.9	8.4	—	—	58[b]
chloroform[c]	28.9	−0.4	nd	nd	nd

[a] Crystal modification stable at room temperature.
[b] First-order solid–solid transition (vapour pressure of the transition point: 234 mm Hg).
[c] Crystal modification stable at higher temperature.

However, ΔH_B and ΔH_C are affected by errors linked to the not perfectly ideal behaviour of the liquid phase.

Equations 3–6 can be appropriately modified by using the scheme of regular solutions[34,56] as follows:

B–E curve: Equations 9 and 10 instead of Equations 3 and 5:

$$\ln a_B = \ln(\gamma_B x_B) = \ln(\gamma_B(1 - (P/(P_A\gamma_A)))) = -(\Delta H_B/R)(1/T - 1/T_B)$$
(9)

$$\ln x_B = -(\Delta H_B/R)(1/T - 1/T_B) - (w/RT)(1 - x_B)^2$$
(10)

R–M–C–E curve: Equations 11 and 12 instead of 4 and 6:

$$\ln(a_A/a_{0A}) + n\ln(a_B/a_{0B})$$

$$= \ln((n+1)P/(P_A\gamma_{0A})) + n\ln((1 - (P/(P_A\gamma_A)))(n+1)\gamma_B/(n\gamma_{0B}))$$

$$= -(\Delta H_C/R)(1/T - 1/T_C)$$
(11)

$$\ln(x_A/x_{0A}) + n\ln(x_B/x_{0B})$$

$$= -(\Delta H_C/R)(1/T - 1/T_C) - (w/RT)(nx_A^2 + x_B^2 - x_{0B})$$
(12)

where $\ln \gamma_i = (w/RT)(1 - x_i)^2$ and w is an interaction parameter between the two components.

In the case of heptane w is estimated around 960 J mol^{-1}.[34]

The same analysis can be extended to other volatile guests: here only the cases of hexane, cyclohexane, dioxan and $CHCl_3$ shall be examined.

The phase diagrams of the inclusion compound with hexane are very similar to those with heptane except for the fact that point E coincides with or is very close to C, indicating a lower stability of the inclusion compound.[34] If the length of the included hydrocarbon is further reduced, the intersection between the vapour pressure curve of the inclusion compound and the liquidus curves can be expected to occur at values of x_B lower than x_{0B}. In this case the eutectic point would be substituted by a peritectic point, also invariant, but the equations describing the behaviour remain the same. The low stability of the inclusion compound would be indicated by its incongruent decomposition at relatively low temperatures and by the higher vapour pressure (the constant D of Equation 7 is, in this case a function of the pressure and the temperature of the peritectic point). Figures 7 and 8 show the decomposition pressures of the inclusion compounds PHTP–cyclohexane and PHTP–dioxan.[34] The almost horizontal pattern followed by log P/P_A for both compounds indicates a near zero value of ΔH_{dec}, far different from that observed for linear hydrocarbons. This fact may be related to the statistical orientation of the nearly spherical guest molecules inside the channels.

The resemblance between the two inclusion compounds disappears when we examine the equilibria involving the liquid phase. Whilst in cyclohexane the behaviour is almost ideal ($w = 200 \text{ J mol}^{-1}$), dioxan is far from ideal ($w = 3350 \text{ J mol}^{-1}$) and it is necessary to use Equations 9–12 to interpret the phase diagrams. A comparison between these two cyclic guests appears significant in clarifying a feature of PHTP inclusion compounds: their formation is essentially influenced by factors of shape and size, whilst the liquid-phase behaviour predominantly depends on polarity factors. The connection between these two aspects is given by the location of point E (or possibly P) in accordance with what has already been discussed.

An examination of the phase diagrams of the inclusion compound PHTP–CHCl$_3$ shows the value of thermodynamic methods in identifying new crystalline phases.[34] In Figs. 7 and 8 a sudden variation can be seen in the slope of the vapour pressure curves at about 58° C, which can be interpreted as a first-order solid-solid transition. The transition can also be observed by DSC. The structure described in the preceding section corresponds to the low-temperature stable phase.

5. Inclusion compounds with long chain hydrocarbons

The longer the chain of the guest hydrocarbon becomes, the more its vapour pressure falls, until it reaches values low enough for it to be possible to neglect the presence of the vapour phase from many points of view. I shall therefore limit my treatment to an examination of the T-x diagrams, obtained generally by DSC at atmospheric pressure or slightly above.

Two facts are immediately apparent: the increase in congruent melting temperature of the inclusion compound (T_C) as the length of the chain increases (Table 5); and the progressive change in the $T-x$ phase diagram, which is more and more dominated by the liquidus curve of the inclusion compound (Figs. 9 and 10).

Table 5. Melting point of PHTP inclusion compounds with linear hydro-carbons C_nH_{2n+2}

Guest n	Melting point (°C)	Guest n	Melting point (°C)
7	120	15	143
8	124	16	145
9	128.5	18	148
10	131	20	151
11	134	24	152.5
12	136	28	156
13	138	32	160
14	141.5	36	161
		polyethylene	178

As regards the first point, we may observe the asymptotic tendency of the value of T_C which tends to 178° C (the value measured for the compound with polyethylene*[23]), over 50 K higher than that of pure PHTP. This behaviour is totally different from that observed with the inclusion compounds of urea,[57] which undergo an incongruent decomposition below the melting point of urea (132.7° C). In the preceding section we discussed the conditions under which it is possible to observe a peritectic point with incongruent decomposition in PHTP inclusion compounds. Nevertheless, the two instances, must not be confused: with PHTP, incongruent decomposition is found only with very short chain guests (below C_6) and the liquidus curve is, in any case, well represented by Equations 3 and 4; with urea such decomposition is observed even with guests as long as C_{28} and the phase diagram is totally modified by the presence of a practically complete miscibility gap.[27] It is precisely this peculiarity which leads to the lower stability of inclusion compounds with urea compared with those with PHTP.

Figure 9 shows the phase diagram of the inclusion compound PHTP–n-$C_{20}H_{42}$ obtained by DSC.[34] Both the liquidus and the solidus curves have been determined over the whole range of compositions; the latter show the absence of mixed crystals between the inclusion compound and the single constituents.

* It is worth mentioning that in the earliest publications[13,14,24] the T_C values were slightly overestimated (by 1–4 K). The discrepancy is essentially due to the different method of observation (hot-plate polarizing microscope instead of DSC).

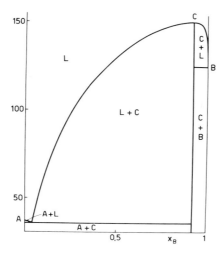

Fig. 9. *T-x* phase diagram of the binary system PHTP–*n*-C$_{20}$H$_{42}$.

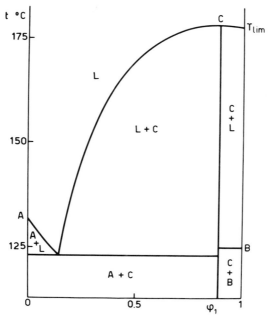

Fig. 10. *T-x* Phase diagram of the binary system PHTP–polyethylene.

Even from a qualitative examination it is possible to observe the greater stability of this inclusion compound compared with that of PHTP–*n*-heptane; in particular, the eutectic point is shifted towards the right-hand extremity of the diagram and its composition is very close to $x_B = 1$. Equation 4 interprets of the experimental points very well indeed: the value of ΔH_C determined by the liquidus curve (303 kJ mol^{-1} guest, or 27.0 kJ mol^{-1} PHTP) is coincident with the calorimetric value (302 kJ mol^{-1} guest), demonstrating the ideal behaviour of the liquid phase PHTP–*n*-$C_{20}H_{42}$.

Although Equation 4 is adequate to describe the phase diagrams between PHTP and *n*-alkanes having a relatively long chain, it is not satisfactory in interpreting the behaviour of the PHTP–polyethylene system, or, in general, the behaviour of inclusion compounds having a macromolecular component. This drawback arises from the non-ideal behaviour of macromolecular solutions, even in cases in which ΔH of mixing is nil. The Flory–Huggins theory states, in fact, that the entropy of mixing is greater than that which can be computed by Raoult's law, because of the different molecular volume of the two components.[58]

The liquidus curve of the PHTP–polyethylene inclusion compound (Fig. 10) can be calculated by introducing into the formula of non-ideal solutions suitable expressions of the activity of the polymer and of the solvent.[23] Equation 11 is thus converted into Equation 13:

$$\ln(\phi_1/\phi_{01}) + 1 - (\phi_1/\phi_{01}) + (\chi/\phi_{01})(\phi_{01} - \phi_1)^2$$
$$= -(\Delta H/R)(1/T - 1/T_C) \tag{13}$$

where ϕ_1 and ϕ_{01} are the volume fractions of PHTP in solution and in the inclusion compound, χ is the polymer–PHTP interaction parameter, T_C is the melting temperature and ΔH the melting enthalpy of the inclusion compound (this value being referred to 1 mole of PHTP). When χ is negligible, Equation 13 can be simplified to Equation 14:

$$\ln(\phi_1/\phi_{01}) + 1 - (\phi_1/\phi_{01}) = -(\Delta H/R)(1/T - 1/T_C) \tag{14}$$

It must be observed that this equation is not symmetrical: in fact, for ϕ_1 tending to 0, T tends to 0 K, whilst for ϕ_1 tending to 1 a limit temperature (T_{lim}) is reached which is close to T_C (451 K). It follows that, since T_{lim} is higher than the melting temperature of PHTP, no eutectic can exist in the range of compositions between ϕ_{01} and 1. The addition of minimal amounts of polymer into the molten PHTP leads to the formation of the inclusion compound, insoluble in the excess of PHTP below T_{lim}. On the contrary, a eutectic point is observed in the high polymer concentration zone (123° C, $\phi_1 = 0.146$).[23]

Agreement between theory and experimental findings is excellent both as regards the fit of the experimental points along the liquidus curve, and as regards the values of ΔH ($=26.5$ kJ mol^{-1} PHTP, determined according to Equation 14; 27.9 determined calorimetrically by means of DSC).

Once again we must underline the versatility of PHTP as a host for the formation of inclusion compounds with hydrocarbons: making use of various experimental techniques but with the same theoretical approach it was possible to delineate a homogeneous picture of the behaviour throughout a wide range of guest structures running from butadiene through to polyethylene. Over and above its intrinsic importance, this study has decisively thrown light on many aspects of inclusion polymerization. In particular, it has made it possible to establish the nature of the initial and final states of the process, to define the limiting conditions of polymerization and to determine the thermal balance of the reaction.[27]

This subject will be dealt with more fully in Chapter 10, Volume 3.

Acknowledgements

I have particular pleasure in expressing my thanks to Prof. Allegra and his research group, and in particular to Prof. A. Colombo and Prof. A. Immirzi. The determination of the crystal structures given in this chapter was carried out by them. I am no less grateful to Dr. G. Audisio and Dr. G. Di Silvestro who at various times have made decisive contributions to this research, the former in the earlier investigations and in the resolution of PHTP into optical antipodes, the latter in the study of phase diagrams. The research work carried out at the University of Milan was financed partly by grants from the Italian National Research Council (CNR), Rome, and by the Italian Ministry of Education, Rome.

References

1. M. Farina, *Tetrahedron Lett.*, 1963, 2097.
2. G. Natta and M. Farina, *Tetrahedron Lett.*, 1963, 703.
3. M. Farina, M. Peraldo and G. Natta, *Angew. Chem.*, 1965, 77, 149; *Angew. Chem., Int. Ed. Engl.*, 1965, **4**, 107.
4. R. S. Cahn, C. K. Ingold and V. Prelog, *Angew. Chem.*, 1966, **78**, 413; *Angew. Chem., Int. Ed. Engl.*, 1966, **5**, 385.
5. C. Mannich, *Chem. Ber.*, 1907, **40**, 153.
6. cf. W. Reppe, *Justus Liebigs Ann. Chem.*, 1955, **596**, 134.

7. M. Farina, G. Audisio and P. Bergomi Bianchi, *Chim. Ind. (Milan)*, 1968, **50**, 446.
8. M. Farina and G. Audisio, *Tetrahedron*, 1970, **26**, 1827.
9. M. Farina and G. Audisio, *Tetrahedron Lett.*, 1967, 1285.
10. M. Farina and G. Audisio, *Tetrahedron*, 1970, **26**, 1839.
11. M. Farina, G. Allegra, G. Logiudice and U. Pedretti, *Tetrahedron Lett.*, 1969, 551.
12. G. Allegra, M. Farina, A. Immirzi and R. Broggi, *Tetrahedron Lett.*, 1964, 1975.
13. M. Farina, G. Allegra and G. Natta, *J. Am. Chem. Soc.*, 1964, **86**, 516.
14. G. Allegra, M. Farina, A. Immirzi, A. Colombo, U. Rossi, R. Broggi and G. Natta, *J. Chem. Soc. (B)*, 1967, 1020.
15. G. Allegra, M. Farina, A. Colombo, G. Casagrande-Tettamanti, U. Rossi and G. Natta, *J. Chem. Soc. (B)*, 1967, 1028.
16. J. Parrod and A. Kohler, *C. R. Hebd. Séances Acad. Sci.*, 1958, **246**, 1046; and *J. Polym. Sci.*, 1960, **48**, 457.
17. A. Kohler, G. Hild and J. Parrod, *C. R. Hebd. Séances Acad. Sci.*, 1962, **255**, 276.
18. J. Parrod, A. Kohler and G. Hild, *Makromol. Chem.*, 1964, **75**, 52.
19. H. Tadokoro, T. Yoshihara, Y. Chatani and S. Murahashi, *Polym. Lett.*, 1964, **2**, 363.
20. J. F. Brown, Jr. and D. M. White, *J. Am. Chem. Soc.*, 1960, **82**, 5671.
21. D. M. White, *J. Am. Chem. Soc.*, 1960, **82**, 5678.
22. K. Monobe and F. Yokoyama, *J. Macromol. Sci., Phys.*, 1973, **8**, 295.
23. M. Farina, G. Di Silvestro and M. Grassi, *Makromol. Chem.*, 1979, **180**, 1041.
24. M. Farina, G. Natta, G. Allegra and M. Löffelholz, *J. Polym. Sci.*, 1967, **C16**—2517.
25. M. Farina, in *Proceedings of the International Symposium on Macromolecules, Rio de Janeiro, 1974*, (ed. E. B. Mano) Elsevier, Amsterdam, 1975, p.21.
26. M. Farina, *Makromol. Chem. Suppl.*, 1981, **4**, 21.
27. M. Farina and G. Di Silvestro, *Gazz. Chim. Ital.*, 1982, **112**, 91.
28. M. Farina, G. Audisio and M. T. Gramegna, *Macromolecules*, 1971, **4**, 265.
29. M. Farina, G. Audisio and M. T. Gramegna, *Macromolecules*, 1972, **5**, 617.
30. M. Farina, U. Pedretti, M. T. Gramegna and G. Audisio, *Macromolecules*, 1970, **3**, 475.
31. M. Farina and G. Di Silvestro, *J. Chem. Soc., Perkin Trans. 2*, 1980, 1406.
32. M. Farina, G. Di Silvestro and P. Sozzani, *Mol. Cryst. Liq. Cryst.*, 1983, **93**, 169.
33. P. Sozzani, G. Di Silvestro and M. Farina, Preprints of the 2nd International Symposium on Clathrate Compounds and Molecular Inclusion Phenomena, Parma, 1982, p. 97.
34. M. Farina, G. Di Silvestro, M. Grassi and P. Sozzani, to be published.
35. A. Colombo and G. Allegra, *Rend. Accad. Naz. Lincei*, 1967, **43**, (8), 41.
36. A. Immirzi and G. Allegra, *Rend. Accad. Naz. Lincei*, 1967, **43**, (8), 57.
37. A. Immirzi and G. Allegra, *Rend. Accad. Naz. Lincei*, 1967, **43**, (8), 181.
38. A. Colombo, E. Torti and G. Allegra, *Rend. Accad. Naz. Lincei*, 1967, **43**, (8), 196.
39. A. E. Smith, *Acta Crystallogr.*, 1952, **5**, 224.
40. F. Laves, N. Nicolaides and K. C. Peng, *Z. Kristallogr.*, 1965, **121**, 258.
41. M. Farina, G. Audisio and G. Natta, *J. Am. Chem. Soc.*, 1967, **89**, 5071.
42. J. Jacques, A. Collet and S. H. Wilen, *Enantiomers, Racemates and Resolutions*, J. Wiley, New York, 1981.
43. A. Colombo and G. Allegra, *Macromolecules*, 1971, **4**, 579.
44. S. Iwaianapi, I. Sekurai, T. Sekurai and T. Seto, *Rep. Progr. Polym. Phys. Jpn.*, 1967, **70**, 167.
45. G. Natta and P. Corradini, *Suppl. Nuovo Cimento, series X*, 1960, **15**, 9.

46. H. Tadokoro, Y. Chartani, T. Yoshihara, S. Tahara and S. Murahashi, *Makromol. Chem.*, 1964, **73**, 109.
47. H. Rheinboldt, *Justus Liebigs Ann. Chem.*, 1927, **451**, 256.
48. H. Rheinboldt, *Z. Physiol. Chem.*, 1929, **180**, 180.
49. W. Schlenk Jr., *Justus Liebigs Ann. Chem.*, 1949, **565**, 204.
50. J. H. van der Waals and J. C. Platteeuw, *Nature (London)*, 1959, **183**, 462.
51. H. W. Roozeboom, *Rec. Trav. Chim. Pays-Bas*, 1884, **3**, 28; 1885, **4**, 65, 108, 331; 1886, **5**, 335, 393; 1887, **6**, 262, 304; 1889, **8**, 1.
52. J. E. Ricci, *The Phase Rule and Heterogeneous Equilibrium*, Van Nostrand, New York, 1951.
53. I. Prigogine and R. Defay, *Chemical Thermodynamics* (transl. by D. H. Everett) Longmans Green, London, 1954.
54. R. Haase, *Thermodynamik der Mischphasen*, Springer Verlag, Berlin, 1956.
55. A. Riesman, *Phase Equilibria* Academic Press, New York, 1970.
56. A. S. Jordan, *Metall. Trans.*, 1971, **2**, 1959.
57. H. G. McAdie, *Can. J. Chem.*, 1962, **40**, 2195.
58. P. Flory, *Principles of Polymer Chemistry*, Cornell University Press, Ithaca, 1953.

4 · INCLUSION COMPOUNDS OF CYCLOTRIVERATRYLENE AND RELATED HOST LATTICES

A. COLLET
Collège de France, Paris, France

1. Introduction

The acid-catalysed reaction of veratrole and formaldehyde has been known since 1915 to produce in good yield a high-melting crystalline solid (m.p. *c.* 226–234° C) of general formula $(C_9H_{10}O_2)_n$.[1] This compound was first considered by G. M. Robinson to be 2,3,6,7-tetramethoxy-9,10-dihydroanthracene ($n = 2$), a suggestion widely accepted in the literature up to the 1950's, when a hexameric structure ($n = 6$) was proposed by Italian chemists, on the basis of chemical studies and X-ray diffraction data.[2–6]

That Robinson's compound was in fact a trimer ($n = 3$) was demonstrated in 1963–65 by the work of Lindsey,[7,8] Erdtman *et al.*,[9] and Goldup *et al.*,[10] based on direct molecular weight determinations, chemistry, mass spectroscopy, re-examination of the cell dimensions by X-ray diffraction, and, finally, by NMR spectroscopy. The compound, named *cyclotriveratrylene* (CTV) for convenience,* was shown at the same time, by variable temperature NMR experiments,[11] to adopt a rigid crown conformation having (C_{3v})

---

* Also named 10,15-dihydro-2,3,7,8,12,13-hexamethoxy-5*H*-tribenzo[*a,d,g*]cyclononene.

INCLUSION COMPOUNDS 2
ISBN 0-12-067101-8

symmetry. It follows that suitably substituted CTV derivatives are chiral, and can be resolved into stable enantiomers. Since the first successful resolution, by Lüttringhaus and Peters[12] in 1966, a relatively large number of optically active (C₃)-CTV derivatives have been described.

The ability of CTV to form crystalline inclusion compounds was first mentioned by Bhagwat *et al.*[13] in 1931. This behaviour was studied in greater detail by Caglioti *et al.*[6] some twenty five years later, but the first crystal structure of a CTV inclusion compound has been reported only recently.[14] The hexademethylated derivative *cyclotricatechylene* has been found to be more efficient in forming crystalline complexes than CTV itself,[15] and several other analogues exhibiting this property have been described.

The exploitation of the bowl-shaped structure of CTV for the design of new complex-forming ligands is currently the object of increasing interest. Compounds in which the six methyl groups of CTV have been replaced by chains of suitable structure,[16,17,50] and macrocages containing the CTV subunit,[33,52] actually exhibit interesting complexing properties.

Although there now exist about 50 references concerning cyclotriveratrylene and its analogues, these topics do not seem to have been reviewed as yet; it is the purpose of the present article to provide such a survey. Accordingly, the first section of this Chapter is devoted to the chemistry of CTV, including the synthetic and conformational aspects, and the access to optically active compounds. The second section treats the inclusion properties, and is itself divided into two main parts: the crystalline inclusion compounds, and the molecular complexes.

2. Chemistry of cyclotriveratrylene and related compounds

CTV and its analogues can be prepared by using various synthetic routes, that are discussed in this section, together with the conformational behaviour of these molecules.

2.1. Synthesis of cyclotriveratrylene

CTV (**1**), appears to be the major condensation product of the veratryl cation, which itself can be generated from a variety of precursors, usually in the presence of strong acids. Accordingly, CTV can be conveniently prepared by using one of the various procedures assembled in Table 1. Incidentally, these reactions usually afford as the main by-product the cyclic

Table 1. Selected procedures for the preparation of CTV and CTTV

Starting materials	Catalyst	Solvent	Temp.	Isolated yields		Refs.
				CTV	CTTV	
3 + aqueous HCHO	70% H_2SO_4	none	r.t.	70		1, 10
3 + aqueous HCHO	70% H_2SO_4	none	0°C	21		8
3 + aqueous HCHO	70% H_2SO_4	none	0°C	~68		18
3 + aqueous HCHO	60% $HClO_4$	none	r.t.	70	~16	20
3 + paraformaldehyde	6 M HCl	none	reflux	"good"		11
3 + paraformaldehyde	conc. HCl	none	r.t.	45		24
3 + trioxymethylene	$HCl/ZnCl_2$	none	−10° C	unspecified		26, 27
4	conc. HCl	none	r.t.	unspecified		26, 27
4	H_2SO_4	acetic acid	90° C	68	16	18
4	H_2SO_4	acetic acid	warm	87		8
4	60% $HClO_4$	none	r.t.	35		20
5a–d	60% $HClO_4$	none	r.t.	80–89		19, 20
5a	BF_3 ether.	benzene	r.t.	45	26	20
5a	BF_3 ether.	benzene	reflux	35	30	20
5a	TsOH	benzene	reflux	56	21	20

tetramer *cyclotetraveratrylene* (CTTV) (**2**), which also has a potential, though as yet unstudied, host structure.

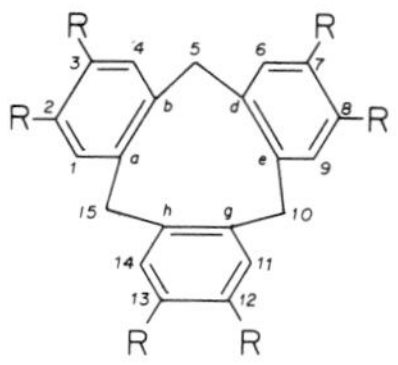

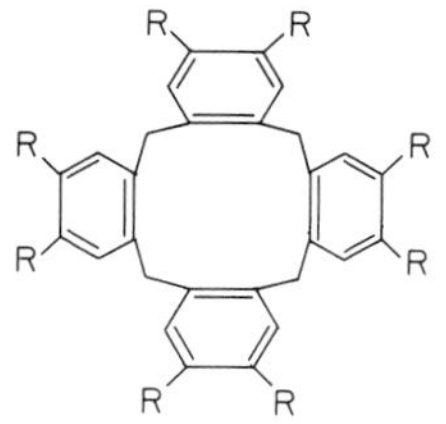

1   R = OMe   2

Although veratrole (**3**) and veratryl alcohol (**4**) have proven to be the most suitable starting materials for preparing CTV, alternative syntheses have been reported[19,20] in which the intermediate veratryl cation is generated from *N*-tosyl veratrylamine derivatives, **5a–d**. These compounds have been shown to afford CTV in virtually quantitative yield in the presence of 60% perchloric acid. Moreover, relatively good yields of CTTV (**2**) have been obtained from **5a** by using BF_3 etherate as the catalyst in benzene solution.[20]

The undesired or unexpected formation of CTV was also reported on several occasions: as the result of the self-condensation of veratryl chloride on standing in the presence of moisture,[26,27] during the course of the attempted distillation of veratryl alcohol in the presence of traces of nickel,[28] and in the oxidation of laudanosine **6**.[13]

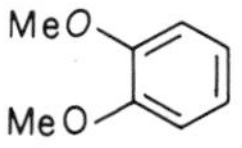

3 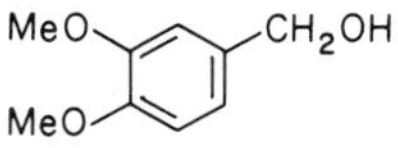   4 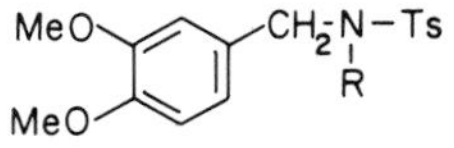   5

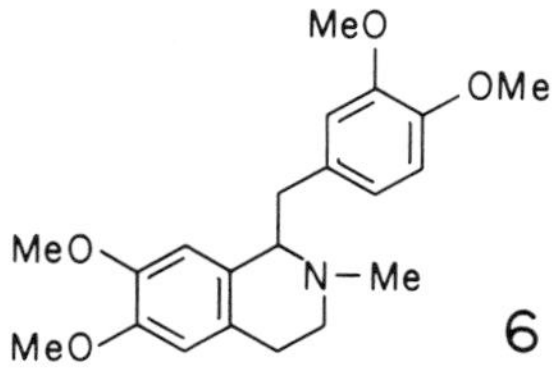

a   R = CH_2CH_2OH

b   R = Et

c   R = Me

d   R = H

6

As yet it is not entirely clear whether the ease of formation of CTV in these reactions reflects a thermodynamic equilibrium in favour of the cyclic trimer (possibly displaced by crystallization), or is due to a kinetically controlled process. Moreover, different mechanisms can be expected, depending on whether veratrole and formaldehyde, or a veratryl alcohol

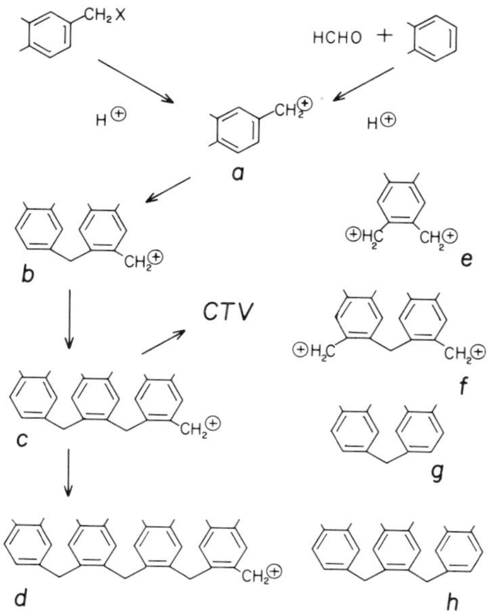

Scheme 1

derivative, are employed as starting materials (Scheme 1). In the latter case, the reaction could, in principle, proceed stepwise via the mono-, di-, and trimeric cations (a), (b), and (c). Intermediate (c) can either cyclize to CTV, or react with (a) to give (d), a precursor of CTTV and the higher polymers.

On the other hand, starting from veratrole and formaldehyde, we can expect the presence of additional intermediates, including (e), (f), (g), and (h). Reaction of (e) with (g), of (f) with veratrole, or of (h) with formaldehyde, would lead to (c) which in turn can cyclize to CTV.

Experimentally, the presence of (e) and (f) can be inferred from the fact that the bis(chloromethyl) derivatives **7** and **8** have been isolated during the course of a temperature controlled chloromethylation of veratrole. Compound **8** has been shown to react with veratrole in refluxing acetic acid to give CTV.[8] The formation of the tetramethoxydiphenylmethane (g) has been observed at the onset of the same reaction carried out at 263 K.[27] The behaviour of this intermediate has been the object of controversial results. According to Robinson,[1] (g) condenses with formaldehyde in sulphuric acid to give CTV; this finding, which led to the conclusion that CTV should be a dimer, was contested by Lindsey,[8] and subsequently confirmed by Umezawa et al.,[20] and Arcoleo et al.[29] The formation of CTV from (g) requires that the latter be first cleaved back to intermediates (a) or (e), and therefore suggests that some among intermediates (a)–(h) could be reversibly formed (however, see below).

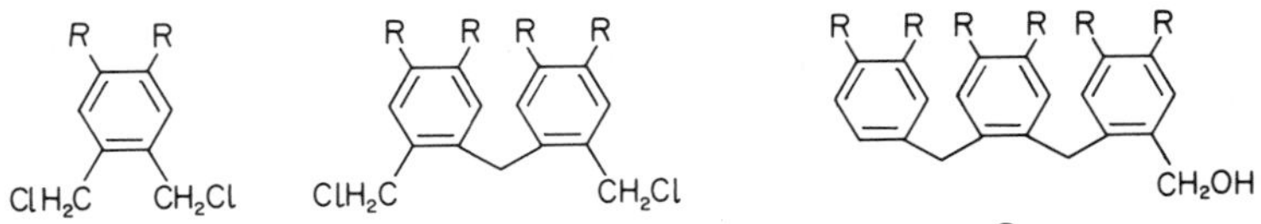

$$(R = OMe)$$

Compound (h) has also been isolated and subsequently transformed into CTV, by reaction with formaldehyde.[2] Finally, the ease with which the cyclization of intermediate (c) can occur is attested by the fast and quantitative conversion of the benzylic alcohol (9) into CTV, in the presence of perchloric acid at 3° C.[20]

2.2. Preparation of analogues

Several general methods are available for the synthesis of CTV analogues: (1) the condensation of suitably substituted benzenes with formaldehyde in the presence of strong acids, or the acid-catalyzed trimerization of appropriate benzyl alcohols or halides; (2) the condensation of 6,6'-dihalomethyl-, or 6,6'-dihydroxymethyldiphenylmethanes with suitable benzene derivatives, and (3) the chemical modification of the CTV structure itself.

2.2.1. The trimerization route
The utilization of the acid-catalyzed condensation of aromatic compounds with formaldehyde to obtain CTV analogues is restricted, in practice, to 1,2-disubstituted benzenes bearing electron-donating groups, such as the *catechol ethers* (10). However, this reaction is expected to produce a mixture of two racemic regioisomers, having (C_1) and (C_3) symmetry, unless a symmetrical ether is involved. For example, methylenedioxybenzene (10a), and 1,2-diethoxybenzene (10b), afford the achiral (C_{3v})-trimers 11a (cyclotripiperonylene), and 11b, respectively,[4,5] whereas the dissymmetrical ether 10c furnishes a mixture of the (C_1) and (C_3) isomers 11c.[53]

In contrast to the lack of regioselectivity in the reaction of non-symmetrical catechol ethers, the condensations of suitably substituted *benzyl alcohol derivatives* such as 12,[30–33] 13,[37] and others,[54,55] have been reported to afford preferentially, if not exclusively, the corresponding (C_3)-trimers. The latter method, rather than the condensation of catechol ethers with formaldehyde, should therefore be recommended for the synthesis of

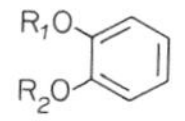

R$_1$O— / R$_2$O— benzene ring **10**

a R$_1$, R$_2$ = CH$_2$<

b R$_1$ = R$_2$ = Et

c R$_1$ = Me R$_2$ = CH$_2$CO$_2$H

R$_1$O— / R$_2$O— CH$_2$OH ring **12**

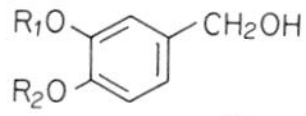

R$_1$ = Me , Et

R$_2$ = CH$_2$CO$_2$H , CH(Me)CO$_2$H

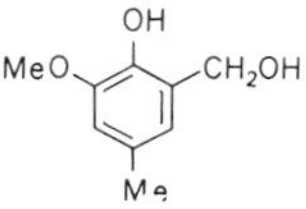

MeO— OH / —CH$_2$OH **13**

Me

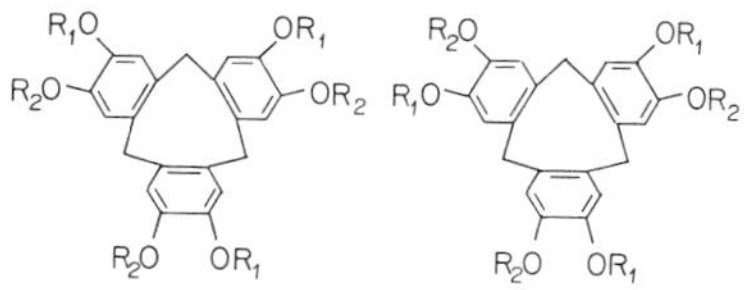

C_1 C_3

11

a R$_1$, R$_2$ = CH$_2$<

b R$_1$ = R$_2$ = Et

c R$_1$ = Me R$_2$ = CH$_2$CO$_2$H

analogues having C_3 symmetry.[55] Incidentally, we note that the apparent regiospecificity of the trimerization in this case raises an interesting question, with regard to the mechanism discussed in Section 2.1 above and, specifically, on the extent of reversibility of the reaction. In fact, these results probably indicate that the stepwise condensation (a) → (b) → (c) → (C_3)-trimer (Scheme 1) occurs faster than the cleavage and equilibration of the intermediates or of the final product.

2.2.2. *The diphenylmethane route*

The condensation of the bis(chloromethyl)diphenylmethane (**8**) with 1,2-disubstituted benzenes to give analogues of structures **14a–c** was originally described by Lindsey.[8] The first optically active CTV derivative (**15**) was prepared by this method, and was partially resolved by chromatography over cellulose acetate into stable enantiomers exhibiting $[\alpha]_{405} +7.5°$ and $-5°$.[12]

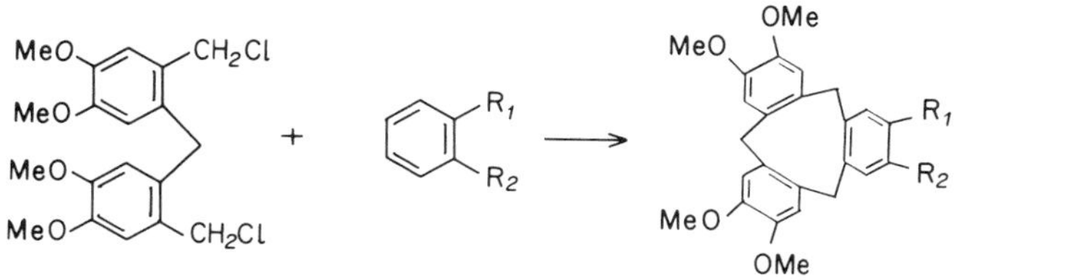

MeO— / —CH$_2$Cl
MeO—
MeO—
MeO— / —CH$_2$Cl

8

+

benzene ring —R$_1$ / —R$_2$

→

product structure **14**

a R$_1$ = OMe R$_2$ = OEt

b R$_1$ = OMe R$_2$ = Me

c R$_1$ = R$_2$ = OEt

15 R$_1$ = OMe R$_2$ = OCH$_2$Ph

Sato *et al.*[23–25] have generalized this reaction to the synthesis of analogues devoid of methoxy substituents (**19–21**), by condensation of diols **16–18** with benzene in the presence of sulphuric acid under high dilution conditions. The hydrocarbon *cyclotribenzylene* (**19**) adopts the same rigid crown conformation as CTV. This is no longer the case for the oxa- and thia-derivatives **20** and **21**: we will return to this point in Section 2.4.

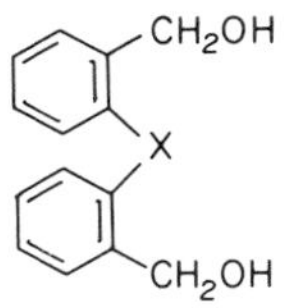

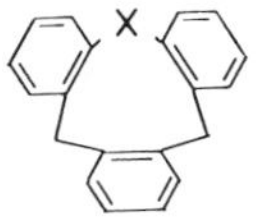

16	X = CH₂	**19**
17	X = O	**20**
18	X = S	**21**

(formula labels: X = CH$_2$, X = O, X = S)

2.2.3. Chemical modification of the ring system

Only few reactions have been reported which allow the functionalization or the derivatization of CTV, without destroying the nine-membered ring. Nitration, as well as bromination and chlorination, only afford ring cleavage products (**22**) which are further cleaved to diphenylmethane derivatives.[21,22,24,25] This result should probably be ascribed to the high electron density induced in the vicinity of the nine-membered ring by the six methoxy substituents, rather than to ring strain; in effect the parent hydrocarbon **19** gives on nitration the mononitro derivative **23**, with no indication of ring cleavage.[24]

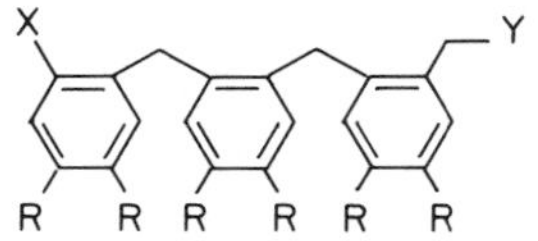

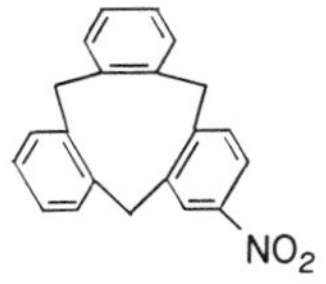

22   X = NO₂, Cl, Br
          Y = OH, Cl, Br
          R = OMe

23

CTV and the parent ring system **19** can be oxidized in good yield to the corresponding monoketones **24**[7,34,35] and **25**[23,25], which exhibit flexible conformations (Section 2.4). Further oxidation of **24** leads to a compound which was first considered[35] to be the triketone **26**, and which in fact[36] is the rearranged lactone **27**.

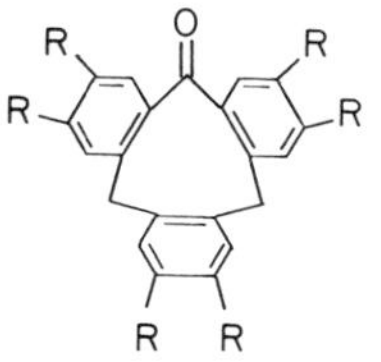

24   R = OMe   **26**   **27**

25   R = H

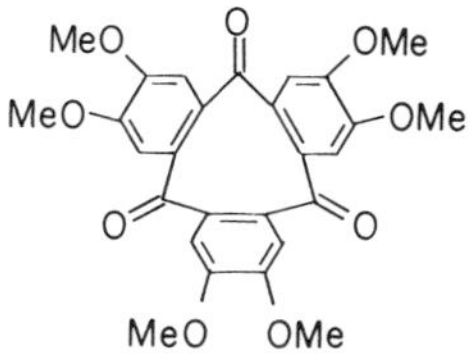

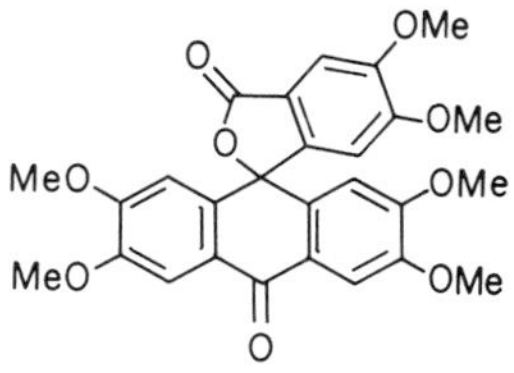

The carefully controlled reduction of ketone **24** with sodium borohydride or lithium aluminium hydride exclusively affords the unstable, conformationally flexible, alcohol **28a**, which in turn can be readily and irreversibly converted on heating into the stable isomer **29a**, having a rigid crown conformation, with the OH group equatorial.[34,35] It is interesting that the labile alcohol **28a** can be virtually instantly etherified by alcohols such as ethanol in the presence of traces of mineral acid, to give the corresponding flexible ether **28b**, which on heating transforms into the stable ether **29b**. The reactivity of the unstable alcohol **28a** might be due to a better stabilization of the cationic transition state (S_N1 process) in the flexible conformer rather than in the crown. The sequence of reactions **24→28a→28b→29b** therefore provides a very convenient route for the functionalization of CTV at the methylene bridge positions.

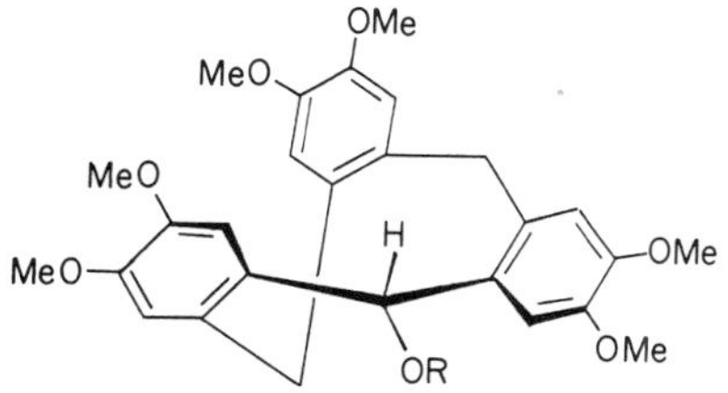

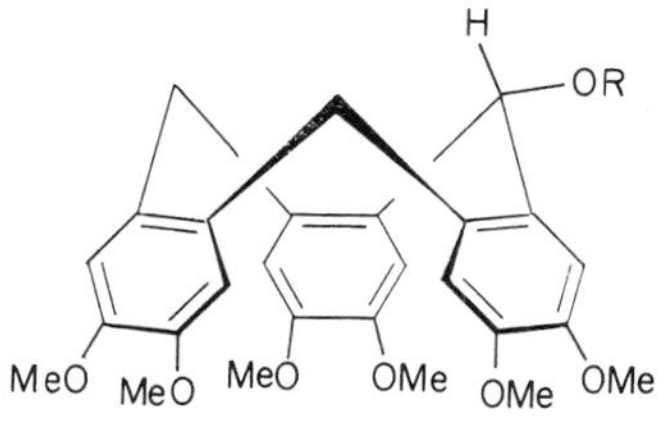

28   *a*   R = H   **29**
                b   R = Et

Another attractive route for derivatization of CTV consists in its complete demethylation, which can be satisfactorily achieved with boron tribromide.[8] The resulting hexaphenol, *cyclotricatechylene* (**30**),[15] has been utilized for the synthesis of various hexa(O-alkylated) derivatives described in Section 3.

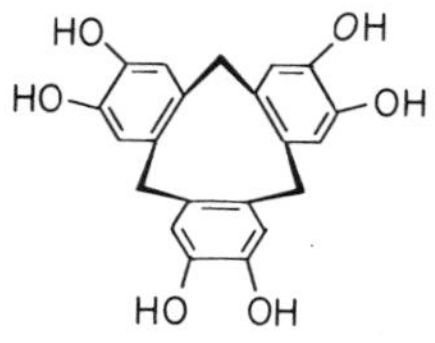

30

Recently, an efficient process allowing stepwise deoxygenation of *cyclo-triguaiacylene* (**34**) into (**19**), via *cyclotrianisylene* (2,7,12-trimethoxycyclo-tribenzylene), had been described.[56]

2.3. Optically active analogues

To date the interest in optically active CTV analogues has been focused principally on the problem of their synthesis, and on the studies of their intrinsic properties. More recently, their utilization in the design of new chiral hosts has been envisaged, and one example of such a compound was reported;[33] this is the reason why a brief account of the work in this direction is warranted here.

With the exception of the monobenzyloxy derivative **15**, having C_1 symmetry, which was resolved by Lüttringhaus and Peters[12] in 1966, virtually all optically active analogues known so far belong to the C_3 point group, and have been described by Collet *et al.*[30–33,38,39,56] The preparation of these

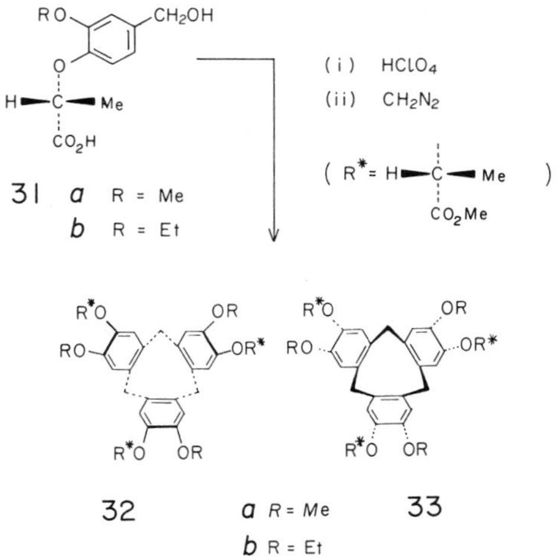

Scheme 2

compounds is based on the condensation of an optically active vanillyl alcohol derivative, e.g. R(+)-**31a**, in the presence of perchloric acid.[30] This reaction affords a mixture of diastereoisomeric (C_3)-trimers, **32a** and **33a**, which differ only in the configuration (M or P)† of the crown structure (Scheme 2). The chromatographic separation of these diastereoisomers is followed by the cleavage of the chiral auxiliary groups R*, to give the optically active triphenols P(−)-**34** and M(+)-**34**, $[\alpha]_D \pm 253°$ (in chloroform), which can be further alkylated,[31] or deoxygenated,[56] thus giving access to a variety of optically active (C_3) analogues of CTV.

   The same sequence of reactions has been followed starting from R(+)-**31b**, and the structure and absolute configuration of diastereoisomer **32b** have been established by single-crystal X-ray crystallography.[31] The absolute configurations of enantiomers **34–40** have been derived from that of **32b** by chemical correlation, in agreement with independent assignments based on an exciton analysis of their circular dichroism.[38–39,57]

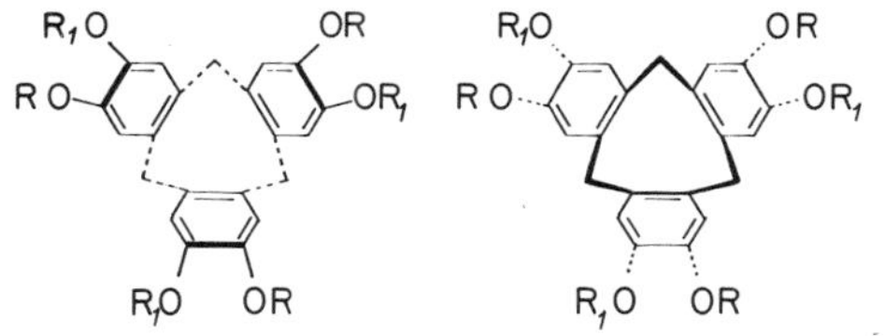

	R	R₁	
P (−) **34**	Me	H	M (+) **34**
M (−) **35**	Me	CD₃	P (+) **35**
M (−) **36**	Me	Ac	P (+) **36**
P (−) **37**	Et	H	M (+) **37**
M (−) **38**	Et	i-Pr	P (+) **38**
M (−) **39**	Et	Ac	P (+) **39**
P (−) **40**	Et	Me	M (+) **40**

2.4. Geometry and conformational behaviour

CTV has been shown by NMR spectroscopy to possess a stable crown conformation.[8–11] The geometry of the crown system (Fig. 1) may be defined[31] by the angle $\Phi = 43 \pm 2°$ between the plane of each benzene ring and the C_3 axis, and by the distance $d = 4.8$ Å between their centres. The aromatic hydrogen atoms, e.g., H(1) and H(14), are virtually at contact

† The specification of the chirality of CTV analogues by means of the P and M descriptors is described in ref. 31.

distance $(2.5 \pm 0.1$ Å between their centres)*. The pseudo-axial hydrogens of the methylene bridges (H_a), separated by only 2 Å, experience a severe steric compression and, as a consequence, resonate 1.2 ppm downfield with respect to their pseudo-equatorial (H_e) counterparts. The resulting AB (nearly AX) quartet shown in Fig. 1 does not change over a temperature range up to 200° C, indicating a high resistance of the nine-membered ring to inversion.

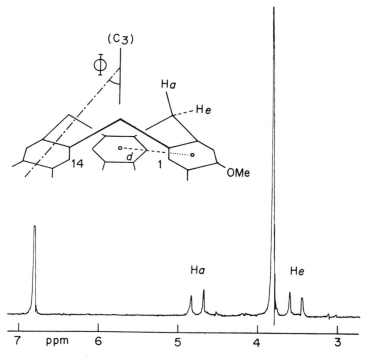

Fig. 1. Geometry and ^{1}H NMR spectrum of cyclotriveratrylene in CDCl$_3$ solution (δ from internal TMS, 90 MHz).

The energy barrier for the crown-to-crown interconversion process in CTV, which is too high to be measured by NMR techniques, has been found[32] to be 110.9 kJ mol^{-1}, from the racemization rates of optically active (C_3)-CTV-d_9 (**35**). Barriers of comparable magnitude have been measured by the same method in the case of (C_3)-*cyclotriguaiacylene* (**34**), and its triacetate (**36**).[38] The activation parameters for ring inversion in **34–36** are assembled in Table 2.

---

* Calculated from the X-ray atomic co-ordinates of H-atoms in cyclotricatechylene (ref. 15).

Table 2. *Activation parameters for crown inversion in* **34–36**

	Ea (kJ mol^{-1})	A s^{-1}	$\Delta H^{\neq}$ (kJ mol^{-1})	$\Delta S^{\neq}$ (J mol^{-1} K^{-1})	$\Delta G^{\neq}_{298}$ (KJ mol^{-1})
34	113.4	0.6×10^{13}	110.5	−9.6	113.4
35	110.9	0.7×10^{13}	108.4	−7.9	110.9
36	130.0	1.4×10^{13}	110.5	−2.1	110.9

The rate constants for crown inversion in these compounds are close to 10^{-7} s^{-1} at 20° C; this value indicates a half-life of *c.* 10^6–10^7 s for a given crown conformer at room temperature. The calculated lifetimes ($t_{1/2}$) for the racemization of **35** are of the order of 960 days, 36 days, and 3 minutes at 0, 20, and 100° C respectively. Complete racemization is expected to occur within a few seconds at 200° C; this is not inconsistent with the invariance of the NMR spectrum on heating, which is due to the large frequency difference of the exchanging sites $H_a \rightleftharpoons H_e$ ($\Delta \nu \sim 75$ Hz at 60 MHz) with respect to the rate constant of the inversion at 200° C ($k \sim 4$ s^{-1}).

The energy barrier for ring inversion in CTV is about 50 kJ mol^{-1} higher than in the parent ring system *cis, cis, cis*-1,4,7-cyclononatriene **41**.[40–42] The

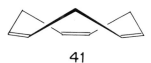

41

mechanism of the conformational interconversion in these compounds has been discussed by Sato and Uno[25] and by Dale.[43] Detailed energy calculations on the tribenzocyclononatriene system (**19**) have been reported by Ermer.[44] It is assumed that inversion of the crown occurs through a flexible conformer readily pseudorotating, via *twist* forms, among six equivalent *saddle* forms, or among three equivalent pairs of enantiomeric saddles in the case of chiral (C$_3$) compounds such as **34–36** (Scheme 3), or three non-equivalent pairs in the case of (C$_1$) compounds such as **15**. The measured energy barrier, corresponding to the rate-determining step, certainly represents the passage from a crown to a saddle form, a process which might involve either the flipping of one phenyl ring (of one *cis* double bond in **41**), or the flipping of one methylene bridge; the latter mechanism has been favoured by Ermer[44] on the basis of force-field calculations, whereas the former has been preferred by Sato,[25] and by Dale.[43]

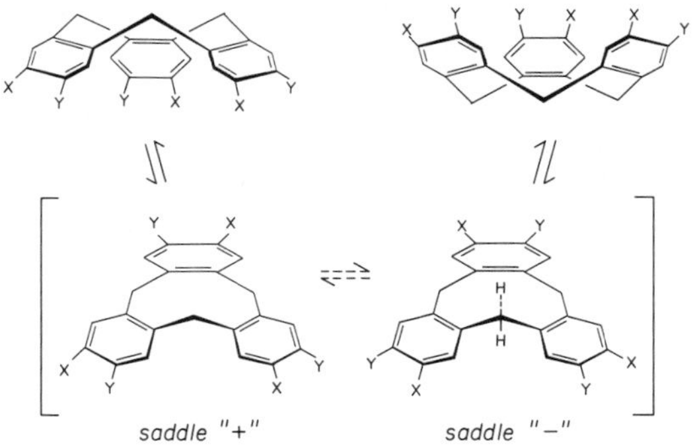

A. Collet

saddle "+"   saddle "−"

Scheme 3

The intermediate saddle/twist form, which is not observable in CTV itself, should therefore be disfavoured with respect to the crown by at least $12–17 \, kJ \, mol^{-1}$, which might be due, in part, to a repulsive interaction between the inward hydrogen atom of the methylene bridge pointing into the ring, and the opposite phenyl group. The flexible conformer may become populated only when suitable structural changes in the molecule leads either to its stabilization, or to the destabilization of the crown form.

The former reason probably holds in the monoketones **24** and **25** which exist exclusively in a rapidly interconverting twist form, allowing better conjugation of the carbonyl group with the adjacent aromatic rings. When the ketone **24** is reduced, the conjugation energy is lost, and the resulting, metastable, flexible alcohol **28b** rapidly returns to the crown form. In the exocyclic methylene analogue **42**, both flexible and crown forms are in equilibrium, and the barrier for the flexible to crown transformation has been estimated at $c. \, 92 \, kJ \, mol^{-1}$.[35]

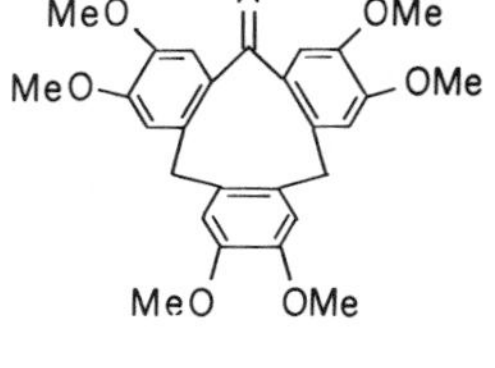

24	X = O	
42	X = CH₂	
43	X = Me, OH	

Destabilization of the crown form usually arises from steric hindrance created either by geminal substitution of one methylene group, as in **43**,[35] or by substitution at the aromatic positions *ortho* to the nine-membered ring. This is the case, for example, in the tribromo derivative **44**, a naturally-occurring substance isolated from red algae *Halopytis pinastroides*,[45] and in compound **45**.[37]

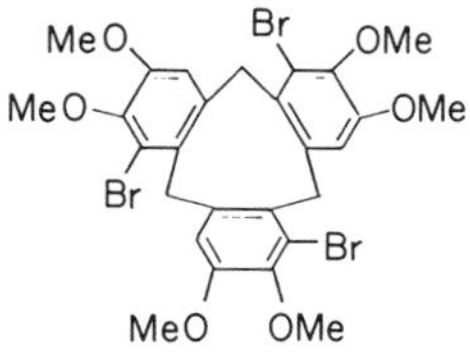

The replacement of one methylene bridge by heteroatoms such as oxygen or sulphur also has conformational consequences. The oxonin (**20**) is a flexible molecule, which probably undergoes repeated interconversion between saddle and twist forms,[43] rather than a fast exchange between crown forms, as was originally suggested.[25] The preferred conformation of **20**, which becomes observable by NMR at 183 K, might be a saddle form with the ether oxygen pointing into the ring. In contrast, both the sulphide **21**, and the sulphone **46**, are conformationally locked. Although a rigid crown conformation was proposed,[25] a rigid saddle form, with the sulphur atom pointing into the ring, cannot be ruled out on the basis of the reported NMR data.

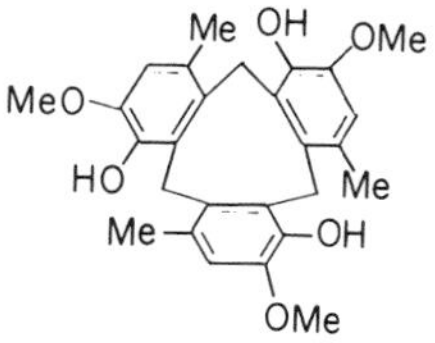

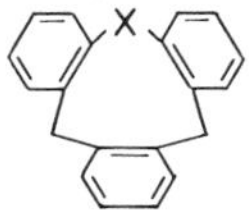

20	X = O
21	X = S
46	X = SO_2

Finally, the conformational behaviour of cyclotetraveratrylene **2** has been studied by White and Gesner,[18] and has been later discussed by Dale.[43] CTTV, which is related to the (unknown) *cis,cis,cis,cis*-1,4,7,10-cyclo-dodecatetraene,[44] has been found by variable temperature NMR experiments to be a flexible molecule. The preferred conformation, observable at 183 K, is a sofa (C_{2h}), exchanging among 8 equivalent forms over a barrier of 54 kJ mol^{-1}.

3. Inclusion compounds

CTV and many of its analogues, like many other compounds having trigonal symmetry (see Chapter 5, Volume 2), form crystals which are not close packed and which can thus accommodate suitable guest molecules within their lattice. This property, as we shall see, is not always associated with a crown conformation. On the contrary, the existence of a rigid, bowl-shaped geometry is probably of greater importance for the formation of molecular complexes.

3.1. Crystalline inclusion compounds

The compounds discussed here form crystalline inclusion compounds and, with the exception of trithia-CTV (Section 3.1.3), probably cannot form stable molecular complexes in solution.

3.1.1. Cyclotriveratrylene

The ability of CTV to form crystalline solvates (with water or benzene) was first observed by Bhagwat *et al.*[13] Caglioti *et al.*[6] subsequently reported the formation of inclusion compounds with benzene, chlorobenzene, toluene, chloroform, acetone, carbon disulphide, butyric acid, acetic acid, thiophene, and decalin. From infrared and X-ray measurements (determination of the cell dimensions) carried out on a number of these complexes, two types of monoclinic phases (α and β) were identified, depending on whether bulky molecules (C_6H_6, $CHCl_3$) or thread-like molecules (CS_2, butyric acid) were included. However, at that time, the host was still considered to be a hexamer, and the host:guest stoichiometries indicated by Caglioti should therefore be modified accordingly. The corrected values have been listed in Table 3, together with those of other complexes described by Hyatt *et al.*[15]

The crystal structure of the benzene compound, which in fact is a CTV-benzene-water complex with a 1:0.5:1 ratio, was determined in 1979 by Cerrini *et al.*[14] and simultaneously by Cesario *et al.*[52] The structure, which was solved in the monoclinic space group C2/c, consists of columns of CTV molecules, juxtaposed parallel to a crystallographic axis *b*, as shown in Fig. 2. The arrangement of the columns, projected onto the *ac* plane, defines a nearly rectangular network (Fig. 2a) which is not amenable to

Table 3. *Cyclotriveratrylene inclusion compounds*

Guest molecules	Type	b (Å)	Ratio CTV:guest	Refs.
Benzene	α	9.61	1:0.6	6
Benzene-water	α	9.629	1:0.5:1	14
Chlorobenzene	α	9.64	1:0.55	6
Toluene	α	9.73	1:0.1	6
Chloroform	α	9.78	1:1.46	6
Acetone	β	8.39	1:0.27	6
Carbon disulfide	β	8.28	1:0.48	6
Butyric acid	β	8.07	1:1	6
Ethyl acetate			1:1.6	15
Methyl ethyl ketone			1:3.2	15
Ethanol			1:1.5	15

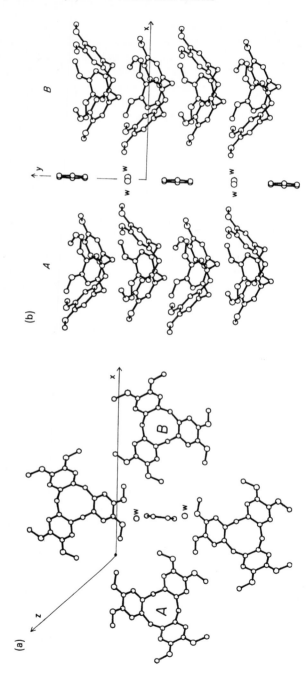

Fig. 2. Crystal structure of the CTV-benzene-water inclusion compound: (a) view along *b*, showing the cross section of the channel; (b) projection on the *ab* plane showing the packing of columns A and B and of the benzene and water (W) molecules within the channel (after Cesario *et al.*[52]).

close packing, and provides channels parallel to *b*, in which the benzene
and water molecules are accommodated. The channels have an approximately oval section, with a maximum extension almost perpendicular to *a*,
and are constricted every 9.63 Å by waists formed by methyl groups. A
channel can thus be described as a succession of cages communicating with
each other. The same structure probably holds for the other compounds
listed in Table 3, since *b*, which represents the stacking of two molecules
in a column, is nearly constant.

A 1:0.5 CTV-guest ratio requires that there is one guest molecule in each
cavity, and this seems to be the case with the benzene, chlorobenzene,
toluene, and carbon disulphide compounds. The data in Table 3 also suggest
the presence of two molecules of butyric acid, or of three molecules of
chloroform, ethyl acetate, or ethanol, per cage. Not all the cages seem to
be occupied in the toluene and acetone compounds, either for steric reasons
(toluene), or because the molecule can easily leave the crystal by moving
throughout the channels (acetone). A surprisingly high ratio (*c.* six
molecules/cage) has been reported in the methyl ethyl ketone adduct.

The stability of the CTV inclusion compounds is not very high, and these
can be desolvated relatively easily by heating under vacuum. In the same
way, when the benzene complex is immersed in chlorobenzene, in which it
is scarcely soluble, the benzene molecules leave the crystal and are replaced
by those of chlorobenzene.[14]

3.1.2. Cyclotricatechylene

According to Hyatt *et al.*,[15] the hexaphenol cyclotricatechylene (CTC) (**30**),
forms a wider range of well defined inclusion compounds than does CTV
itself (Table 4). It is interesting that several molecules, ranging in size from

Table 4. Inclusion compounds of cyclotricatechylene[a]

Guest molecules	CTC:guest ratio
N,N-dimethylformamide	1:3.1
N-methylpyrrolidone	1:3.0
N,N-dimethylacetamide	1:3.1
dimethylsulfoxide	1:3.0
Water	1:3.0
HMPA	1:3.0
Acetone	1:2.0
2-propanol	1:2.0

[a] Reproduced from J. A. Hyatt, E. N. Duesler, D. Y. Curtin
and I. C. Paul, *J. Org. Chem.*, 1980, **45**, 5074, by permission of
the publisher; © 1980, The American Chemical Society.

water to hexamethylphosphoric triamide, give compounds having a 1:3 host-guest ratio, whereas other molecules, although of comparable size, give well-defined 1:2 compounds. Also surprising is the fact that, although water, and 2-propanol, form inclusion compounds, ethanol does not. The compounds listed in Table 4 are indefinitely stable in air, and release the guest molecules only upon heating under vacuum.

The crystal structure of the 1:2 CTC/2-propanol compound has been determined; the crystals belong to the triclinic system, space group $P\bar{1}$, with two CTC molecules and four guest molecules in the unit cell. The packing consists of rows of molecules, parallel to the b axis, that provide channels running in the same direction (Fig. 3). Adjacent rows translationally related along a form the "walls" of the channels (Fig. 3a), whereas centrosymmetrically related rows form the "sides" (Fig. 3b). The channels are constricted by several OH groups, so as to give a series of cavities, each of which contains two guest molecules. The structure is held by hydrogen bonding involving the six OH groups of each CTC molecule as sketched in Fig. 3b.

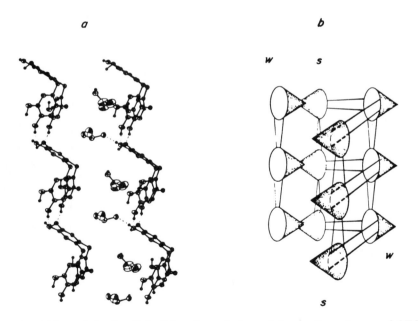

a   *b*

Fig. 3. (a) View of the "walls" of the channel, formed from adjacent rows of CTC molecules; (b) schematic drawing showing the "walls" (W), the "sides" (S) and the hydrogen bonding pattern between the host molecules. Reproduced from J. A. Hyatt, E. N. Duesler, D. Y. Curtin, and I. C. Paul, *J. Org. Chem.*, 1980, **45**, 5074, by permission. © 1980, The American Chemical Society.

The two guest molecules located in each pocket of the channel are not equivalent. One, which exhibits disorder, has its 2-propyl group sitting in the concave cavity of a CTC molecule, while the other, the ordered one, is located at the junction of four CTC molecules.

3.1.3. Trithiacyclotriveratrylene

Trithia-CTV (47) has been prepared by the diphenylmethane route (see Section 2.2.2), by reacting the dithiol **46** with 4,5-dibromoveratrole in the presence of Cu_2O.[46] According to von Deuten *et al.*[47,48] **47** exists in a temperature and solvent dependent equilibrium of the crown and the saddle forms. It is interesting that both conformers can be isolated in a pure state by a suitable choice of solvents, from which they form crystalline inclusion compounds. Thus, the crown form crystallizes from chloroform as a 1:2 adduct, and the saddle form is obtained from benzene as a 1:0.5 adduct, the crystal structure of which has been determined.[48]

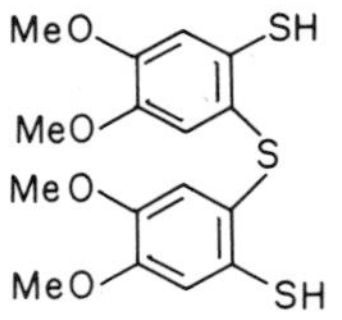

46 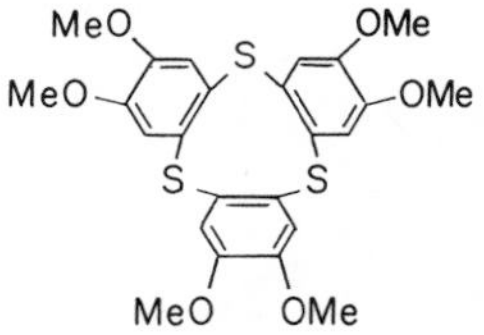   **47**

Trithia-CTV also forms metal complexes, such as [47-CuBr-H_2O-acetone], in which the copper is tetrahedral, [47-Rh(NO_3)$_3$-(dimethylacetamide)$_3$], in which the rhodium is octahedrally co-ordinated[47] and [47-PtCl$_2$-(dimethyl-acetamide)$_{3/2}$] in which the Pt has a distorted square pyramidal coordination.[47]

3.1.4. Oxocyclotriveratrylene

The ketone **24** provides an example of a conformationally flexible CTV derivative forming 1:1 crystalline inclusion compounds with benzene, chloroform, and other solvents.[53] The structure of these compounds is not known.

3.2. Molecular complexes

Two main types of CTV derivatives capable of (or designed for) complexation in solution have been described: *octopus molecules*, in which suitable chains have been attached to the six phenolic oxygens, and *macrocages* incorporating one or two CTV subunits so as to provide a molecular cavity of well defined structure.

3.2.1. Octopus molecules

The first octopus molecules in the cyclotriveratrylene series have been described by Hyatt.[16] Compounds **48–53**, bearing linear poly(oxyethylene) chains of various length (1 to 4 units) have been prepared by reacting the hexaphenol **30** with the tosylates of appropriate mono-, di-, tri-, and tetraethyleneglycol monoethers. As shown by the NMR spectra, the crown conformation of the CTV ring is retained in these compounds.

48	n = 1	R = Me
49	n = 2	R = Me
50	n = 2	R = Et
51	n = 2	R = n-Bu
52	n = 3	R = Et
53	n = 4	R = Me

Their ability to solvate alkali metal salts in aprotic solvents was qualitatively examined by means of the methanol-toluene procedure of Htay and Meth-Cohn.[51] With the exception of the short-armed derivative **48**, which is inactive, the solubilizing power of **49–53** is found to be relatively uniform and comparable to that of 18-crown-6. Namely, Na^+, K^+, NH_4^+, and Cs^+ are strongly, albeit non-selectively, complexed, while Ba^{2+} and Mg^{2+} interact only weakly.

Frensch and Vögtle have described two compounds, **56** and **57**, in which crown ethers are attached to the phenolic oxygens of CTV, in order to provide specific complexation sites for alkali metal ions.[49] Reaction of benzo[15]crown-5 (**54**) or of benzo[18]crown-6 (**55**) with formaldehyde in the presence of concentrated HCl afforded the desired ligands **56** and **57** in 4 and 32% isolated yields, respectively.

Although the complexing ability of these compounds in solution was not described, both ligands gave 1:3 crystalline complexes with NaSCN (**56**), and KSCN (**57**), and **57** was also reported to crystallize with two molecules of water. The crystal structures of these complexes are not known as yet. Attempts to obtain crystalline complexes with triacid salts such as sodium aminotriacetate, or silver mesitoate, were unsuccessful.

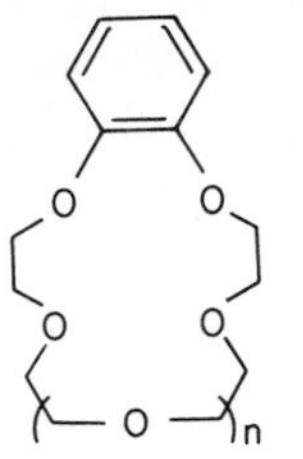

54 n = 1

55 n = 2

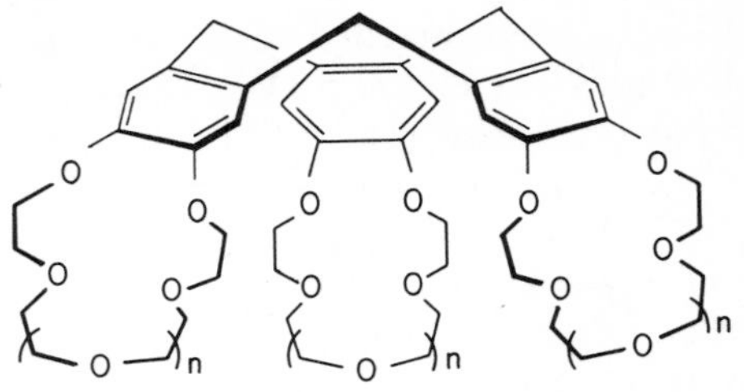

56 n = 1

57 n = 2

The compounds described above are capable of solubilizing mineral cations in aprotic solvents; the inverse behaviour, that is, the solubilization of organic molecules in polar protic solvents such as water, is of particular interest. The hexa-10-carboxydecyl ether of cyclotricatechylene (**58**) synthetized by Menger *et al.*,[17] appears to be the first example of a CTV-octopus molecule that exhibits this property.

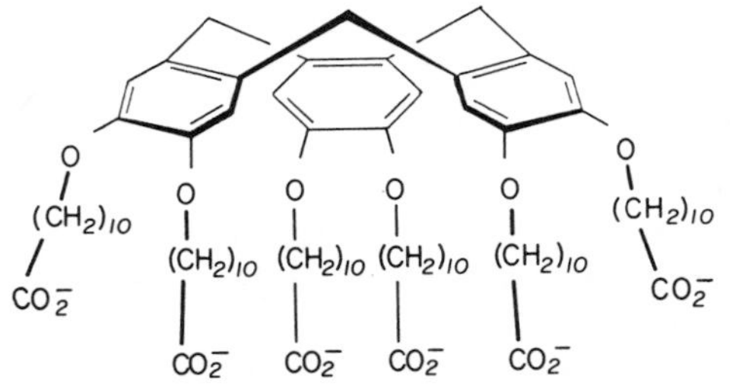

58

The six carboxylate groups in **58** solubilize the compound (>0.04M) in midly basic water in which it forms aggregates, even at concentrations as low as 1×10^{-5} M at pH 9.5 (compared to 1×10^{-2} M for a C_{12} surfactant). Each aggregate is formed from 9 ± 1 molecules of **58** and thus contains about 54 chains, which corresponds to a typical aggregation number in micelles of single-chained surfactants. On the other hand, **58** exhibits much less surface activity than fatty acid salts, presumably owing to the difficulty of placing above the water both the hydrocarbon chains and the bulky aromatic part of the molecule.

Host **58**, in 0.01 M aqueous solution at pH 9.5, is an effective, non selective complexing agent for a variety of organic molecules. The water-soluble dye phenol blue is strongly bound, with an association constant of

$1 \times 10^4 \, M^{-1}$; **58** solubilizes naphthalene (host-guest ratio 2.5:1) and *p*-nitroaniline (1:1), and slightly enhances the solubility of cholesterol in water. Finally, it binds *p*-nitrophenyl butyrate and inhibits its base-catalysed hydrolysis. Whether the complexing properties of **58** are due to the molecule, or to the micellar (or aggregate) structure is not established.

3.2.2. Macrocages

The macrocage **60**, containing two CTV subunits, has been prepared by intramolecular "replicative" cyclization of the tris(vanillyl alcohol) derivative (±)-**59**, in formic acid solution (0.8×10^{-3} M) at 90° C.[33] The reaction is stereospecific, affording exclusively the racemic (D_3)-isomer, in which the new ring is formed with the same chirality as the parent (C_3)-ring. The same reaction, starting from the optically active precursor (−)-**59** ($[\alpha]_D - 65°$ in chloroform), furnishes the active cage (+)-**60** ($[\alpha]_D + 180°$). The macrocage **60** forms a crystalline 1:1 complex with chloroform, the structure of which is not known.

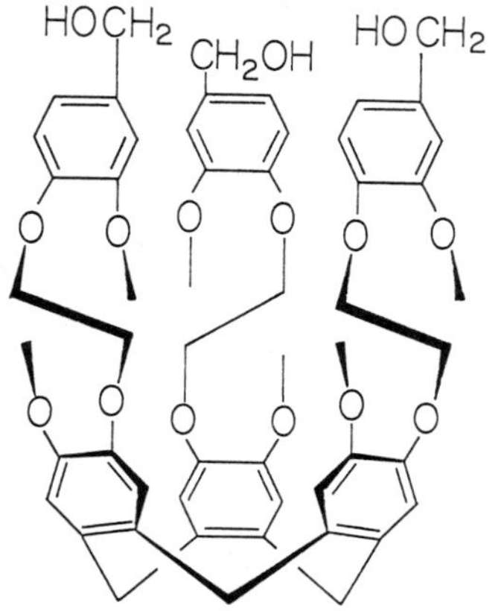

(−) 59 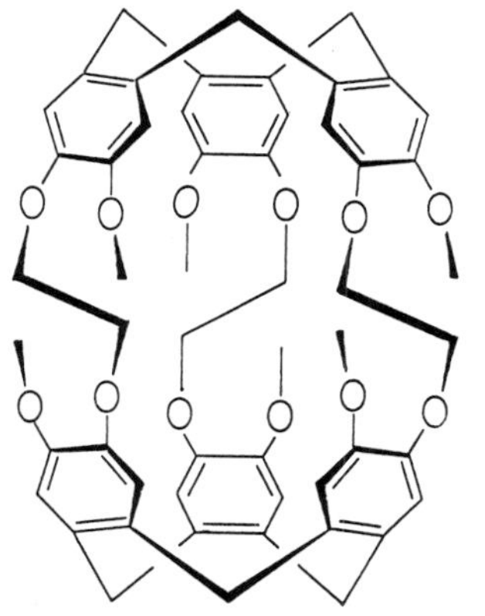   (+) 60

Condensation of the triacid chloride **61** (obtained from triacid **11c**)[33] with the [18]-N_3O_3 crown ether **62** under high dilution conditions, followed by the reduction of the resulting triamide, gives the new macrocage **63**,[50] which combines both a lipophilic cavity, and a binding site for small cations. This compound actually binds strongly and specifically the methylammonium cation in $CDCl_3/CH_3OD$ solution, with the methyl group of the guest being located inside the molecular cavity, as shown from NMR experiments. The slightly larger ethylammonium cation is not complexed by **63**.

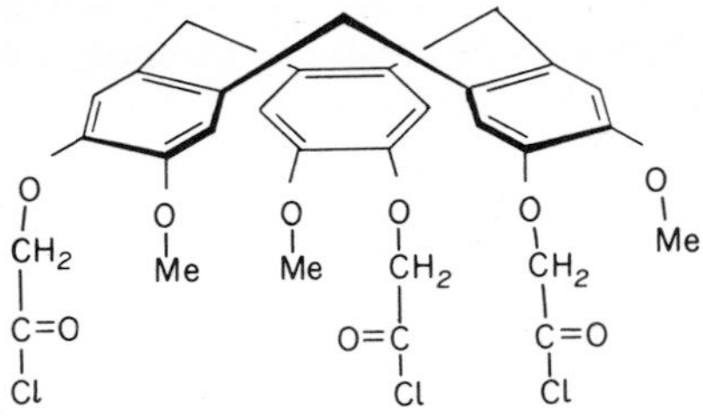

61

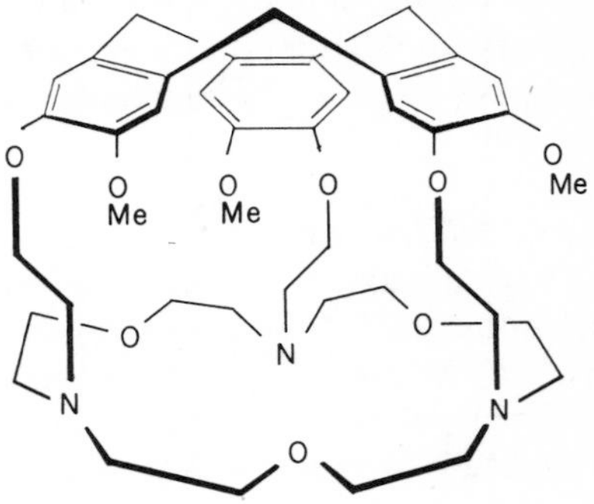

62

63

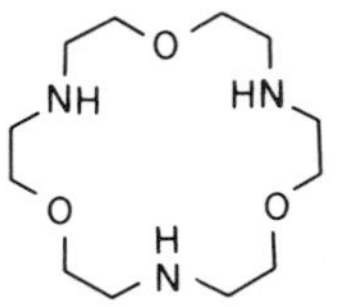

References

1. G. M. Robinson, *J. Chem. Soc.*, 1915, 267.
2. A. Oliverio and C. Casinovi, *Ann. Chim.* (*Rome*), 1952, **42**, 168.
3. C. Casinovi and A. Oliverio, *Ann. Chim.* (*Rome*), 1956, **46**, 929.
4. T. Garofano and A. Oliverio, *Ann. Chim.* (*Rome*), 1957, **47**, 896.
5. T. Garofano, *Ann. Chim.* (*Rome*), 1958, **48**, 125.
6. V. Caglioti, A. M. Liquori, N. Gallo, E. Giglio and M. Scrocco, *Ric. Sci. Suppl.*, 1958, **28**, 3; and *J. Inorg. Nucl. Chem.*, 1958, **8**, 572.
7. A. S. Lindsey, *Chem. Ind.* (*London*), 1963, 823.
8. A. S. Lindsey, *J. Chem. Soc.*, 1965, 1685.
9. H. Erdtman, F. Haglid and R. Ryhage, *Acta Chem. Scand.*, 1964, **18**, 1249.
10. A. Goldup, A. B. Morrison and G. W. Smith, *J. Chem. Soc.*, 1965, 3864.
11. B. Miller and B. D. Gesner, *Tetrahedron Lett.*, 1965, 3351.
12. A. Lüttringhaus and K. C. Peters, *Angew. Chem.*, 1966, **78**, 603; *Angew. Chem., Int. Ed. Engl.*, 1966, **5**, 593.
13. V. K. Bhagwat, D. K. Moore and F. L. Pyman, *J. Chem. Soc.*, 1931, 443.
14. S. Cerrini, E. Giglio, F. Mazza and N. V. Pavel, *Acta Crystallogr.*, 1979, **B35**, 2605.
15. J. A. Hyatt, E. N. Duesler, D. Y. Curtin and I. C. Paul, *J. Org. Chem.*, 1980, **45**, 5074.
16. J. A. Hyatt, *J. Org. Chem.*, 1978, **43**, 1808.
17. F. M. Menger, M. Takeshita and J. F. Chow, *J. Am. Chem. Soc.*, 1981, **103**, 5938.
18. J. D. White and B. D. Gesner, *Tetrahedron Lett.*, 1968, 1591.

19. B. Umezawa, O. Hoshino, H. Hara and J. Sakakibara, *Chem. Pharm. Bull.*, 1968, **16**, 177.
20. B. Umezawa, O. Hoshino, H. Hara, K. Ohyama, S. Mitsubayashi and J. Sakakibara, *Chem. Pharm. Bull.*, 1969, **17**, 2240.
21. B. Umezawa, O. Hoshino, H. Hara and S. Mitsubayashi, *J. Chem. Soc.* (C), 1970, 465.
22. T. Sato, T. Akima, S. Akabori, H. Ochi and K. Hata, *Tetrahedron Lett.*, 1969, 1767.
23. T. Sato, K. Uno and M. Kainosho, *J. Chem. Soc., Chem. Commun.*, 1972, 579.
24. T. Sato, T. Akima and K. Uno, *J. Chem. Soc., Perkin Trans. 1*, 1973, 891.
25. T. Sato and K. Uno, *J. Chem. Soc., Perkin Trans. 1*, 1973, 895.
26. P. Carré and D. Libermann, *C.R. Hebd. Séances Acad. Sci.*, 1934, **199**, 791.
27. P. Carré and D. Libermann, *Bull. Soc. Chim. Fr.*, 1935, 5ᵉ série, **2**, 291.
28. G. Tsatsas, *C.R. Hebd. Séances Acad. Sci.*, 1951, **232**, 530.
29. A. Arcoleo, G. Giammona and G. Fontana, *Chem. Ind. (London)*, 1976, 853.
30. A. Collet and J. Jacques, *Tetrahedron Lett.*, 1978, 1265.
31. A. Collet, J. Gabard, J. Jacques, M. Cesario, J. Guilhem and C. Pascard, *J. Chem. Soc., Perkin Trans. 1*, 1981, 1630.
32. A. Collet and J. Gabard, *J. Org. Chem.*, 1980, **45**, 5400.
33. J. Gabard and A. Collet, *J. Chem. Soc., Chem. Commun.*, 1981, 1137.
34. N. K. Anand, R. C. Cookson, B. Halton and I. D. R. Stevens, *J. Am. Chem. Soc.*, 1966, **88**, 370.
35. R. C. Cookson, B. Halton and I. D. R. Stevens, *J. Chem. Soc. (B)*, 1968, 767.
36. J. E. Baldwin and D. P. Kelly, *J. Chem. Soc., Chem. Commun.*, 1968, 1664.
37. J. F. Manville and G. E. Troughton, *J. Org. Chem.*, 1973, **38**, 4278.
38. A. Collet and G. Gottarelli, *J. Am. Chem. Soc.*, 1981, **103**, 204.
39. A. Collet and G. Gottarelli, *J. Am. Chem. Soc.*, 1981, **103**, 5912.
40. P. Radlick and S. Winstein, *J. Am. Chem. Soc.*, 1963, **85**, 344.
41. K. G. Untch and R. J. Kurland, *J. Am. Chem. Soc.*, 1963, **85**, 346.
42. W. R. Roth, *Justus Liebigs Ann. Chem.*, 1964, **671**, 10.
43. J. Dale, *Topics Stereochem.*, 1976, **9**, 199.
44. O. Ermer, *Aspekte von Kraftfeldrechnungen*, Wolfgang Baur Verlag, Munich, 1981, pp. 368–379.
45. G. Combaut, J-M. Chantraine, J. Teste, J. Soulier and K-W. Glombitza, *Tetrahedron Lett.*, 1978, 1699.
46. T. Weiss, G. Klar, *Z. Naturforsch. Teil B*, 1979, **34**, 448.
47. (a) K. von Deuten, J. Kopf and G. Klar, *Cryst. Struct. Commun.*, 1979, **8**, 721.
    (b) J. Kopf, K. von Deuten and G. Klar, *Cryst. Struct. Commun.*, 1979, **8**, 1011.
    (c) K. von Deuten and G. Klar, *Cryst. Struct. Commun.*, 1981, **10**, 757.
48. K. von Deuten, J. Kopf and G. Klar, *Cryst. Struct. Commun.*, 1979, **8**, 569.
49. K. Frensch and F. Vögtle, *Justus Liebigs Ann. Chem.*, 1979, 2121.
50. J. Canceill, A. Collet, J. Gabard, F. Kotzyba-Hibert and J-M. Lehn, *Helv. Chim. Acta*, 1982, **65**, 1894.
51. M. M. Htay and O. Meth-Cohn, *Tetrahedron Lett.*, 1976, 469.
52. M. Cesario, J. Guilhem and C. Pascard, unpublished work, see ref. 31.
53. J. Canceill and A. Collet, unpublished work.
54. J. Bosch, J. Canals and R. Granados, *An. Quim.*, 1976, **72**, 709.
55. J. Canceill, J. Gabard and A. Collet, *J. Chem. Soc., Chem. Commun.*, 1983, 122.
56. J. Canceill and A. Collet, *J. Chem. Soc., Chem. Commun.*, 1983, 1145.
57. A. Collet and G. Gottarelli, *J. Am. Chem. Soc.*, 1982, **104**, 7383.

5 · STRUCTURE AND DESIGN OF INCLUSION COMPOUNDS: THE HEXA-HOSTS AND SYMMETRY CONSIDERATIONS

D. D. MACNICOL

University of Glasgow, Glasgow, Scotland

1. Introduction

A commonly held view, persisting well into the 1960s, was that new crystalline multimolecular inclusion compounds could only be found by chance. Appearing to consolidate this notion, perhaps, was the knowledge that in the vast majority of organic molecular crystals, in order to minimize the potential energy of the system, the molecules pack together efficiently leaving no significant intermolecular gaps available for guest accommodation. Confronted with this situation, which would imply a very serious barrier to any reasonable rate of progress in this field, the writer, towards the end of the 1960s, started serious consideration of the following questions.

(1) Can new multimolecular hosts be found, otherwise, than by chance?
(2) If so, can fundamentally new cage shapes be produced, ideally can these be produced by design?

And a very fundamental question,

INCLUSION COMPOUNDS 2
ISBN 0-12-067101-8

(3) Can any common features be found from experience which will
suggest a principle or strategy, or indeed strategies, for the synthesis
of new hosts whose molecular structures are not related to any
previously known?

One of many reasons for our enthusiasm in trying to produce new hosts
with *controlled cavity geometry* was the possibility of studying the conformation, reactivity, and other properties of a given guest molecule in a series
of stereocontrolled environments. All very fine, but what method, or
methods, could be used to find new hosts? Well, in the initial studies (the
first strategy) we successfully employed the idea that judicious modification
of a known host might lead to new hosts with properties significantly
different from those of the parent. As described already in Chapter 1, Volume
2, both cavity shape and selective inclusion behaviour have been fundamentally altered for molecules related to Dianin's compound. In the present
chapter, however, different approaches are described. The second strategy,
the hexa-host analogy (*vide infra*), leads to the successful design and
synthesis of new hosts with no *direct* structural relationship to any known
host, while the third strategy, incorporating only a *symmetry* relationship
to known hosts, also provides a new principle for the synthesis of novel
host molecules.

2. The hexa-host analogy, hexa-hosts and related systems

2.1. Nature of the concept

As discussed in Chapter 1, Volume 2, a common feature of the clathrates
formed by hydroquinone, phenol, Dianin's compound, and related systems
is the linking of the OH groups of six host molecules by a network of
hydrogen bonds such that the oxygen atoms form a hexagonal arrangement
(A), shown in Fig. 1. This hydrogen-bonded grouping, which plays a key
role in maintaining the 'open' clathrate structure, is however, clearly of a
temporary nature subject to collapse as R is varied, and indeed most phenols
exhibit denser packing and do not include solvent on crystallization. We
were, however, struck by the parallel between this temporary unit (A) and
the permanent consolidated structure of a suitably hexasubstituted benzene
(B), and reasoned that molecules of the latter type might possess a greatly
increased tendency to crystallize forming non-close-packed structures with
interesting inclusion properties. That is, in this strategy, aptly termed the
hexa-host analogy,[1,2] we hoped to retain the particularly favourable
geometry for guest inclusion, while exploiting the permanent nature of (B)

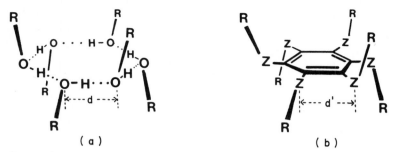

(a)   (b)

Fig. 1. Comparison of (a) hydrogen-bonded hexamer unit with (b) hexasubstituted benzene analogue.

to circumvent hexamer collapse which normally provides an efficient escape route to closer-packed structures. It may be noted that unit (B) corresponds to (A) both in terms of overall geometric aspects and "hexamer" dimensions (cf. distances d' and d in Fig. 1, where Z denotes a general atom or group attached directly to the central benzene ring). For the moment we assume the likely disposition of groups alternately above and below the central benzene ring plane in (B), a point returned to later.

2.2. Synthesis of prospective hexa-hosts and initial observations

In view of the above considerations, the preparation of a number of suitably hexasubstituted benzenes appeared particularly attractive. A literature search revealed that a few of these compounds had been prepared previously and encouragingly, a number of these materials had been found to retain certain solvents, though at the time of publication the nature of inclusion compounds was less well understood. Also, at the time of our initial studies on this strategy in the mid-70's, a number of "octopus" molecules based on a hexasubstituted benzene nucleus, were independently shown to be capable of binding metal ions *in solution.*[3]

Compound **1** was initially prepared in moderate yield by reaction of hexachlorobenzene with PhSCu, as described in a literature method.[4] (An efficient and general synthesis of **1** and related compounds is discussed in a subsequent section). Other hexasubstituted benzenes were prepared[2] by the action of the appropriate phenol, thiol, selenol, amine, or alcohol on hexakisbromomethylbenzene, $C_6(CH_2Br)_6$ in the presence of base.

We now reach the moment of truth: an analogy may have its interest, but the central questions always are, of course, whether the analogy has any validity and, particularly crucial, whether it leads anywhere useful.

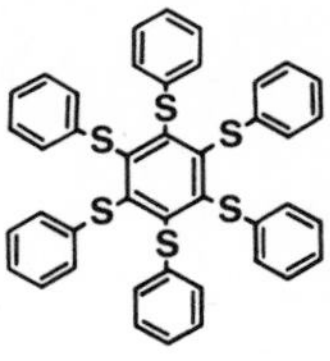

(1)

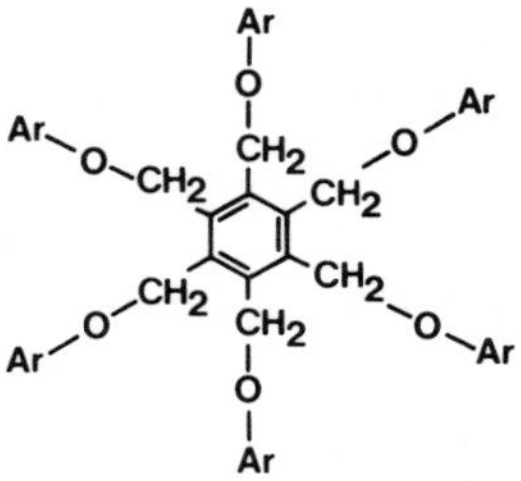

(2) Ar = —⟨phenyl⟩

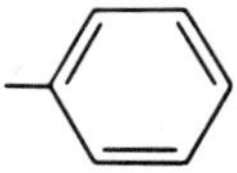

(3) Ar = —⟨phenyl with Me (ortho)⟩

(4) Ar = —⟨phenyl with Me (meta)⟩

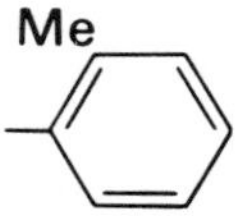

(5) Ar = —⟨phenyl with Me (para)⟩

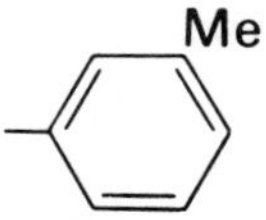

(6) Ar = —⟨phenyl with Pr^i (ortho)⟩

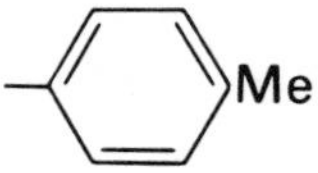

(7) Ar = —⟨phenyl with Pr^i (para)⟩

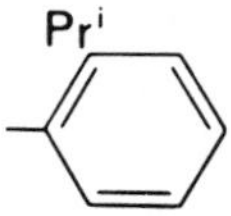

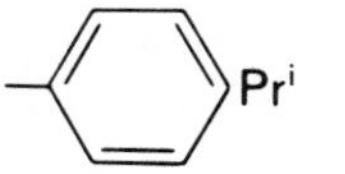

(8) Ar = —⟨phenyl with Bu^t (para)⟩

(9) Ar = —⟨phenyl with (1-adamantyl) (para)⟩

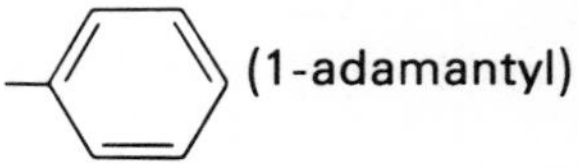

(10) Ar = —⟨phenyl with OMe (para)⟩

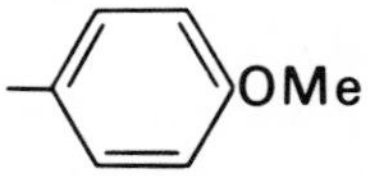

(11) Ar = —⟨phenyl with Me, Me⟩

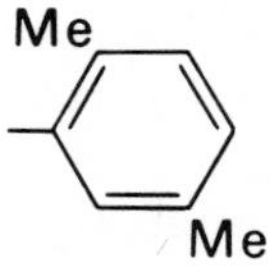

(12) Ar = —⟨phenyl with Me, Me⟩

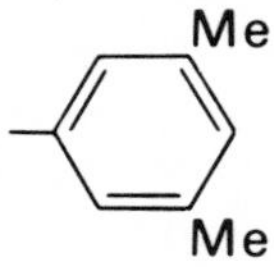

(13) Ar = —⟨phenyl with Pr^i, Me⟩

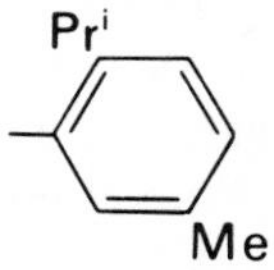

Well, in the present case we were left in no doubt as to the essential correctness and value of the analogy, since a vast range of new inclusion behaviour was immediately uncovered. Among the first compounds studied,[1,2] hexakis(phenylthio)benzene (1) was found to include CCl_4, CCl_3CH_3, and other chlorine-containing guest molecules (Table 1). Hexakis(phenoxymethyl)benzene (2) and its sulphur analogue 14 showed interesting host properties, though no parallel behaviour has yet been found for the related selenium compound 20. Hexakis(*p*-t-butylphenylthio-methyl)benzene (15), further discussed later, turns out to be a particularly

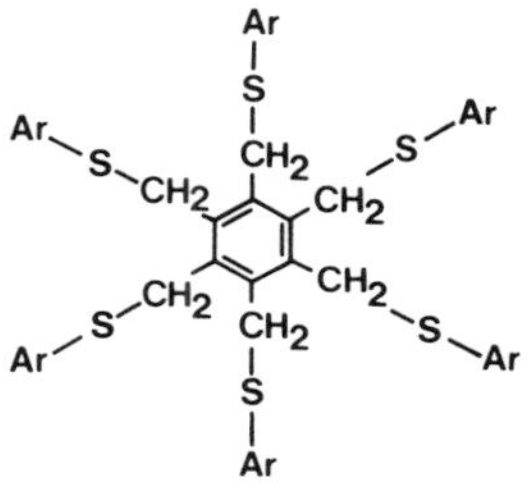

(14) Ar =

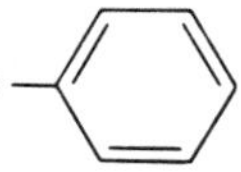

(15) Ar = Bu^t

(16) Ar = (1-adamantyl)

(17) Ar = OH

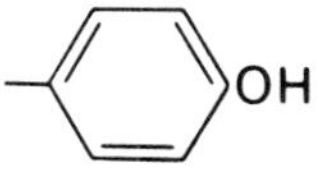

(18) Ar = NH$_2$

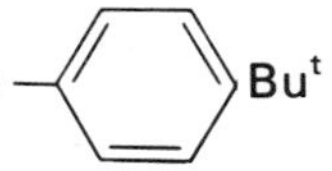

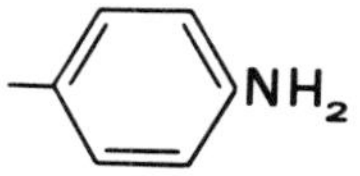

(19) Ar =

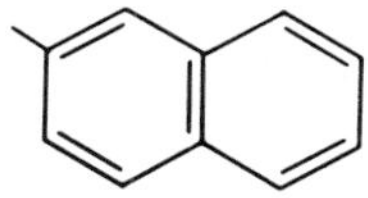

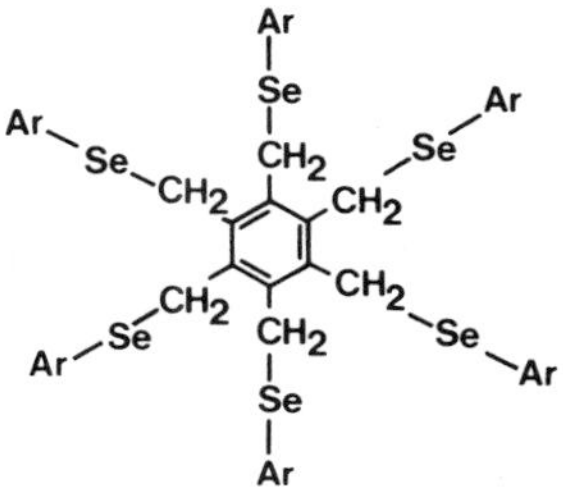

(20) Ar =

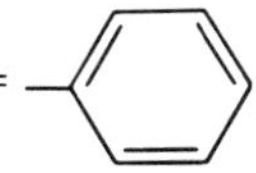

(21) Ar = Bu^t

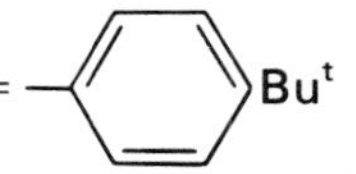

Table 1. Representative inclusion compounds formed by some hexa-host molecules

Host	Guest	Mole ratio of host: guest[a]	Method of analysis	Ref.
1	CCl_4	1:2	b	2, 1, 5
	CCl_3CH_3	1:2	c	2
	CCl_3Br	1:1	b	2, 1
	CCl_3SCl	1:2	b	2
	CCl_3NO_2	1:1	b	2
2	Toluene	1:2	c	2, 1
	1,4-Dioxan	1:3	c	2, 1
	Tetrahydrothiophen	1:1	c	2
3	1,4-Dioxan[e]	1:2	c	20
4	Toluene	f	c	20
	Benzene	f	c	20
6	Toluene	1:1	c	20
	Chloroform	1:2	b	20
	p-Chlorotoluene	1:1	c	20
	Acetonitrile	1:2	c	20
	1,4-Dioxan	1:2	c	20
12	1,4-Dioxan	1:2	c	20
	Benzene	1:2	c	20
13	Acetonitrile	1:1	c	20
	Diethyl ether	1:1	c	20
	Toluene	1:1	c	20
	p-Xylene	1:1	c	20
	p-Chlorotoluene	1:1	c	20
14	Toluene	1:1	c	2, 1
	1,4-Dioxan	1:2	c	2, 1
15	Cyclohexane	1:2	c	2
	Cycloheptane	1:2	c	2, 1
	Cyclooctane	1:2	c	2, 1
	Toluene	1:2	c	2, 1
	Iodobenzene	1:2	b	2, 1
	Phenylacetylene	1:2	c	2, 1
	1-Methylnaphthalene	1:2	c	2
	2-Methylnaphthalene	1:2	c	2
	Bromoform	1:2	b	2, 1
	Hexamethyldisilane	2:1	c	2
	Squalene	2:1	c	2, 12
16	Toluene[e]	1:1	c	2
	1,4-Dioxan	1:1	c	2
	o-Xylene	1:2	c	2
17	Acetone	1:3	c	2
	Methanol	1:4	c	2
18	1,4-Dioxan	1:3	c	2
19	Toluene[e]	1:1	c	2
	o-Xylene	2:1	c	2

Table 1—continued

Host	Guest	Mole ratio of host:guest[a]	Method of analysis	Ref.
21	Cyclooctane	2:3	c	2
	Toluene	2:3	c	2
	1,4-Dioxan	2:3	c	2
22	Cyclohexane	1:1	c	2, 1
	Toluene	1:1	c	2, 1
	1,4-Dioxan	1:1	c	2, 1
	Acetone	1:2	c	2, 1
	1,1,1-Trichloroethane	1:1	c	6
	Ethyl acetate	1:1	c	6
	p-Chlorotoluene	1:1	c	6
	Benzene	1:1	c	6
23	1,1,1-Trichloroethane	f	c	6
24	1,4-Dioxan	1:1	c	6
	Benzene	1:1	c	6
	Acetone	1:2	c	6
	Cyclohexane	1:1	c	6
	p-Xylene	1:1	c	6
	o-Xylene	1:1	c	6
	Ethylbenzene	1:1	c	6
	Anisole	1:1	c	6
	Furan	1:2	c	6
	Ethyl acetate	1:1	c	6
25	1,4-Dioxan	1:1	c	6
	Benzene	1:1	c	6
26	Benzene	1:1	c	6
	Cyclohexane	1:2	c	6
	1,1,1-Trichloroethane	1:2	c	6
27	Anisole	1:1	c	6
	Toluene	1:1	c	6
28	Toluene	1:1	c	6
	Benzene	1:1	c	6
	Anisole	1:1	c	6
28	1,4-Dioxan	1:2	c	6
	Ethyl acetate	1:2	c	6
	Nitromethane	1:2	c	6
	Acetone	1:2	c	6
29	1,4-Dioxan	1:4	c	6
30	Benzene	1:2	c	6
	1,4-Dioxan	2:3	c	6
31	Nitromethane	1:3	c	6
32	Cyclohexane	1:1	c	6
	Benzene	1:1	c	6
	1,1,1-Trichloroethane	1:1	c	6
	1,4-Dioxan	1:1	c	6

Table 1—continued

Host	Guest	Mole ratio of host : guest[a]	Method of analysis	Ref.
	Toluene	1 : 1	c	6
	Ethyl acetate	1 : 1	c	6
33	1,4-Dioxan	*c.* 1 : 6[f]	c	6
35	Chloroform	1 : 1	d	2
	1,1,1-Trichloroethane	1 : 1	d	2
36	Benzene	2 : 3	c	2
	Toluene	2 : 3	c	2
	Methyl acetate	2 : 3	c	2
	o-Xylene	2 : 3	c	2
	1,4-Dioxan	2 : 3	c	2
37	1,4-Dioxan	1 : 1	c	7, 8
	Benzene	1 : 2	c	7
	Toluene	1 : 2	c	7
	Tetrahydropyran	2 : 3	c	7
	Fluorobenzene	1 : 2	c	7
	Chlorobenzene	1 : 2	c	7
	Bromobenzene	2 : 3	b, c	7
	Iodobenzene	1 : 1	b	7
	1,1,1-Trichloroethane	1 : 1	b, c	7
52	Acetic acid	1 : 4	c	22
57	Nitromethane	1 : 2	c	32
	Tetramethylurea	1 : 2	c	32
	N,N-Dimethylformamide	1 : 2	c	32
	N,N-Dimethylacetamide	1 : 2	c	32
	N-n-butyl-*N*-methyl-formamide	1 : 2	c	32
69	1,4-Dioxan	1 : 1	c	35
	N,N-Dimethylpiperazine	1 : 1	c	g
	N,N-Dimethylformamide	1 : 1	c	g
70	Carbon tetrachloride	1 : 2	b	35
	Carbon tetrabromide	1 : 2	b	35
72	Acetonitrile	1 : 1	c	41
	Nitromethane	1 : 1	c	41
	Acetone	2 : 1	c	41

[a] Given to nearest integral ratio.
[b] Mole ratio of host : guest determined by microanalysis for halogen.
[c] Mole ratio of host : guest determined by multiple integration of the ^{1}H NMR spectrum employing $CDCl_3$, CCl_4, or $[^2H_6]$ acetone as solvent.
[d] Mole ratio of host : guest determined by weight loss.
[e] Unsolvated material has also been obtained from this solvent.
[f] Variable ratios have been found.
[g] D. D. MacNicol and A. Murphy, unpublished results.

Table 2. *Guest selectivity properties of specified hosts*

Host	Recrystallization solvent mixture	Respective mole percentage of guest included[a]			Overall host:guest ratio[b]	Ref.
2	50:50; o-Xylene–p-xylene	85		15[c]	1:2	2, g
	50:50; o-Xylene–p-xylene	90		10[d]	2:1	2
	50:50; Mesitylene–*pseudo*cumene	65		35	1:2	2
14	50:50; o-Xylene–p-xylene	90		10	1:2	2, g
	50:50; o-Xylene–m-xylene	45		55	1:1	2
15	50:50; o-Xylene–p-xylene	95		5[c,e]	1:2	2, g
	50:50; o-Xylene–m-xylene	85		15	2:3	2, g
	50:50; Mesitylene–*pseudo*cumene	10		90	1:2	2, g
	33:33:33; Cyclopentane–cyclohexane–cycloheptane	20	45	35	1:2	2
16	50:50; o-Xylene–m-xylene	70		30[c]	1:2	2, g
19	50:50; o-Xylene–p-xylene	20		80[e]	1:1	2, g
21	50:50; o-Xylene–p-xylene	90		10[e]	1:1	2, g
	50:50; o-Xylene–m-xylene	80		20[e]	1:2	2, g
	33:33:33; Cyclopentane–cyclohexane–cycloheptane	20	45	35	1:2	2
22	50:50; o-Xylene–p-xylene	25		75	1:1	2, g
24	50:50; o-Xylene–p-xylene	15		85	1:1	6
28	50:50; o-Xylene–p-xylene	80		20	1:1[f]	6
29	50:50; o-Xylene–p-xylene	70		30	2:3	6
30	50:50; o-Xylene–p-xylene	70		30	2:3	6
	50:50; o-Xylene–p-xylene	80		20	2:3	6
36	50:50; o-Xylene–m-xylene	80		20	2:3	2

(a) Measured by multiple ^{1}H NMR integration and given to nearest 5%.
(b) Given to nearest integral ratio.
(c) Average for solvated material formed in two experiments.
(d) For sorption experiment in which 200 mg of host **2** was stirred at room temperature for 24 h in 8 ml solvent mixture.
(e) Shows duality of crystallization behaviour: on one occasion, unsolvated material was deposited.
(f) Variable host:guest ratios have been obtained in this case, owing to rapid guest loss *in vacuo*.
(g) Initial report: D. D. MacNicol and D. R. Wilson, *Chem. Ind. (London)*, 1977, 84.

Table 3. Selected crystal data for some hexa-host molecules and other aromatic systems

Compound	Space group	Lattice parameters[a]	Guest	Mole ratio host : guest	Ref.
1	R$\bar{3}$	$a = 14.263$; $c = 20.717$ Å, $Z = 3$(host)	CCl$_4$	1:2	5, 2
22	P2$_1$/c	$a = 10.542$, $b = 20.863$, $c = 12.496$ Å, $\beta = 95.48°$, $Z = 2$(host)	1,4-dioxan	1:1	6
24	P2$_1$/c	$a = 9.62$, $b = 15.45$, $c = 22.72$ Å, $\beta = 111.0°$, $Z = 2$(host)	p-o-xylene[b]	1:1	6
25	P2$_1$/c	$a = 8.66$, $b = 23.69$, $c = 13.50$ Å, $\beta = 92.9°$, $Z = 2$	None[c]	—	6
25	Pcab	$a = 18.67$, $b = 14.18$, $c = 23.22$ Å, $Z = 4$(host)	1,4-dioxan	1:1	6
37	P$\bar{1}$	$a = 9.623(2)$, $b = 12.520(4)$, $c = 12.763(3)$ Å, $\alpha = 92.39(3)$, $\beta = 97.19(2)$, $\gamma = 103.15(2)°$, $Z = 1$(host)	1,4-dioxan	1:1	8
52	P2$_1$	$a = 16.320(5)$, $b = 13.870(4)$, $c = 16.731(7)$ Å, $\beta = 106.47(3)°$, $Z = 2$(host)	acetic acid	1:4	22
15	P$\bar{1}$	$a = 14.710(5)$, $b = 15.773(6)$, $c = 20.417(5)$ Å, $\alpha = 107.40(2)$, $\beta = 113.90(3)$, $\gamma = 81.93(3)°$, $Z = 2$(host)	squalene	2:1	12
54	Pbca	$a = 12.49$, $b = 17.13$, $c = 13.78$ Å, $Z = 4$	d	—	27
63	C2/c	$a = 39.521(6)$, $b = 5.538(1)$, $c = 26.701(6)$ Å, $\beta = 115.92(2)°$ $Z = 8$	d	—	30
57	P$\bar{1}$	$a = 12.511(3)$, $b = 12.781(3)$, $c = 13.224(3)$ Å, $\alpha = 70.94(1)$, $\beta = 66.04(2)$, $\gamma = 75.27(2)°$, $Z = 1$(host)	DMF	1:2	32

			KSCN, H₂O	1:3:1[(e)]	32
61	P2/c	$a = 15.249(2)$, $b = 10.087(2)$, $c = 30.328(6)$ Å, $\beta = 95.10(2)°$, $Z = 2(\text{host})$			
69	P$\bar{1}$	$a = 10.118(2)$, $b = 15.172(2)$, $c = 20.379(5)$ Å, $\alpha = 75.05(2)$, $\beta = 82.64(2)$, $\gamma = 68.45(1)°$, $Z = 2(\text{host})$	1,4-dioxan	1:1	35
71	P$\bar{1}$	$a = 9.585(2)$, $b = 9.610(1)$, $c = 19.171(3)$ Å, $\alpha = 99.65(1)$, $\beta = 91.95(2)$, $\gamma = 114.67(2)°$, $Z = 2$	d	—	41
72	P$\bar{1}$	$a = 14.505(3)$, $b = 14.663(3)$, $c = 14.869(4)$ Å, $\alpha = 117.46(2)$, $\beta = 109.43(2)$, $\gamma = 62.95(2)°$, $Z = 2$	acetonitrile	1:1	41
75	P2₁/n	$a = 13.534(2)$, $b = 10.379(1)$, $c = 17.838(2)$ Å, $\beta = 104.35(1)°$, $Z = 4$	None	—	41
75	P2₁/c	$a = 5.653(3)$, $b = 17.052(4)$, $c = 28.202(7)$ Å, $\beta = 86.23(4)°$, $Z = 4(\text{host})$	Chlorobenzene	2:1	41
76, yellow form	I2/c	$a = 20.469(3)$, $b = 10.567(2)$, $c = 25.415(3)$ Å, $\beta = 118.13(2)°$, $Z = 4$	None	—	42
76, red form	P$\bar{1}$	$a = 9.149(2)$, $b = 11.473(3)$, $c = 12.466(2)$ Å, $\alpha = 101.42(2)$, $\beta = 96.37(1)$, $\gamma = 109.67(2)°$, $Z = 1$	None	—	42

(a) For R$\bar{3}$, host **1**, the values of a and c are referred to a hexagonal unit cell ($\alpha = \beta = 90°$, $\gamma = 120°$); for other space groups unspecified angles are 90°.

(b) Crystals prepared by recrystallization from an equimolar mixture of o- and p-xylene; a marked selectivity for the *para* isomer is found (see Table 2).

(c) Unsolvated material produced by recrystallization from cyclohexane (solvated crystals have also been obtained from this solvent).

(d) No inclusion behaviour found to date.

(e) Molar ratio of host **61** to potassium thiocyanate to water.

versatile host forming beautifully crystalline adducts with a very wide range of guest species. In this case, the direct selenium counterpart **21** retains host properties. Notably, compounds **35** and **36** which have saturated rather than aromatic "side-chain" rings[2] also possess the ability to form inclusion compounds. However, not only did many of the molecules show interesting inclusion behaviour on recrystallization from single-component solvents, but a number also exhibited remarkable selective inclusion behaviour from solvent mixtures, see Table 2. For example, on recrystallization from an equimolar mixture of *o*- and *p*-xylene, hosts **2, 14, 15**, and **21** favour inclusion of the *ortho*-isomer; and for hexakis(*p*-t-butylphenylthiomethyl)benzene (**15**) the ratio of *o*- to *p*-isomer included was as high as 95% : 5%. On the other hand, a marked preference for the *para*-isomer was soon discovered for hexakis(2-naphthylthiomethyl)benzene (**19**) and hexakis(benzyl-thiomethyl)benzene (**22**), structural modification of the last-named host having subsequently led, as described below, to improved selectivity for *p*-xylene. Satisfyingly, many more new hosts were rapidly discovered; and at this stage it became essential to elucidate the nature and void geometry of inclusion compounds formed by these novel host molecules. Accordingly, detailed X-ray investigations were undertaken, and these will be described below. Where systematic structural modification of a given hexa-host molecule has been carried out, as for hexakis(benzylthiomethyl)benzene (**22**), for example, this will also be described.

2.3. Crystal structure of the carbon tetrachloride inclusion compound of hexakis(phenylthio)benzene 1

The first inclusion compound of a hexa-host studied[5,2] by X-ray methods was the CCl_4 compound of **1**. To our delight the CCl_4 compound crystallized in the familiar "clathrate" space group R$\bar{3}$ (see, for example, Chapter 1, Volume 2), with three host molecules and six guest molecules in the hexagonal unit cell, Table 3. Accordingly, the host molecule **1** is centred on a point of $\bar{3}$ symmetry, so that the crystallographically equivalent "legs" of the molecule point alternately above and below the plane of the central benzene ring, Fig. 2. The relative disposition of the central and side-chain aromatic rings may be appreciated by consideration of the torsion angles $C(1^*)C(1)S(2)C(3)$ and $C(1)S(2)C(3)C(4)$, which are 56° and 28° respectively. The true clathrate nature of this adduct may be seen from the packing diagram in Fig. 3, which is a view onto the *ac* plane; the host molecules which complete the cage (which lie above and below the cavity for the view shown) have been omitted for clarity. In the long closed cavity, effective length *c*. 17 Å, two CCl_4 guest molecules are accommodated, such that a

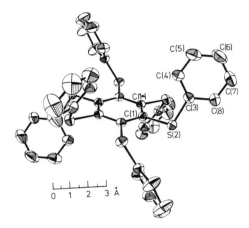

Fig. 2. An ORTEP drawing showing a general view of the molecular structure of host molecule hexakis(phenylthio)benzene (**1**) in the crystal of the CCl$_4$ clathrate. (The hydrogen atoms are not shown.) (Reprinted by permission from *Nature*, Vol. 266, No. 5603, pp. 611–612 Copyright © 1977 Macmillan Journals Limited.)

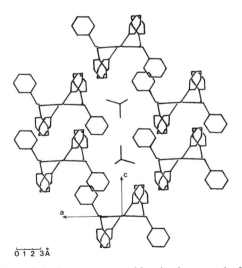

Fig. 3. An illustration of the host–guest packing in the crystal of the CCl$_4$ clathrate of hexakis(phenylthio)benzene (**1**), as viewed onto the *ac* plane. Two host molecules which lie above and below the cavity as viewed in this direction have been excluded to show the guest molecules more clearly. (Reprinted by permission from *Nature*, Vol. 266, No. 5603, pp. 611–612 Copyright © 1977 Macmillan Journals Limited.)

C–Cl bond of each is collinear with the c-axis. The space available for guest accommodation may be appreciated from the contours shown in Fig. 4, which are at various levels parallel to the ab plane and represent only half the centrosymmetric cage.

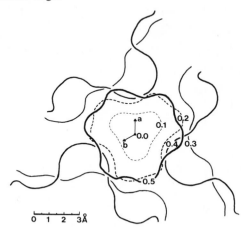

Fig. 4. The space available in the cage of **1** in cross section at various fractional levels parallel to the ab plane. The contours shown represent only half the centrosymmetric cage. (Reproduced, by permission, from ref. 2).

2.4. Structure of the 1,4-dioxan inclusion compound of hexakis(benzylthiomethyl)benzene (22) and systematic structural modification of host 22

The effect of increasing the length of the linkage between each outer ring and the central aromatic core of the hexa-host molecule clearly merits investigation.[6] In hexakis(benzylthiomethyl)benzene (**22**), as well as its structurally related analogues **23–33**, one now has a three-atom CH_2–S–CH_2 inter-ring linkage, as opposed to the one-atom, sulphur, link in **1**. The crystals of the 1,4-dioxan adduct of **22** are monoclinic (Table 3) with two host and two guest molecules in the unit cell. Unlike $C_6(SPh)_6$ in its CCl_4 clathrate, molecule **22** does not possess exact three-fold symmetry, though it retains a centre of symmetry (Fig. 5). Consideration[6] of torsion angles in **22** reveals approximate three-fold "core" symmetry, though deviations from three-fold symmetry become more marked for the remainder of the molecule. The sulphur atoms in the "legs" of **22** are situated alternately above and below the plane of the central benzene ring, and in each of the three independent terminal phenyl groups, half the ring atoms lie above the plane

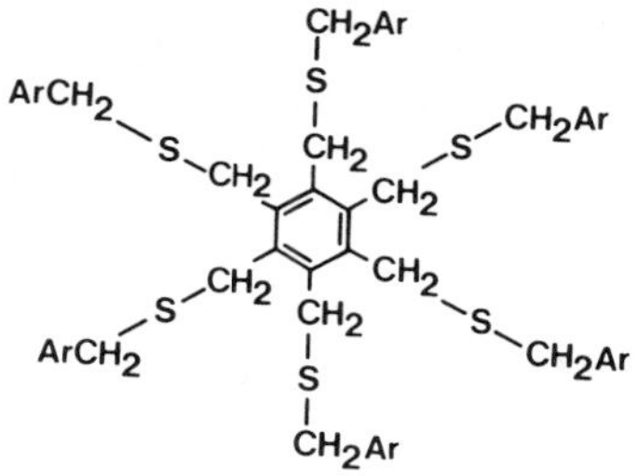

(22) Ar =

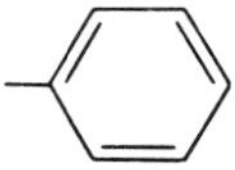

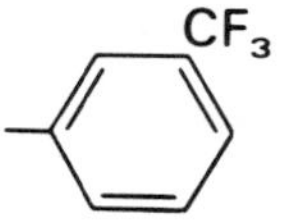

(28) Ar =

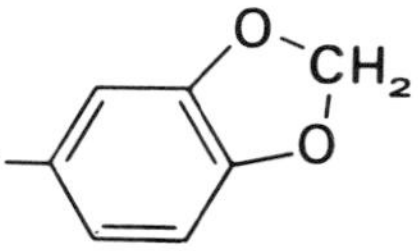

(23) Ar =

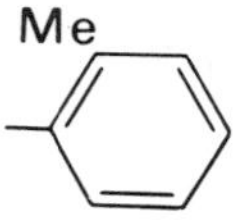

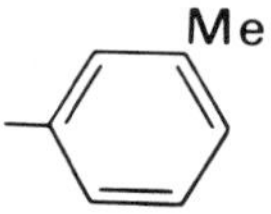

(29) Ar =

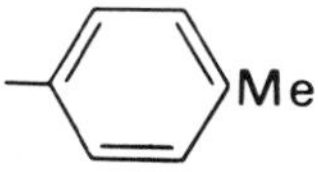

(24) Ar =

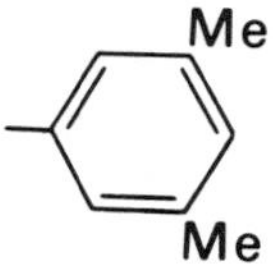

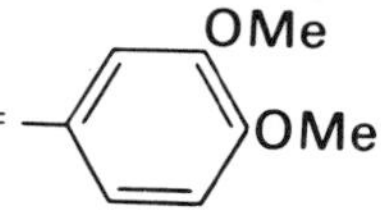

(30) Ar =

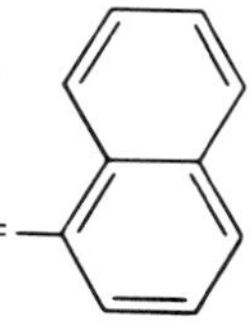

(25) Ar =

(31) Ar =

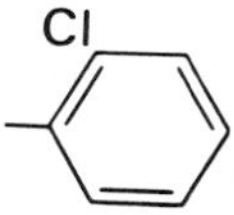

(26) Ar =

(32) Ar =

(27) Ar =

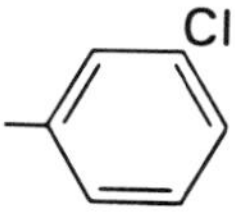

(33) Ar =

of the central benzene ring, while the other half lie below. The host-guest packing arrangement is shown in Fig. 6, the host and guest molecules above and below those shown being directly superposed in this view. All the host and guest molecules are situated at crystallographic centres of symmetry

$$C_6(CH_2OCH_2Ph)_6$$

(34)

$$C_6(CH_2XR)_6$$

34038   **(35)** X = O, R = cyclopentyl
34039   **(36)** X = S, R = cyclohexyl

on the *bc* plane. The acute angle between the normal to the central benzene ring plane and the *c*-axis is 36°. The adduct is of the true clathrate type, the chair-shaped 1,4-dioxan guest molecules being accommodated in effectively closed cages. The contours shown in Fig. 7 (representing only half the centrosymmetric cage) are at various fractional levels perpendicular to the *c*-axis and show the space available for guest accommodation.

A study of the inclusion properties of eleven structurally related compounds **23–33** has been carried out.[6] Substitution of each of the six aromatic rings of **22** with a single methyl group to give compounds **23–25** or a single chlorine atom to give **31–33** produces in each case a new host, the most general inclusion properties being found for the *meta*-analogues **24** and **32**, Table 1. Interestingly, the trifluoromethyl-analogue **27**, which closely resembles host **24** in structure, shape, and molecular volume, does not match the general inclusion properties of **24**, suggesting that the size and polarity

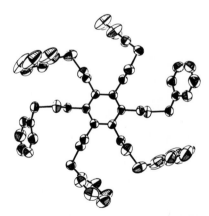

Fig. 5. An ORTEP drawing with perspective normal to the plane of the central benzene ring of the host molecule hexakis(benzylthiomethyl)benzene (**22**), in its 1,4-dioxan clathrate. The hydrogen atoms have been omitted for clarity. (Reproduced, by permission, from ref. 6.)

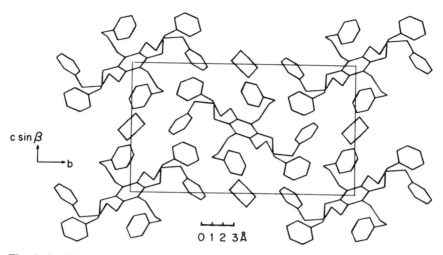

Fig. 6. An illustration of the host-to-guest packing in the crystal of the 1,4-dioxan clathrate of hexakis(benzylthiomethyl)benzene (**22**), as viewed along the *a*-axis. (Reproduced, by permission, from ref. 6.)

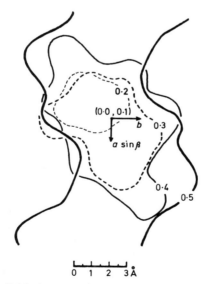

Fig. 7. The space available in the cage of **22** in cross section at various fractional levels perpendicular to the *c*-axis, in the 1,4-dioxan clathrate. The contours shown represent only half the centrosymmetric cage. (Reproduced, by permission, from ref. 6.)

of the *meta*-substituent are important. Methyl substitution at both *meta*-positions of **22** gives **26**, which retains inclusion properties; while fusion of a methylenedioxy-unit onto each outer aromatic ring, as in **28**, leads to a new host with more general inclusion properties than have been found for the related molecule **29**. Consideration of inclusion compounds formed by hosts **23–33** shows that, in this series, quite a wide range of structural change at the outer rings is allowed, without eliminating inclusion properties. Notably, *meta*-substituted host **24** shows a significantly greater selectivity for *p*-xylene, from an equimolar *ortho*- and *para*-xylene mixture, than does parent **22** itself (see Table 2). In contrast to the monoclinic adducts of **22** and **24**, Table 3, compound **25** forms solvated orthorhombic crystals from 1,4-dioxan, the unit cell volume being approximately doubled.[6]

2.5. Effect of further inter-ring chain elongation and symmetrical removal of three hexa-host "legs"

Detailed design considerations have led to a study[7] of chain elongation beyond three members, and hexakis(2-phenylethylthiomethyl)benzene (**37**) and hexakis(3-phenylpropylthiomethyl)benzene (**38**), respectively possessing four- and five-atom links, have been prepared. Recrystallization of **37**

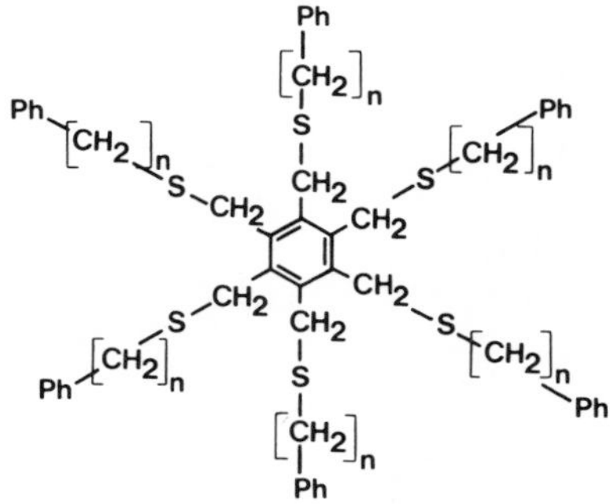

(**14**) n = 0
(**22**) n = 1
(**37**) n = 2
(**38**) n = 3

from a number of solvents[7] established inclusion behaviour for this hexa-host molecule, the first with a four-atom-link to be reported, Table 1. Interestingly, compound **38**, which corresponds to further chain extension compared to **37**, has given no evidence of inclusion behaviour to date, unsolvated material being obtained from cyclohexane, nitromethane, and acetone.[7] We have carried out a detailed X-ray study[8] of the 1,4-dioxan adduct of hexakis(2-phenylethylthiomethyl)benzene (**37**). By contrast to the monoclinic unit cell of its three-atom-link counterpart **22**, the 1,4-dioxan adduct of **37** is now triclinic, with one host and one guest molecule in the unit cell, Table 3. As found for **22**, host molecule **37** is located on a point of $\bar{1}$ symmetry and is therefore constrained to be centrosymmetric; again the sulphur atoms project alternately above and below the central benzene ring plane with the C–S bonds lying in planes approximately normal to the central ring (Fig. 8). Overall, however, **37** deviates more from three-fold symmetry than its shorter-chain counterpart **22**. As can be seen from Fig. 8, which gives relevant torsion angles for **37**, all three independent "legs" have an antiperiplanar conformation around the C–S bonds closest to the central ring, and the connected bonds S(3)–C(4), S(14)–C(15), and S(25)–C(26) all have a synclinal arrangement; however, while bonds C(4)–C(5) and C(15)–C(16) in two of the "legs" have an antiperiplanar conformation,

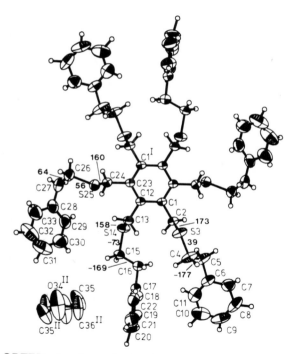

Fig. 8. An ORTEP drawing of the host molecule hexakis(2-phenylethyl-thiomethyl)benzene (**37**) and the 1,4-dioxan guest in the clathrate, viewed normal to the plane of the central benzene ring of **37**. (Reproduced, by permission, from ref. 8.)

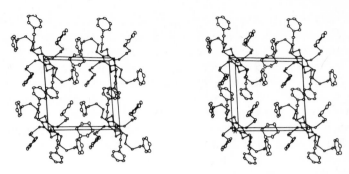

Fig. 9. A stereoview looking down the *a*-axis illustrating the host-to-guest packing in the 1,4-dioxan clathrate of the host hexakis(2-phenylethylthiomethyl)benzene (**37**). (Reproduced, by permission, from ref. 8.)

in the third "leg" a synclinal arrangement is found about the corresponding bond C(26)–C(27). Figure 9 shows a stereoview of the host-guest packing in **37**, looking down the *a*-axis. The 1,4-dioxan guest molecule is sandwiched between two centrosymmetric host molecules a *b* translation apart, a similar feature to that found in **22** where the sandwiching hosts are a *c*-translation apart (cf. Fig. 6). In **37** the normal to the central benzene ring is inclined to the *b*-axis at an angle of 47°. The centrosymmetric 1,4-dioxan guest, located at $(0, \frac{1}{2}, 0)$, has a chair conformation, and is accommodated in an effectively closed cage,[8] establishing that this adduct is also of the true clathrate type.

In order to study the effect of symmetrical removal of three of the hexa-host "legs" we have synthesized the trigonal molecules **39–42**. These were prepared by the action of the appropriate thiols on the known[9]

$$
\begin{array}{c}
\text{Ph} \\
| \\
[\text{CH}_2]_n \\
| \\
\text{S} \\
| \\
\text{CH}_2
\end{array}
$$

(39) n = 0
(40) n = 1
(41) n = 2
(42) n = 3

compound 1,3,5-tris(bromomethyl)benzene in presence of base.[7] None of
these molecules has given any evidence of inclusion behaviour (the longer
"chain" compounds **41** and **42** were obtained as oils). The apparent lack
of inclusion ability of these and other[10,11] 1,3,5-trisubstituted benzenes
further amplifies the significance of sixfold substitution in the hexa-host
analogy, and highlights the care required in the design of new trigonal host
molecules (*vide infra*).

2.6. Channel-type inclusion compound of hexakis(*p*-t-butylphenylthiomethyl)benzene (15) with squalene as guest; conformational selection

So far we have discussed only X-ray studies of clathrates containing fairly
small guest molecules. We will now consider the structure[12] of the adduct
of hexakis (*p*-t-butylphenylthiomethyl)benzene (**15**) with the large guest
species squalene (**43**), $C_{30}H_{50}$. This highly crystalline adduct of **15** with its

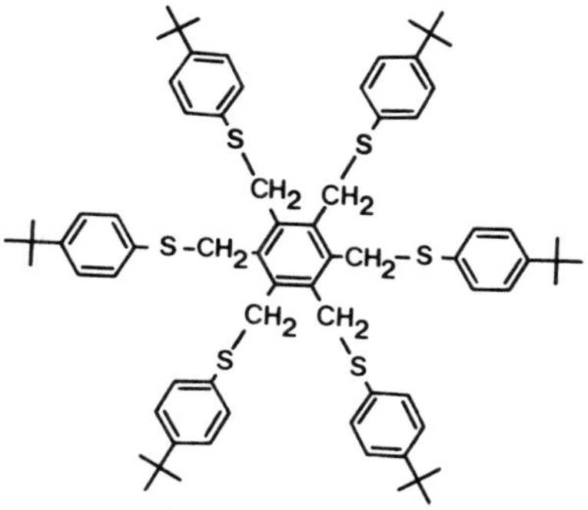

(15)

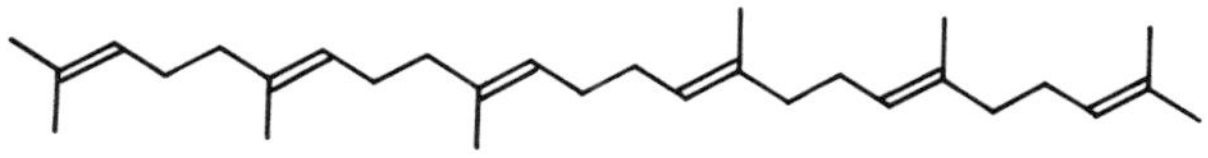

(43)

triterpene guest appeared particularly attractive because of the exact 2.0:1
host–guest ratio (determined by multiple ^{1}H NMR integration of a $CDCl_3$
solution of the adduct) which suggested a possible high degree of order
between host and guest components. The adduct is triclinic, space group
$P\bar{1}$, with a unit cell containing two host molecules and one squalene guest
molecule. Figure 10 shows a general view of the host molecule **15** which
occupies a general position in the unit cell with a centroid at approximately

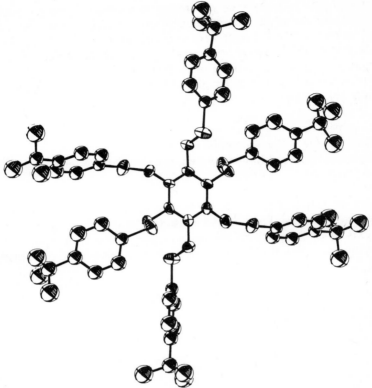

Fig. 10. An ORTEP drawing showing a general view of the molecular structure of the host hexakis (*p*-t-butylphenylthiomethyl)benzene (**15**) in the crystal of the squalene adduct. (Reproduced by permission from *Tetrahedron Letters*[12].)

$(0, \frac{1}{4}, \frac{3}{4})$; the sulphur atoms of the two-atom links point alternately above and below the plane of the central benzene ring. The host–guest packing arrangement is shown in Fig. 11, the squalene guests being accommodated in continuous channels running through the crystal. By virtue of location of the centre of either of the double bonds adjacent to the molecular centre of **43** at the crystallographic centre $(0, \frac{1}{2}, \frac{1}{2})$, the disordered squalene exists as a pair of enantiomeric conformations belonging to the point group C_1, and appears as a continuous chain in Fig. 11. The successful resolution of the guest disorder[12] allows comparison of the guest squalene conformation with that found[13] in the molecular crystal of pure squalene itself. As can be seen from Fig. 12, these conformations differ markedly, the new squalene conformation being closely controlled by host–guest interactions. So here

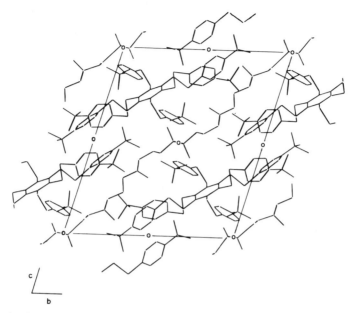

Fig. 11. A view looking onto the *bc* plane illustrating the host-to-guest packing in the adduct of hexakis(*p*-t-butylphenylthiomethyl)benzene (**15**) with squalene. (Reproduced by permission from *Tetrahedron Letters*[12]).

one has a novel way of producing and observing directly by X-ray methods conformations not present in the normal molecular crystal. This suggests the exciting possibility, currently under investigation, of specific reaction of guests which have been subjected to "conformational selection". In the case of squalene, because of low rotational energy barriers, the usual distribution of solution conformations is expected when the guest is released by dissolving the adduct. In recent work Anet and coworkers,[14,15] employing an elegant cryogenic deposition technique, have observed the *axial*-methyl form of methylcyclohexane and the twist-boat form of cyclohexane itself. In this method the gaseous form of the substance under study is rapidly condensed on to a cold surface, trapping the conformer proportions corresponding to the high temperature equilibrium, typically at 500–800° C. Conformational selection by clathration followed by guest release, a method potentially applicable to thermally labile compounds, appeared to us an attractive way of producing conformers strongly disfavoured in the normal solution equilibrium.[16] A promising guest candidate was 3,3,6,6-tetramethyl-*s*-tetrathiane, for here existing dynamic NMR studies[17] indicated that, under suitable conditions of guest release (*vide infra*), conformational integrity

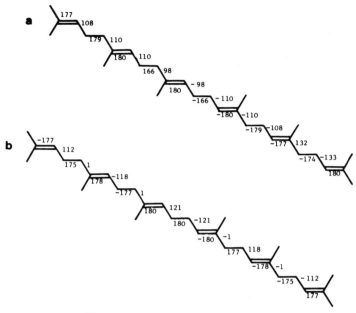

Fig. 12. A comparison[12] of main chain torsion angles for (a) squalene in its inclusion compound with host **15**; and (b) squalene in its molecular crystal at −110° C.[13]

would be maintained. The detailed studies of Bushweller[17] have established that the twist-boat form **44** of 3,3,6,6-tetramethyl-s-tetrathiane is substantially favoured over the less stable chair conformer **45**; and in the tetragonal molecular crystal this tetrathiane is found to have the twist-boat conformation with exact D_2 symmetry.[18] Also, calculations[19] by the molecular mechanics method indicate that the chair form **45** is 2.9 kJ mol^{-1} less stable than its D_2 twist-boat counterpart. The symmetry distinction between the centrosymmetric chair from **45** and the chiral twist-boat **44** suggested that

(44)   **(45)**

guest location at a crystallographic centre of inversion in versatile host hexakis(p-t-butylphenylthiomethyl)benzene (**15**) might constrain the guest to be centrosymmetric, hence yielding **45**. The appropriate adduct of **15** was prepared,[16] employing mesitylene as solvent, since the tetrathiane is a solid. On dissolving a sample of the adduct in a suitable precooled solvent mixture in an NMR tube, and quickly transferring this to the precooled probe of the spectrometer at $-90°$ C, the guest proton spectrum in Fig. 13a was observed.[16] This shows the predominance of chair form **45** with resonances from the non-equivalent methyl groups at 1.59δ and 2.00δ, a low proportion (8%) of twist-boat giving rise to the signal at 1.69δ. For comparison, Fig. 13b shows the normal predominance of the twist-boat form, observed in a similar sample warmed to room temperature, then cooled to $-80°$ C. In fact by heating samples of the adduct *in vacuo* to remove or minimise any "free" tetrathiane which may have co-crystallized during the preparation, it was possible to obtain the normally disfavoured chair form of 3,3,6,6-tetramethyl-*s*-tetrathiane (**45**) routinely *in solution* with a purity of $\geqslant 95\%$. (The related question of *configurational* selection will be considered later.)

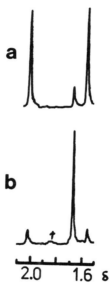

Fig. 13. (a) ^{1}H NMR spectrum of guest 3,3,6,6-tetramethyl-*s*-tetrathiane released from its adduct with host **15** showing the predominance of chair form **45** ($-90°$ C, solvent CS_2/CD_2Cl_2, *c.* 4:1, v/v); (b) corresponding spectrum for a sample warmed to room temperature, and recooled (to $-80°$ C), illustrating the normal predominance of the twist-boat form **44**. (Peak denoted † is a spinning side-band of the host t-butyl resonance.) Reproduced by permission from *Tetrahedron Letters.*[16])

At this point we shall discuss briefly the inclusion properties of a number of two-atom-link molecules conforming to the general formula $C_6(CH_2OAr)_6$. In addition to results for the parent hexakis(phenoxymethyl)benzene (**2**), already mentioned, related compounds **3–13** have been prepared and a preliminary survey carried out.[2,20] Although introduction of methyl, isopropyl, *t*-butyl, 1-adamantyl, or methoxy *para* substituents into each of the phenyl groups of hexakis(phenoxymethyl)benzene **2** to give compounds **5** and **7–10** respectively, has yielded to date no evidence of inclusion behaviour, well characterized adducts (Table 1) have been found for hosts **6, 12,** and **13** derived from *o*-isopropylphenol, 3,5-dimethylphenol, and thymol respectively. Compound **11,** unlike its isomer **12,** has so far yielded no adducts.[20] Further work on molecules belonging to this class is currently underway in Glasgow, and independent work carried out in Bonn[11] has established the ability of related hosts **46–49** to form crystalline inclusion compounds with polar guest molecules such as tetramethylurea, 2,6-dimethylpyridine, 2,4,6-trimethylpyridine, dimethylacetamide, hexamethyl-phosphoramide, chloroacetonitrile, 2,2,2-trichloroethanol, and diglyme; **46, 47,** and **49** even form complexes[11] with crown ethers as guests.

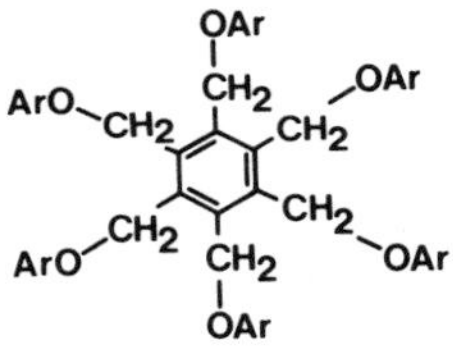

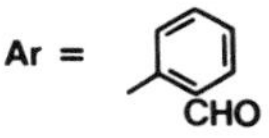

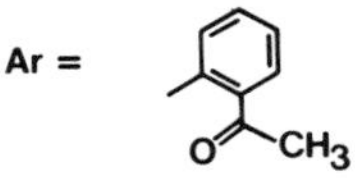

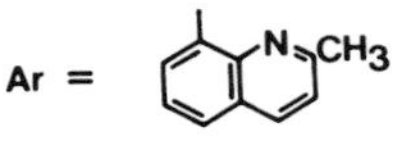

2.7. Synthesis of the chiral hexa-host molecule hexakis(R-α-phenylethylsulphonylmethyl)benzene (52) and the X-ray crystal structure of its acetic acid inclusion compound

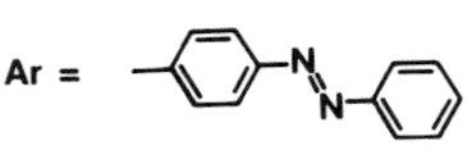

In the last ten years or so there has been a renaissance of interest in chiral phenomena, both with respect to spectroscopic investigations and to the synthesis of novel chiral systems. It was therefore of obvious interest to try to prepare a chiral hexa-host molecule. To do this a uniform chiral array of "legs" must be introduced, avoiding possible complications due to racemization. We selected (+)-1-phenylethanethiol (**50**) as a suitable leg-forming component, and this compound was prepared by an excellent literature route.[21] We assigned[22] to this thiol (**50**) the R-(+)-configuration from the Raman CID spectrum, this assignment being concordant with chemical correlation.[23] Six-fold substitution of hexakis(bromomethyl)ben-zene by the thiolate anion of **50** generated in NaOEt/EtOH occurred

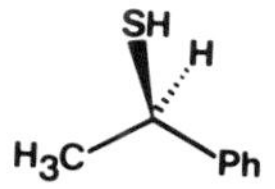

(**50**)

smoothly[22] leading to a high yield of optically pure hexakis(R-α-phenyl-ethylthiomethyl)benzene (**51**), as bullet-shaped, non-centrosymmetric crys-tals, m.p. 137–138° C. Oxidation of **51** with excess hydrogen peroxide in acetic acid gave in high yield the hexa-sulphone **52**, the first chiral hexa-host molecule. The potential use of hexakis(R-α-phenylethylsulphonylmethyl) benzene (**52**), and related chiral molecules, is currently under study. Although no X-ray information is yet available about this chiral host containing a chiral guest, we shall briefly discuss the interesting monoclinic acetic acid adduct of **52**, which has a host-guest ratio of 1:4 (Table 3).

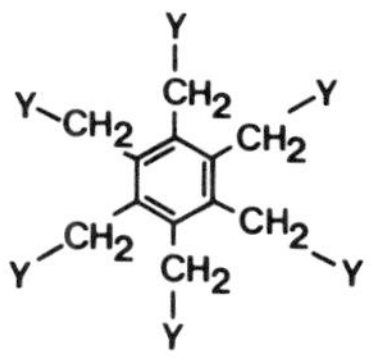

(**51**) $Y = -S-\overset{\overset{\displaystyle CH_3}{|}}{\underset{\underset{\displaystyle H}{|}}{C}}-Ph$

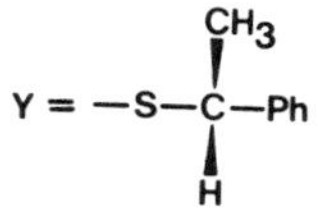

(**52**) $Y = -\overset{\overset{\displaystyle O}{\|}}{\underset{\underset{\displaystyle O}{\|}}{S}}-\overset{\overset{\displaystyle CH_3}{|}}{\underset{\underset{\displaystyle H}{|}}{C}}-Ph$

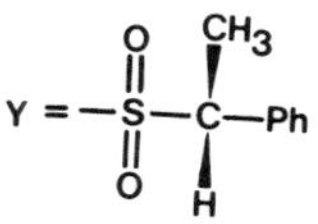

Figure 14 shows a view of the molecular structure of host **52**, and also illustrates the acetic acid guest molecules. In the host molecule, which has approximate three-fold symmetry, the $-SO_2-$ moieties in the inter-ring chains are situated alternately above and below the mean plane of the central

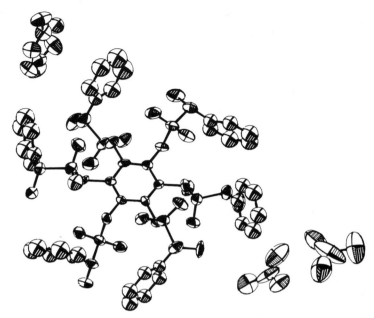

Fig. 14. A view illustrating the hexakis(*R*-α-phenylethylsulphonylmethyl)benzene (**52**) host molecule, and the two crystallographically independent acetic acid guest dimers. Hydrogen atoms have been omitted for clarity. (Reproduced by permission from *Tetrahedron Letters.*[22])

benzene ring. The four acetic acid molecules appear as hydrogen-bonded dimeric pairs which are crystallographically independent. Each dimer consists of two planar acetic acid molecules whose mean planes intersect at an angle of approximately 18 degrees. The host–guest packing is shown in Fig. 15, and as can be seen one dimer is situated directly between two host molecules a *b* translation apart, while the second dimer is situated in a different type of site, on the *b* axis. It is noteworthy that this work[22] allowed the first direct observation of dimeric acetic acid, infinite chains of hydrogen-bonded molecules having been previously found in the crystal structure of acetic acid[24,25] and also in the 1:1 complex of deoxycholic acid and acetic acid.[26]

2.8. Nitrogen-based hexa-hosts and related aromatic molecules

Reaction of hexakis(bromomethyl)benzene $C_6(CH_2Br)_6$ with aniline occurs in the expected way, giving as product[2] hexakis(phenylaminomethyl)-

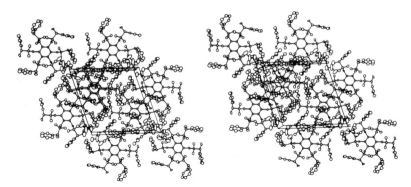

Fig. 15. A stereoview looking on the *ac* plane illustrating the host–guest packing in the acetic acid adduct of hexakis(R-α-phenylethylsulphonylmethyl)benzene (**52**). The space group is P2$_1$ with two host, and eight acetic acid guest molecules in the monoclinic unit cell.

benzene (**53**). On the other hand, reaction of the same hexabromide with morpholine yields[27] the novel bis-quaternary dispiro system **54**, while structurally-related molecules are produced by reaction with piperidine and *N*-methylpiperazine.[28] A view of the structure of **54** in the crystal is shown

$$C_6(CH_2\overset{H}{N}Ph)_6$$

(53)

(54)

in Fig. 16, the centrosymmetric molecule being located on a point of $\bar{1}$ symmetry. The two independent morpholine rings both have chair conformations, and the quaternary nitrogen atoms are displaced (±0.47 Å) above and below the plane of the central benzene ring.[27]

In the context of inclusion chemistry, reaction of C$_6$(CH$_2$Br)$_6$ with benzylamine is important. With nitromethane as solvent this reaction leads to the ready isolation of **62** and **63**, products corresponding to the incorporation of four[29] and three[30] molecules of benzylamine respectively. The structure of **63** has been determined by X-ray methods[30] and this triamine has been

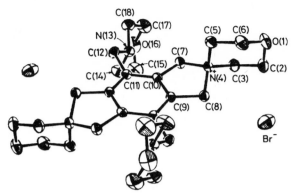

Fig. 16. A view of the molecular structure of **54** in the crystal looking onto the *ab* plane. (Reproduced, by permission, from ref. 27.)

converted into $NN'N''$-tribenzylbenzo[1,2-c:3,4-c':5,6-c'']tripyrrole (**64**), the first example of the trisazahexaradialene ring system.[31] If hexane, instead of nitromethane, is used as solvent in the reaction of $C_6(CH_2Br)_6$ with excess benzylamine, a good yield of the six-fold-substituted product **55** is easily obtained.[32] This hexa-amine, by virtue of the possibility of structural elaboration at the nitrogen atoms, is a useful precursor for the synthesis of new hexa-host molecules; though compound **55** itself, like its structurally-related hexa-ether counterpart **34**, hexakis(benzyloxymethyl)benzene,[2] has given no sign of guest trapping to date. The hexa-amides **56–61** have been prepared[32] by refluxing **55** with an excess of the appropriate acid anhydride, either neat or with ethyl acetate as solvent. For these, the formulated structures were confirmed spectroscopically, and, for **57** and **61**, also by X-ray methods,

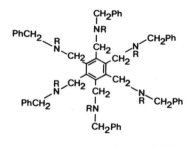

(**55**) R = H
(**56**) R = COCH₃
(**57**) R = COCF₃
(**58**) R = COCF₂CF₃

(**59**) R = COCF₂CF₂CF₃
(**60**) R = COCH₂OCH₃
(**61**) R = COCH₂OCH₂CH₂OCH₃

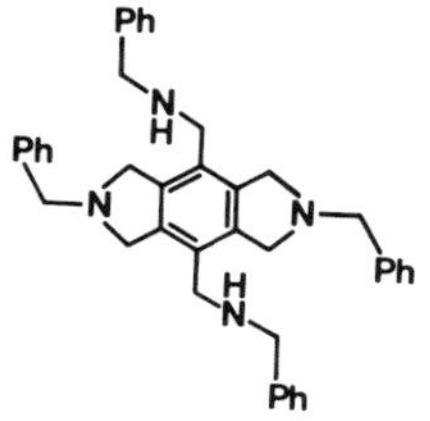

(62)

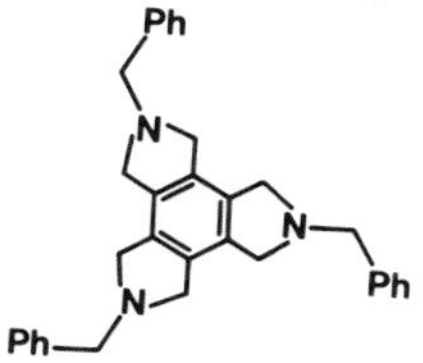

(63)

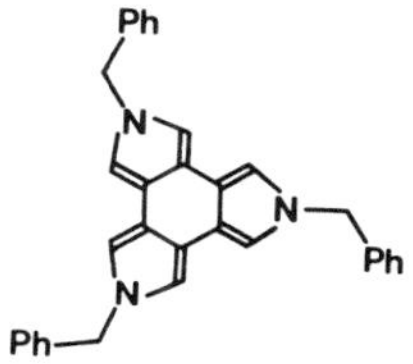

(64)

as described below. Compound **57** forms adducts with a host-guest ratio of 1:2 from *N,N*-dimethylformamide (DMF), *N-n*-butyl-*N*-methylformamide (**65**), and other solvents specified in Table 1. In contrast, the structurally related hexa-amide **56** has shown no guest inclusion to date. The inclusion of guest **65** by host **57** is particularly noteworthy since this occurs with

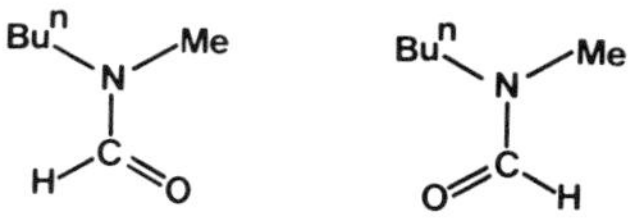

(65)

effectively complete *configurational selection* of the thermodynamically less stable[33] *Z* form. This remarkable selectivity was established[32] by dissolving a specimen of the adduct (previously washed with cold ether) in CDCl$_3$ at −40° C, and monitoring by ^{1}H NMR at −20° C. The absence of a methyl resonance from the *E*-form, known to occur at 2.87δ at −20° C, allowed a *minimum* purity of *c.* 97% to be estimated for the *Z*-form, the configuration of **65** disfavoured in the normal solution equilibrium.

The DMF adduct of host **57** is triclinic, space group P$\bar{1}$, with one host and two DMF guest molecules in the unit cell.[32] The host molecule **57**, shown in Fig. 17a, is centrosymmetric and alternation of near planar amide nitrogens above and below the plane of the central benzene ring is observed in the structure which deviates markedly from 3-fold symmetry. In **57**, each

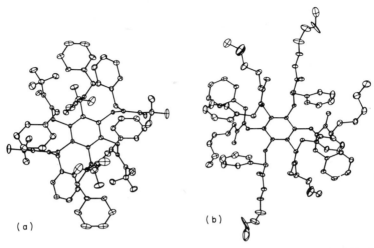

Fig. 17. ORTEP drawings showing general views of (a) host molecule **57** in its DMF adduct; and (b) host molecule **61** in its 3KSCN.H₂O complex. (Reproduced, by permission, from ref. 32.)

of the three independent CF₃ groups has a C–F bond almost eclipsing the adjacent C=O bond. The DMF guest molecule occupies a general position in **57** and undergoes considerable thermal motion; no evidence has been found to suggest that the guest deviates greatly from planarity.

The X-ray study[32] of the complex of **61** with 3KSCN.H₂O is important in that it provided the first definitive structural information concerning a hexa-host inclusion compound with an ionic guest present. This adduct is monoclinic, space group P2/c, with two host molecules and two formula units of (3KSCN.H₂O) as guest per unit cell. As can be seen from Fig. 17b, the host molecule **61** is again centrosymmetric, deviates significantly from 3-fold symmetry, and like **57** shows a regular alternation of nitrogen atoms with respect to the central ring. A common feature of both **61** and **57** is that each carbonyl oxygen atom is *trans* to the corresponding benzyl group. In **61**, two independent sites have been found for the potassium ions, one corresponding to a general position in the unit cell, the other located on a two-fold rotation axis. The potassium ion in the first site is co-ordinated to 4 ether oxygens [K⁺ . . . O; 2.70(1), 2.81(1), 2.96(1), 3.08(1) Å] and 2 carbonyl oxygens [1.76(1), 2.84(1) Å] belonging to nonequivalent "legs" of adjacent host molecules, a K⁺ . . . N contact of 3.03(1) Å to a thiocyanate ion in a general position also being found. The K⁺ ion on the two-fold rotation axis (coordinated to a water molecule) is situated between two equivalent host molecules, though here severe disorder of the two equivalent 2-methoxyethoxy units involved in co-ordination reduces the accuracy with which contact distances can be measured. The corresponding thiocyanate ion, not co-ordinated to K^+, has its nitrogen on a two-fold rotation axis. It may be noted that the theoretically possible mode of binding of K^+ by three "legs" of a single host molecule is not employed at either site in **61**. A $3CoCl_2.6H_2O$ complex of molecule **66** has also been reported,[11] though here the binding geometry in the crystal has not yet been elucidated.

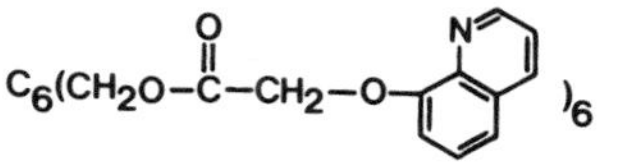

(66)

2.9. Synthesis of the hexakis(arylthio)benzenes **68** and **69** and the hexa-selenoether **70**; X-ray crystal structure of the 1,4-dioxan inclusion compound of hexakis(β-naphthylthio)benzene (**69**)

So far, with the exception of work on hexakis(phenylthio)benzene (**1**), we have in the hexa-host series been exclusively concerned with molecules with a two or more atom link joining the outer rings to the central aromatic core. The comparative neglect of the one-atomlink class was due to synthetic difficulties, rather than lack of inherent interest. Indeed this latter class is of particular interest owing to its close formal resemblance to the (OH . . . O) hydrogen-bonded hexameric unit found in the clathrates of phenol, hydroquinone, and Dianin's compound and related systems (Chapter 1, Volume 2). Although, as already mentioned, hexakis(phenylthio)benzene (**1**) can be obtained in moderate yield by the action of PhSCu on hexachlorobenzene,[4] the related compound hexakis(p-tolylthio)benzene (**67**) has been obtained[2] only in very low yield by the arylthiocuprate route. However, recently, greatly encouraged by a report[34] by Tiecco and colleagues of

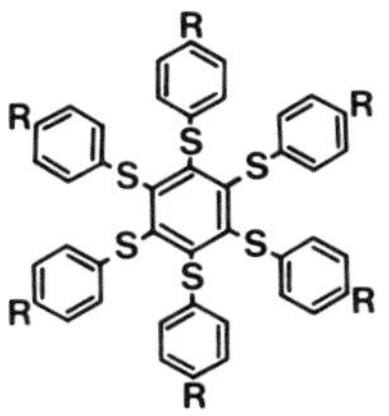

(1) R = H
(67) R = Me
(68) R = t-Bu

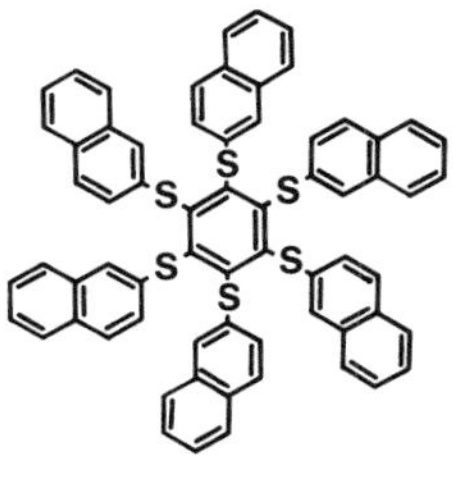

(69)

complete substitution of C_6F_6 and C_6Cl_6 by *i*-PrSNa in hexamethylphos-phoramide (HMPA) to give $C_6(SPr^i)_6$ in high yields, we investigated[35] the action of the arenethiolate salt PhSNa on C_6Cl_6 and found remarkably facile displacement of all the chlorine atoms giving, after appropriate work-up, a non-optimised 90% yield of pure $C_6(SPh)_6$, **1**. In an analogous reaction[35] using the same solvent, HMPA, PhSeNa and C_6Cl_6 gave hexakis(phenylseleno)benzene (**70**), apparently the first molecule with six selenium atoms bonded directly to a benzene ring. Hexa-selenoether (**70**) on recrystallization from pentachloroethane containing excess CBr_4 gave the CBr_4 adduct of **70** as beautiful orange rhombs, the host–guest ratio being 1 : 2, Table 1. This adduct has a true cage structure analogous to that of the CCl_4 clathrate of **1**.[55]

(**70**)

However since HMPA is a known carcinogen[36] we sought an alternative solvent capable of promoting efficient aromatic halogen displacement, and have found that 1,3-dimethyl-2-imidazolidinone, dimethylethyleneurea (DMEU), is a remarkably effective,[37] and probably much less toxic, sub-stitute.[38] With DMEU as solvent, reaction of PhSNa with C_6F_6, C_6Cl_6, or C_6Br_6 gave in each case an excellent yield of pure **1**, and the general nature[35] of this substitution process is suggested by the successful synthesis, employ-ing the sodium salt of the appropriate arenethiol, of hexakis(*p*-t-butylphenylthio)benzene (**68**) and hexakis(*β*-naphthylthio)benzene (**69**), from C_6Cl_6. The last-named compound is of particular interest, showing a new feature of molecular conformation. Compound **69** forms a highly crystalline adduct with 1,4-dioxan as guest, and there are two host and two guest molecules in the triclinic unit cell (Table 3). Figure 18 shows a stereoview of the host molecule **69**, which is located in a general position in the unit cell,[35] and now the new conformational feature is apparent. Contrasting with the trigonal conformation of $C_6(SPh)_6$ which, as will be recalled, is located on a point of $\bar{3}$ symmetry in its CCl_4 clathrate,[2,5] **69** is not constrained to have alternate disposition of side chains above and below the plane of the central ring: indeed, instead of having three-fold symmetry the molecule approximates to the point group C_2, adjacent "legs" being found on the same side of the central ring, with a corresponding pair on

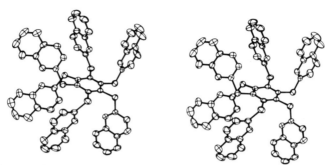

Fig. 18. A stereoview showing the host molecule of hexakis(β-naphthylthio)benzene (**69**) in its 1,4-dioxan channel-type inclusion compound. All hydrogen atoms have been omitted for clarity. The molecule has approximate C₂ symmetry. (Reproduced by permission from *Tetrahedron Letters.*[35])

the opposite side. This situation represents the first example of a Type II hexa-host conformation, Type I corresponding to the regular alternation of "legs" normally encountered. The molecular conformation of **69** is reflected in a significant distortion of the central benzene ring, which has an approximate two-fold rotation axis. The retention of inclusion properties for **69**, whose conformation corresponds to an interchange of two adjacent "legs" away from a trigonal situation (the importance of trigonal symmetry is considered below) is of particular interest in view of the recent consideration by Huang and Mak[39] of the dominant role that two-fold molecular symmetry plays in the architecture of the lattices of many inclusion compounds. Figure 19 shows a stereoview of the host–guest packing: a noteworthy feature is

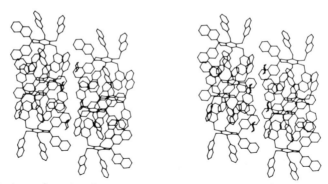

Fig. 19. A stereoview showing the host-guest packing of the adduct of hexakis(β-naphthylthio)benzene (**69**) with 1,4-dioxan as guest. The chair-shaped 1,4-dioxan molecules can be seen to be located in continuous voids in the structure. (Reproduced by permission from *Tetrahedron Letters.*[35])

that neither the host molecule nor the guest molecule is located at a point of inversion in this centrosymmetric structure. The 1,4-dioxan guest molecules have a chair conformation and can be seen to be located in continuous voids running through the crystal.

2.10. Synthesis of the first hexakis(aryloxy)benzenes, (71–73); and X-ray structures of unsolvated hexakis(phenyloxy)benzene (71) and the acetonitrile inclusion compound of hexakis(3,5-dimethylphenyloxy)benzene (72)

The remarkably facile substitution processes just discussed suggested use of the DMEU to achieve synthetic entry to hexakis(aryloxy)benzenes, a new class of molecule. An attractive initial target was hexakis(phenyloxy)benzene 71, the "direct" analogue of the hydrogen-bonded hexameric unit present in phenol clathrates[40] (see also Chapter 1, Volume 2). Compound 71 was readily prepared[35] as colourless needles, m.p. 277–280° C (with sublimation), by reaction of C_6F_6 with 18 molar equivalents of PhONa for 4 days at 120° C, with DMEU as solvent. Hexakis(3,5-dimethylphenyloxy)benzene (72) and hexakis(p-methoxyphenyloxy)benzene (73) were

(71)   (72)   (73)

prepared analogously[41] with DMEU at 120° C, though in these cases a reaction time of several weeks was employed and 0.85 molar equivalent of [2.2.1] cryptand was added: it has not yet been determined whether extended reaction time and added cryptand are strictly necessary for the successful preparation of 72 and 73. Hexa-demethylation of 73, currently being undertaken, is of particular interest since the resulting hexa-phenol has a structure which suggests the theoretical possibility that it may be capable of forming a "hybrid" structure lying between a hexa-host and a β-hydroquinone host lattice; that is, there would be exactly the correct number of hydroxyl groups in relation to the aromatic core to allow half the $[OH]_6$ rings of the β-hydroquinone structure to be replaced by permanent benzene units. The first member of the new series 71 crystallizes unsolvated from DMF, the

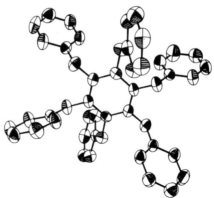

Fig. 20. A general view of hexakis(phenyloxy)benzene (**71**) in the molecular crystal. Reproduced, by permission from *Tetrahedron Letters*.[47]

crystals being triclinic, space group P$\bar{1}$, with two molecules in the unit cell (Table 3). The asymmetric hexakis(phenyloxy)benzene molecule, occupying a general position, is shown in Fig. 20. Although no inclusion behaviour has yet been found for this molecule, its 3,5-dimethyl counterpart **72** readily forms adducts. In the acetonitrile adduct of **72**, which has a host-guest ratio of 1:1 (Tables 1 and 3), there are two crystallographically distinct host molecules and both of these, located at points of $\bar{1}$ symmetry, are constrained to be centrosymmetric. The crystal structure[41] of the acetonitrile clathrate of **72** is illustrated in Fig. 21, crystallographically non-equivalent host

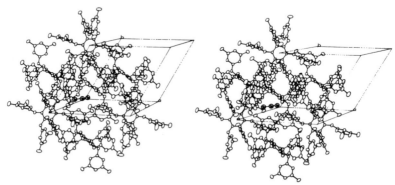

Fig. 21. An illustration of the structure of the (1:1) inclusion compound of hexakis(3,5-dimethylphenyloxy)benzene (**72**) with acetonitrile as guest. Crystallographically nonequivalent host molecules are located at $(0, 0, 0)$ and $(\frac{1}{2}, \frac{1}{2}, \frac{1}{2})$ and there are two $CH_3C\equiv N$ guest molecules in each unit cell, one of which is shown. Reproduced, by permission, from *Tetrahedron Letters*.[41]

molecules being located at $(0, 0, 0)$ and $(\frac{1}{2}, \frac{1}{2}, \frac{1}{2})$; there are two $CH_3C\equiv N$ guest molecules occupying enantiomerically-related general positions in the unit cell, and one of these is shown in the figure. As can be seen, the acetonitrile guest molecule interacts edge-on with a host molecule, methyl . . . O distances of 3.252 (2) Å and 3.366 (3) Å being found.

2.11. Synthesis of pentakis(phenylthio)pyridine (74) and the unexpected formation of 1,3,4-tris(phenylthio)[1]benzothieno[3,2-c]pyridine (75)

Extension of the use of DMEU to perhalogenoheterocycles had interesting consequences.[41] Thus reaction of pentachloropyridine with 10 molar equivalents of PhSNa in DMEU for 5 days at ambient temperature and in normal daylight gave not only the expected, but hitherto uncharacterized, pentakis(phenylthio)pyridine (74), m.p. 133–133.5° C (c. 80% yield), but also a second product, showing an isolated ABMX pattern in the 360.13 MHz ^{1}H NMR spectrum, suggesting a benzothienopyridine structure; the structure 75 was firmly established by X-ray analysis of the unsolvated crystal

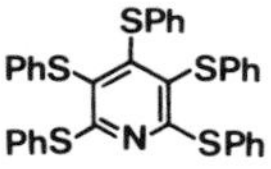

(74) 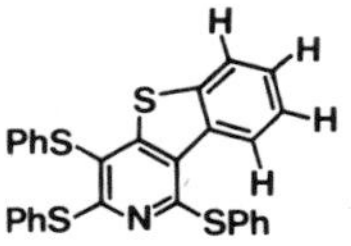   (75)

(from CCl$_4$) (Table 3). The mechanism of formation of 75, currently under study, is of considerable interest: 75 was also formed (in reduced quantity) when the reaction was carried out in the dark, while irradiation by tungsten lamps increased the yield. With pentafluoropyridine, in daylight, PhSNa in DMEU gives only pentakis(phenylthio)pyridine 74. Most interestingly, 1,3,4-tris(phenylthio)[1]benzothieno[3,2-c]pyridine (75) shows unexpected host properties, adducts with a host–guest ratio of 2:1 being formed with toluene, chlorobenzene, bromobenzene, and 1,4-dioxan. The chlorobenzene adduct, like the unsolvated crystal of 75, is monoclinic, Table 3. Figure 22 shows corresponding general views of 75 in the unsolvated crystal and in the chlorobenzene adduct; and these illustrate the significant changes in skeletal torsion angles that occur on guest complexation. The chlorobenzene guest, close to a crystallographic centre in 75, is disordered and is currently under further refinement.

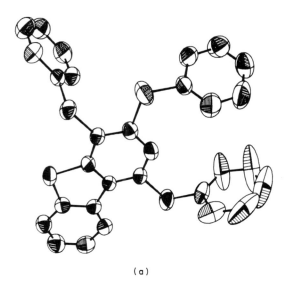

(a)

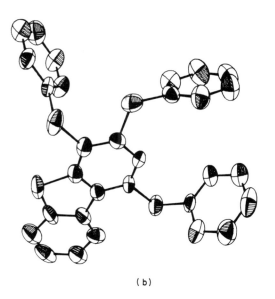

(b)

Fig. 22. Corresponding general views for 1,3,4-tris(phenylthio)[1]benzothieno[3,2-*c*]pyridine (**75**), (a) in the unsolvated crystal; and (b) in the chlorobenzene adduct.

2.12. First synthesis of octakis(arylthio)naphthalenes: X-ray study of the yellow and red forms of (unsolvated) octakis(phenylthio)naphthalene (76) and inclusion properties of 76 and 77

As discussed earlier many suitably hexasubstituted benzenes exhibit the property of forming crystalline inclusion compounds, a possibility originally suggested by the hexa-host analogy.[1,2] The role of three-fold molecular symmetry (discussed below) and two-fold molecular symmetry[39] in host design has also been recently considered. With new hosts, exploiting a higher aromatic core, in mind, the promising candidate molecules 76 and 77 were prepared. This was achieved[42] by reaction of octafluoronaphthalene with 16 molar equivalents of the sodium salt of the appropriate arenethiol in DMEU for 2 days, near quantitative yields being obtained in each case.

(76)   (77)

Significantly, both 76 and 77 form crystalline inclusion compounds on recrystallization from 1,4-dioxan at ambient temperature, thus establishing 76 and 77 as the first members of a new class of eight-legged or "spider" hosts. Two unsolvated forms of 76 have been obtained, a yellow form (from DMF at ambient temperature) and a red form (from anisole at c. 50° C). Intrigued as to the origin of the different colours of these unsolvated forms, and indeed how the chromophoric region of 76 could differ significantly in the two forms, an X-ray study was undertaken. The yellow form is mono-clinic, space group I2/c, while the red form is triclinic, space group PĪ (Table 3). Just how fundamentally the structures differ may be appreciated from Fig. 23. The molecule of 76 in the yellow form is chiral and has exact C_2 symmetry, and approximately D_2 symmetry; both enantiomers are present in the crystal. A very pronounced twist around the bond at the ring fusion for the yellow form can be seen from Fig. 23a, the intra-ring torsion angle corresponding to this bond being −31°. The exact C_2 axis of the molecule runs along the naphthalene's central bond, contrasting with the situation for octabromonaphthalene[43] which also belongs to the point group C_2, but which has the two-fold axis at 90° C to this central bond, and passing

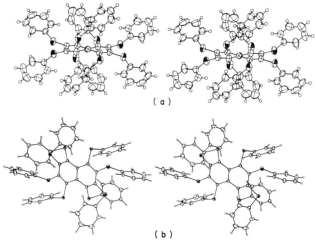

Fig. 23. Stereoviews showing the molecular structure of octakis(phenylthio)naph-thalene (**76**) in (a) the yellow unsolvated crystal; and (b) the red unsolvated crystal. Reproduced, by permission, from ref. 42.

through its mid-point. The structure of **76** in the red form, Fig. 23b, is quite different: the molecule is now achiral with the central region not far from planar; the two six-membered rings comprising the centrosymmetric naph-thalene unit may now be described as enantiomerically-related (distorted) boat conformations. In the yellow form *pairs* of adjacent "legs" alternate above and below the mean plane of the naphthalene ring: taking such a *pair* on one side and relocating it on the other side of the naphthalene, with groups flanking this given pair moving in the opposite sense (to maintain a "balance" of four "legs" up and four down), then gives the achiral leg-distribution of the red form. The individual molecule in the red form occupies 2% less volume than in the yellow form. Fascinatingly, when one presses crystals of the *yellow* form, on for example a glass slide, *red* crystals are produced at the point of pressure application, so that the above apparently formal transformation may well actually occur in the crystal lattice. (The identity of the red form produced in this novel pressure-induced process is currently being investigated.)

3. Design of hosts with trigonal symmetry: the third strategy

Trigonal symmetry is a feature apparent in the molecular structure of a number of important hosts, and families of related host molecules.

Examples[44] of systems of this type shown in Fig. 24 (with cross referencing) are triphenylmethane[45] (**78**), *trans,anti,trans,anti,trans*-perhydrotriphenylene[46] (**79**) cyclotriveratrylene[47] (**80**) (and related systems), the parent tri-*o*-thymotide[48] (**81**), tris(*o*-phenylenedioxy)cyclotriphosphazene[49] (**82**) representing the important class of cyclotriphosphazene hosts, and of course, the (Type I) hexa-hosts, general formula **83**. In the multimolecular inclusion compounds formed by these hosts the surrounding lattice is normally consolidated by van der Waals forces alone, and not by hydrogen-bonding. The individual host molecule does not always attain exact crystallographic three-fold symmetry in these adducts, however, although trigonal (or hexagonal) lattice symmetry is quite often encountered.[50]

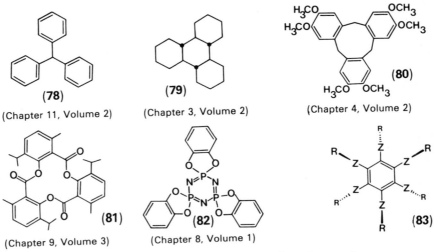

(**78**)
(Chapter 11, Volume 2)

(**79**)
(Chapter 3, Volume 2)

(**80**)
(Chapter 4, Volume 2)

(**81**)
(Chapter 9, Volume 3)

(**82**)
(Chapter 8, Volume 1)

(**83**)

Fig. 24. Some important host molecules possessing trigonal symmetry.

In view of the foregoing, it appeared very attractive to incorporate trigonal symmetry as a key design element in the synthesis of new host molecules. Following this line of inquiry, corresponding to the third strategy, we set out to prepare target molecules **87** and **88**. Bromination of *trans,trans,trans*-cyclododeca-1,5,9-triene by a literature method[51] yields a mixture of the

(**84**)

(**85**)

(**86**)

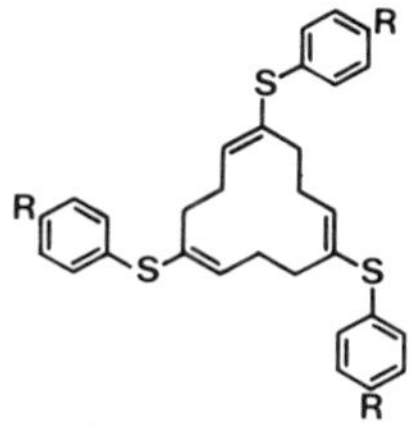

(87) R = H
(88) R = Me

hexabromides **84** and **85**; ^{13}C NMR spectra[52] of the separated isomers, an oil and a solid, establish that the oil has the more symmetrical structure **85** while the solid is **84**. (The structures as here formulated do not, of course, represent the true molecular conformations of **84** and **85**.) Interestingly, the solid isomer itself forms 1:1 adducts with benzene, thiophene, and 1,4-dioxan.[52] Trisdehydrobromination of the isomeric mixture of hexa-bromides[51] gave a mixture of trivinylbromides from which the required, previously well characterized[51,53] symmetrical isomer, 1,5,9-tribromo-*cis,cis-cis*-1,5,9-cyclododecatriene (**86**), could be obtained. Reaction of **86** with PhSCu at 175° C for 6 h in a 10:1 quinoline/pyridine mixture (sealed tube), led to three-fold substitution with complete retention[52] of configuration,

Table 4. *Representative inclusion compounds formed by hosts* **87** *and* **88**

Host	Guest	Mole ratio[a] of host to guest.	Ref.
87	Cyclopentane	2:1	52
	Cyclohexane	2:1	52
	Fluorocyclohexane	2:1	52
	1,4-Dioxan	2:1	52
88	Cyclopentane[b]	2:1	52
	Cyclohexane	2:1	52
	Fluorocyclohexane	2:1	52
	Methylcyclohexane	2:1	52
	Diethyl ether	3:1	52
	Ethyl acetate	4:1	52
	t-Butylacetylene	2:1	52
	2,2-Dimethylbutane	2:1	52
	2,3-Dimethylbutane	2:1	52

[a] The host-guest ratios (given to nearest integer) were determined by multiple integration of the ^{1}H NMR spectrum (in CDCl$_3$ or CS$_2$). All the inclusion compounds were carefully dried *in vacuo* before analysis.
[b] In contrast to cyclopentane, *n*-pentane gives unsolvated material.

yielding the desired product **87**, after appropriate work up. The prospective host **88** was prepared analogously. To our great satisfaction, we found that these compounds were both new hosts, and a representative selection of guests forming stable inclusion compounds with these novel trigonal molecules is shown in Table 4. In each case, the crystalline adduct was obtained by recrystallization of unsolvated material from the appropriate pure solvent. When **88** was recrystallized from an equimolar mixture of cyclopentane, cyclohexane, and cycloheptane, the relative percentages included were 30%, 45%, and 25%, indicating a significant preference for the six-membered cyclic paraffin. It is noteworthy that for host **88**, for example, even such volatile species as *t*-butylacetylene and 2,2- and 2,3-dimethylbutane are tightly retained. An X-ray study of suitable inclusion compounds of **87** and **88**, currently planned, should provide interesting information about the molecular architecture of these hosts as well as elucidating the geometry of the voids available for guest accommodation.

4. Concluding remarks

In the present chapter the writer has hoped to demonstrate that the use of analogy can lead to the discovery of new hosts with no *direct* structural relationship to any known host. The name "hexa-host" selected for these new molecules seems particularly appropriate since it not only suggests the substitution pattern of the central ring, but also reflects the key relationship to the pre-existing hydrogen-bonded hexamer unit commonly encountered in hydroxyaromatic clathrate systems. The hexa-host analogy should not be considered the only useful analogy possible, but simply the first discovered. Indeed, in a recent international symposium held in Parma, Liebau[54] described important work on the relationship between clathrate hydrates and clathrasils, silica-based clathrates; this inorganic analogy has led to the deliberate synthesis of, for example, a previously unknown, direct counterpart of the cubic type II clathrate hydrates in which the water network is replaced by an isostructural silica-based framework. Certainly, other analogies should be sought among organic, inorganic, and organometallic systems.

In the present chapter, host symmetry has also been briefly considered. It is the author's belief that careful host design will lead to the synthesis of many new host molecules possessing trigonal symmetry, and new hosts of importance may also be expected which have a two-fold axis of symmetry. Fortunately, at the time of writing, chemical intuition is still important in host design so that, as in many other areas of inclusion chemistry, chemists have a fascinating and exciting road ahead, stretching as far as the eye can see.

References

1. D. D. MacNicol and D. R. Wilson, *J. Chem. Soc., Chem. Commun.*, 1976, 494.
2. A. D. U. Hardy, D. D. MacNicol and D. R. Wilson, *J. Chem. Soc., Perkin Trans. 2*, 1979, 1011.
3. F. Vögtle and E. Weber, *Angew. Chem., Int. Ed. Engl.*, 1974, **13**, 814.
4. (a) R. Adams, W. Reifschneider and M. D. Mair, *Croat. Chem. Acta*, 1957, **29**, 277. (b) R. Adams and A. Ferretti, *J. Am. Chem. Soc.*, 1959, **81**, 4927.
5. D. D. MacNicol, A. D. U. Hardy and D. R. Wilson, *Nature (London)*, 1977, **266**, 611.
6. A. D. U. Hardy, D. D. MacNicol, S. Swanson and D. R. Wilson, *J. Chem. Soc., Perkin Trans. 2*, 1980, 999.
7. D. D. MacNicol and S. Swanson, *J. Chem. Res. (S)*, 1979, 406–407.
8. K. Burns, C. J. Gilmore, P. R. Mallinson, D. D. MacNicol and S. Swanson, *J. Chem. Res. (S)*, 1981, 30–31 and *J. Chem. Res. (M)*, 1981, 0501–0568.
9. F. Vögtle, M. Zuber and R. G. Lichtenthaler, *Chem. Ber.*, 1973, **106**, 717.
10. D. D. MacNicol and S. Swanson, unpublished results.
11. *Cf.* E. Weber, W. M. Müller and F. Vögtle, *Tetrahedron Lett.*, 1979, 2335.
12. A. Freer, C. J. Gilmore, D. D. MacNicol and D. R. Wilson, *Tetrahedron Lett.*, 1980, 1159.
13. J. Ernst, W. S. Sheldrick and J.-H. Fuhrhop, *Angew. Chem., Int. Ed. Engl.*, 1976, **15**, 778; J. Ernst and J.-H. Fuhrhop, *Justus Liebigs Ann. Chem.*, 1979, 1635.
14. F. A. L. Anet and M. Squillacote, *J. Am. Chem. Soc.*, 1975, **97**, 3243.
15. M. Squillacote, R. S. Sheridan, O. L. Chapman and F. A. L. Anet, *J. Am. Chem. Soc.*, 1975, **97**, 3244.
16. D. D. MacNicol and A. Murphy, *Tetrahedron Lett.*, 1981, **22**, 1131.
17. C. H. Bushweller, *J. Am. Chem. Soc.*, 1969, **91**, 6019; and references therein.
18. J. D. Korp, I. Bernal, S. F. Watkins and F. R. Fronczek, *Tetrahedron Lett.*, 1981, **22**, 4767.
19. N. L. Allinger, M. J. Hickey and J. Kao, *J. Am. Chem. Soc.*, 1976, **98**, 2741.
20. D. D. MacNicol, S. Swanson and A. Murphy, unpublished results.
21. M. Isola, E. Ciuffarin and L. Sagramora, *Synthesis*, 1976, 326.
22. A. A. Freer, C. J. Gilmore, D. D. MacNicol and S. Swanson, *Tetrahedron Lett.*, 1980, **21**, 205.
23. (a) K. Nishihata and M. Nishio, *Chem. Comm.*, 1971, 958. (b) H. M. R. Hoffmann and E. D. Hughes, *J. Chem. Soc.*, 1964, 1244.
24. R. E. Jones and D. H. Templeton, *Acta Crystallogr.*, 1958, **11**, 484.
25. L. Leiserowitz, *Acta Crystallogr.*, 1976, **B32**, 775.
26. B. M. Craven and G. T. DeTitta, *J. Chem. Soc., Chem. Commun.*, 1972, 530.
27. A. D. U. Hardy, D. D. MacNicol and D. R. Wilson, *J. Chem. Soc., Chem. Commun.*, 1977, 525.
28. M. McMath and D. D. MacNicol, unpublished results.
29. J. H. Gall and D. D. MacNicol, unpublished results.

168 D. D. MacNicol

30. J. H. Gall, C. J. Gilmore and D. D. MacNicol, *J. Chem. Soc., Chem. Commun.*, 1979, 927.
31. *Cf.* M. B. Stringer and D. Wege, *Tetrahedron Lett.*, 1980, **21**, 3831; and references therein.
32. A. A. Freer, J. H. Gall and D. D. MacNicol, *J. Chem. Soc. Chem. Commun.*, 1982, 674.
33. L. A. LaPlanche and M. T. Rogers, *J. Am. Chem. Soc.*, 1963, **85**, 3728.
34. L. Testaferri, M. Tingoli and M. Tiecco, *J. Org. Chem.*, 1980, **45**, 4376.
35. D. D. MacNicol, P. R. Mallinson, A. Murphy and G. J. Sym, *Tetrahedron Lett.*, 1982, **23**, 4131.
36. H. Spencer, *Chem. Ind. (London)*, 1979, 728.
37. U.K. Patent application no. 8217510.
38. B. J. Barker, J. Rosenfarb and J. A. Caruso, *Angew. Chem., Int. Ed. Engl.*, 1979, **18**, 503.
39. N. Z. Huang and T. C. W. Mak, *J. Chem. Soc., Chem. Commun.*, 1982, 543; and references therein.
40. (a) M. V. Stackelberg, A. Hoverath and Ch. Scheringer, *Z. Elektrochem.*, 1958, **62**, 123. (b) B. A. Nikitin, *Comp. rend. U.S.S.R.*, 1940, **29**, 571.
41. C. J. Gilmore, D. D. MacNicol, A. Murphy and M. Russell, *Tetrahedron Letters*, 1983, **24**, 3269.
42. R. H. Barbour, A. A. Freer and D. D. MacNicol, *J. Chem. Soc., Chem. Commun.*, 1983, 362.
43. J. H. Brady, A. D. Redhouse and B. J. Wakefield, *J. Chem. Res. (S)*, 1982, 137 and *J. Chem. Res. (M)*, 1982, 1541–1554.
44. L. Manojlović-Muir and K. W. Muir, *J. Chem. Soc., Chem. Commun.*, 1982, 1155.
45. A. Allemand and R. Gerdil, *Acta Crystallogr.*, 1975, **A31**, S130.
46. (a) G. Allegra, M. Farina, A. Immirzi, A. Colombo, U. Rossi, R. Broggi, and G. Natta, *J. Chem. Soc. (B)*, 1967, 1020. (b) G. Allegra, M. Farina, A. Colombo, G. Casagrande-Tettamanti, U. Rossi and G. Natta, *J. Chem. Soc. (B)* 1967, 1028.
47. R. C. Cookson, B. Halton and I. D. R. Stevens, *J. Chem. Soc. (B)*, 1968, 767; and references therein.
48. (a) D. Lawton and H. M. Powell, *J. Chem. Soc.*, 1958, 2339. (b) D. J. Williams and D. Lawton, *Tetrahedron Lett.*, 1975, 111. (c) S. Brunie, A. Navaza, G. Tsoucaris, J. P. Declercq and G. Germain, *Acta Crystallogr.*, 1977, **B33**, 2645. (d) S. Brunie and G. Tsoucaris, *Cryst. Struct. Commun.*, 1974, **3**, 481.
49. H. R. Allcock, R. W. Allen, E. C. Bissell, L. A. Smeltz and M. Teeter, *J. Am. Chem. Soc.*, 1976, **98**, 5120; and references therein.
50. See for example, S. A. Puckett, I. C. Paul and D. Y. Curtin, *J. Chem. Soc., Perkin Trans. 2*, 1976, 1873 (Table 3).
51. K. G. Untch and D. J. Martin, *J. Am. Chem. Soc.*, 1965, **87**, 3518.
52. D. D. MacNicol and S. Swanson, *Tetrahedron Lett.*, 1977, 2969.
53. S. Castellano and K. G. Untch, *J. Am. Chem. Soc.*, 1966, **88**, 4238.
54. (a) F. Liebau, plenary lecture at The Second International Symposium on Clathrate Compounds and Molecular Inclusion Phenomena, Parma, Italy, 1982. (b) H. Gies and F. Liebau, *Acta Crystallogr.*, 1981, **A37(Suppl.)**, C187. (c) H. Gies, F. Liebau and H. Gerke, *Angew. Chem., Int. Ed. Engl.*, 1982, **21**, 206. (d) J. L. Schlenker, F. G. Dwyer, E. E. Jenkins, W. J. Rohrbaugh, G. T. Kokotailo and W. M. Meir, *Nature (London)*, 1981, **294**, 340; H. Gerke, H. Gies and F. Liebau, *Z. Kristallogr.*, 1982, **159**, 51.
55. C. J. Gilmore, D. D. MacNicol, P. R. Mallinson, A. Murphy and M. A. Russel, *J. Incl. Phenom.* (in Press).

6 · THE TRIANTHRANILIDES: A NEW CLASS OF ORGANIC HOSTS

W. D. OLLIS and J. F. STODDART
The University of Sheffield, Sheffield, UK.

1. Historical background

Amongst the known[1-6] trisalicylides **1–8**, only tri-*o*-thymotide (**6**)[5] has been found[5,7-20] to form crystalline inclusion compounds of both the cavity and channel types with a wide range of achiral and chiral guest species (see Chapter 9, Volume 3). Some of these inclusion compounds undergo spontaneous resolution on crystallization as a result of the solvated host molecules adopting asymmetric propeller-like conformations of one chirality or the other. At first sight, the fact that tri-*o*-carvacrotide (**8**),[6] a constitutional isomer of tri-*o*-thymotide (**6**) in which the methyl and isopropyl substituents on the aromatic rings have interchanged their positions, has not been reported to form *any* crystalline inclusion compounds might seem surprising. However, X-ray crystallography of tri-*o*-carvacrotide (**8**) has demonstrated[21] recently that this trisalicylide derivative crystallizes with its molecules existing, half as an enantiomeric pair of propeller-like conformations and, the other half, as an enantiomeric pair of helix-like conformations.

INCLUSION COMPOUNDS 2
ISBN 0-12-067101-8

By contrast, unsolvated tri-*o*-thymotide exists only as a pair of enantiomeric propeller-like conformations in the solid state.[11,22]

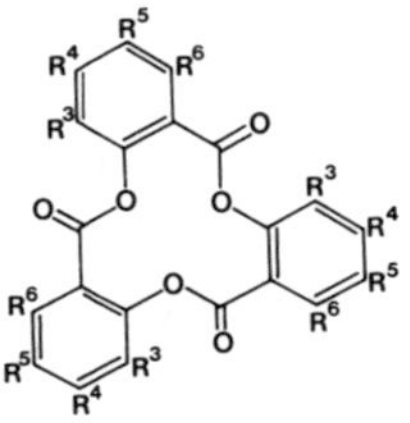

	R³	R⁴	R⁵	R⁶
(1)	H	H	H	H
(2)	Me	H	H	H
(3)	H	Me	H	H
(4)	H	H	H	Me
(5)	Me	H	H	Me
(6)	CHMe₂	H	H	Me
(7)	H	CHMe₂	H	Me
(8)	Me	H	H	CHMe₂

In solution, tri-*o*-thymotide (**6**) and tri-*o*-carvacrotide (**8**) both exist[23-27] as an equilibrium mixture (P⇌H⇌H*⇌P*) of enantiomeric propeller (P and P*) and enantiomeric helical (H and H*) conformations. Only the P and H conformations of tri-*o*-thymotide (**6**) are represented diagramatically in Fig. 1. It will be noted that in both these diastereoisomeric conformations

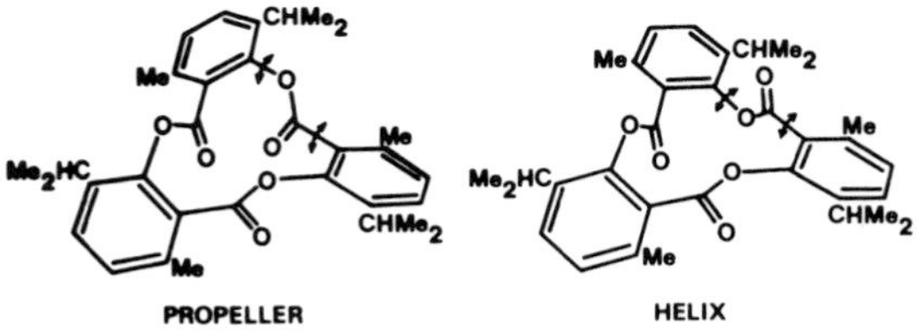

Fig. 1. The propeller and helical conformations of tri-*o*-thymotide (**6**). The double headed arrows indicate the pedalling motion by which a *trans*-ester linkage can change its orientation.

all three ester linkages assume *trans* geometries: in the case of the propeller conformations, the molecular symmetry is that of time-averaged C_3 symmetry, whereas the helical conformations belong to point group C_1. ¹H NMR spectroscopy reveals[26,28] that, in common with tri-*o*-thymotide (**6**) and tri-*o*-carvacrotide (**8**), tri-3,6-dimethylsalicylide (**5**) and tri-*o*-cresotide (**2**) adopt the propeller conformations preferentially (58–90%) in solution. The conformational interconversion and inversion processes are[28] too rapid on the ¹H NMR time scale even at low temperatures to ascertain the conformational preferences for tri-*m*-cresotide (**3**) and trisalicylide (**1**) itself. The highest energy barrier ($\Delta G^{\ddagger} = 89.5$ kJ mol^{-1}) was observed[26] for P → H conformational interconversion in tri-*o*-thymotide (**6**). From our investigations[23-28] on the conformational behaviour of the trisalicylide derivatives **1–3**, **5**, **6**, and **8** in solution, it could be concluded[26-28] that, although the barriers to conformational change are greatly influenced ($\Delta\Delta G^{\ddagger} \simeq 63$ kJ mol^{-1}) by the nature of the *ortho* substituents on the aromatic rings, the positions of the conformational equilibria are affected only slightly ($\Delta\Delta G° \simeq 4$ kJ mol^{-1}).

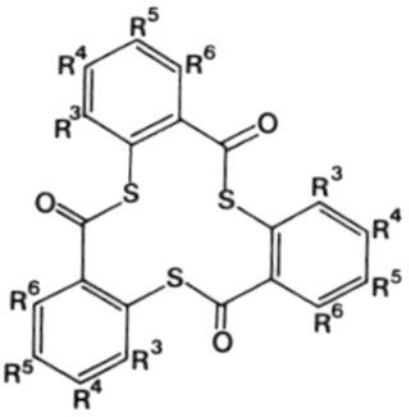

	R^3	R^4	R^5	R^6
(9)	H	H	H	H
(10)	Me	H	H	H
(11)	H	H	Me	H
(12)	H	H	H	Me
(13)	Me	H	H	Me
(14)	CHMe$_2$	H	H	Me

Next, we decided to examine the consequences for solution and solid state properties of replacing the ester linkages in the trisalicylides first by thioester linkages and then subsequently by amide bonds. Since trithiosalicy-lide (9), the parent sulphur analogue of trisalicylide (1), can be prepared[29] by dehydration of thiosalicylic acid in a one-step reaction, we resolved to synthesise some *ortho* mono- and di-substituted trithiosalicylides,[30,31] in addition to tri-5-methylthiosalicylide (11). This derivative may be considered as a suitable model compound (with a convenient ^{1}H NMR probe being provided by the aryl methyl groups) for trithiosalicylide (9), since it lacks substituents in the *ortho* positions of the aromatic rings. Although tri-3-methylthiosalicylide (10), tri-6-methylthiosalicylide (12), and tri-3,6-dimethylthiosalicylide (13) were all obtained and characterized, attempts[30,31] to cyclise thiothymotic acid only afforded di-*o*-thiothymotide; no tri-*o*-thiothymotide (14) was isolated. In view of the supposedly unique complex-ing and chirality properties of tri-*o*-thymotide (6) amongst the known trisalicylide derivatives, the failure to obtain tri-*o*-thiothymotide (14) came as a disappointment.

An investigation of the conformational behaviour of the trithiosalicylide derivatives 10–13 in solution revealed[30,31] that they exist as ring inverting (see Fig. 2) enantiomeric helical conformations with free energy barriers to inversion ranging from 44–100 kJ mol^{-1}. Since *no* propeller conformations were detected by ^{1}H or ^{13}C NMR spectroscopy, it is perhaps hardly surpris-ing that both tri-3-methylthiosalicylide (10) and tri-6-methylthiosalicylide (12)

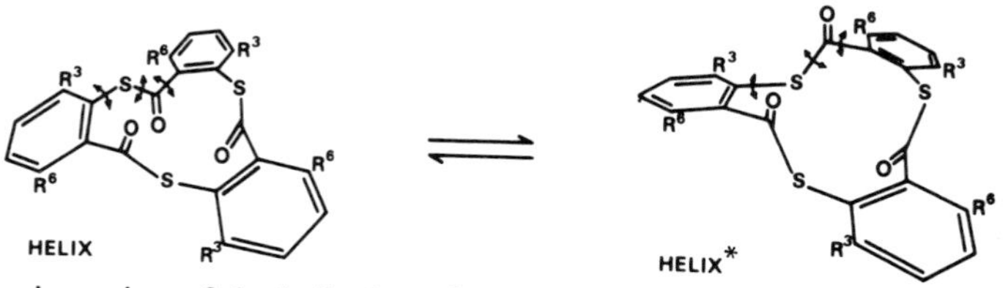

Fig. 2. The inversion of the helical conformation of the trithiosalicylide derivatives 10–13. The double headed arrows signify a process for the reorientation of *trans*-thioester linkages involving *cis* intermediates.

crystallize[21,31] in helical conformations. *None* of the trithiosalicylide derivatives **9–13** has been found to form inclusion compounds although tri-6-
methylthiosalicylide (**12**) undergoes spontaneous resolution on crystallization.[21,31]

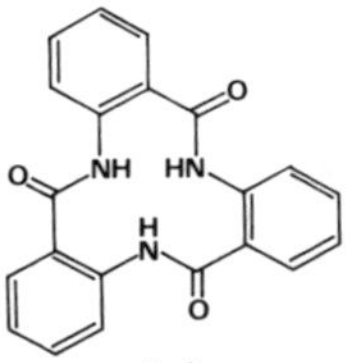

(15) 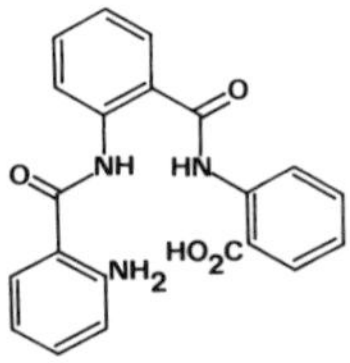   (16)

Despite the relative ease with which salicylic and thiosalicylic acids and
their derivatives can be condensed to form mixtures of cyclic compounds,
which often include dimers, tetramers, and hexamers, as well as the trimers,
all our attempts to prepare trianthranilide (**15**) directly by condensation
from anthranilic acid failed despite the reported[32] isolation of the acyclic
trimer **16** in 1907. It was, therefore, necessary to devise stepwise approaches
to the synthesis of trianthranilide derivatives. This was accomplished[27,33–35]

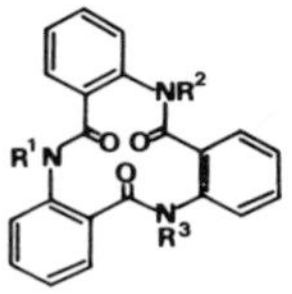

	R^1	R^2	R^3
(17)	Me	Me	H
(18)	Me	CD$_3$	H
(19)	Me	Me	Me
(20)	Me	Me	CD$_3$
(21)	Me	CD$_3$	CD$_3$
(22)	Me	Me	COMe
(23)	Me	Me	COPh
(24)	Me	Me	CH$_2$Ph
(25)	Me	CD$_3$	CH$_2$Ph
(26)	CH$_2$Ph	CH$_2$Ph	H
(27)	CH$_2$Ph	CH$_2$Ph	CH$_2$Ph
(28)	CH$_2$Ph	CH$_2$Ph	Me
(29)	Me	CH$_2$Ph	H
(30)	Me	CH$_2$Ph	Et
(31)	Me	Et	H
(32)	Me	Et	CH$_2$Ph

after much time and effort on the part of numerous research students and associates at Bristol and Sheffield beteween 1951 and 1974. Two different synthetic routes provided the wide range of N,N'-disubstituted and N,N',N''-trisubstituted trianthranilide derivatives represented by the constitutional formulae **17–32**. The following Section describes how these compounds were obtained and, in some cases, characterized as crystalline conformational diastereoisomers.[34,35]

2. Trianthranilides

2.1. Trianthranilide synthesis

Dianthraniloylanthranilic acid (**16**) was identified as a possible intermediate in an attempted synthesis of trianthranilide (**15**). In this way, the first step in our successful synthetic route (see Scheme 1) was based upon the known[36,37] reaction between anthranilic acid and isatoic anhydride to afford anthraniloylanthranilic acid (**33**). Since an attempt to condense **33** with more isatoic anhydride was unsuccessful, a less direct route to dianthraniloylanthranilic acid (**16**)[32] had to be adopted. This involved treatment of **33** under basic conditions with *o*-nitrobenzoyl chloride to afford the *o*-nitrobenzoyl derivative **34** which could then be reduced to give **16**. However, all attempts to cyclize dianthraniloylanthranilic acid (**16**) yielded *only* 8-membered ring dianthranilide derivatives. Hence, we were persuaded to protect both amide linkages in **16** by a methylation procedure which also brought about esterification of the free carboxyl group. The N,N'-dimethylated methyl ester **35** could then be reduced ($NO_2 \rightarrow NH_2$) and de-esterified in a one-pot reaction to give the key amino acid precursor **36**. Treatment of **36** with N,N'-dicyclohexylcarbodiimide (DCCI) afforded the *N*-acylurea derivative **38**—presumably *via* the intermediacy of the *O*-acylisourea derivative **37**—as a crystalline compound. X-ray crystallography confirmed[35,38] the constitutional assignment made to this compound in Scheme 1. Finally, the *N*-acylurea derivative **38** successfully underwent cyclization to afford N,N'-dimethyltrianthranilide (**17**) in refluxing ethanol. Thus, the three aromatic rings in **17** have been derived from anthranilic acid (A), isatoic anhydride (B), and *o*-nitrobenzoyl chloride (C). N,N'-Dibenzyltrianthranilide (**26**) was prepared by a modification to the synthetic route shown in Scheme 1 by which N-[2-(*o*-nitrobenzamido)benzoyl]anthranilic acid was benzylated (PhCH$_2$Br, NaH, THF), reduced ($NO_2 \rightarrow NH_2$ on ring C), de-esterified ($CO_2CH_2Ph \rightarrow CO_2H$ on ring A), and cyclized (DCCI) without isolation of any intermediate *N*-acylurea derivative. Scheme 2 shows that N,N'-dimethyltrianthranilide (**17**) and N,N'-dibenzyltrianthranilide (**26**)

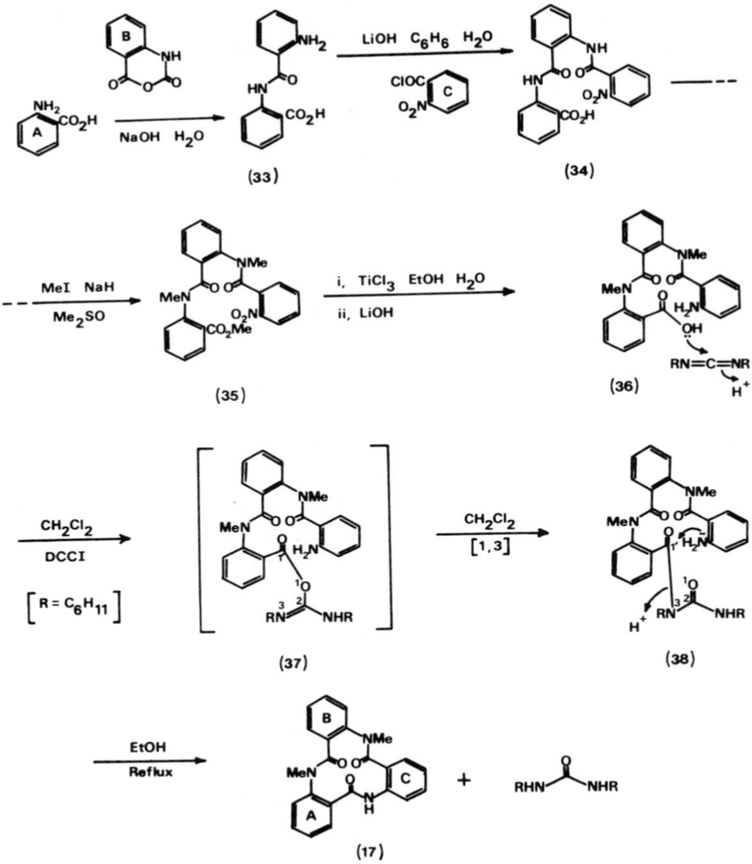

Scheme 1.

can be converted by means of appropriate alkylations (with MeI, CD$_3$I, or PhCH$_2$Br as alkylating agent in either DMSO or THF with NaH as base) into the *N,N',N''*-trisubstituted trianthranilide derivatives **19**, **20**, **24**, **26**, and **27**. Acylation (AcCl, NaH, PhMe) and benzoylation (BzCl, NaH, THF) of *N,N'*-dimethyltrianthranilide (**17**) afforded the *N''*-acetyl **22** and *N''*-benzoyl **23** derivatives. In all the trianthranilide derivatives synthesized *via* the routes summarized in Schemes 1 and 2, the nitrogen substituents R^1 and R^2 are identical. In order to be able to prepare derivatives in which R^1 and R^2 are non-identical, the alternative synthetic route outlined in Scheme 3 had to be adopted. The key compound **42** was built up from *N*-methylanthranilic acid[39] (ring A) and *o*-nitrobenzoyl chloride (rings B and C) by the

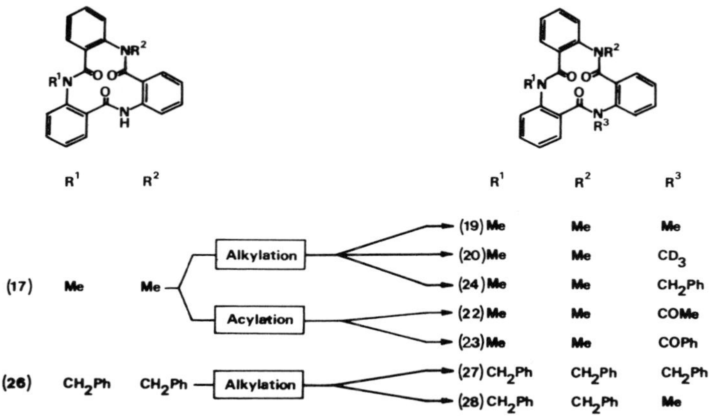

Scheme 2.

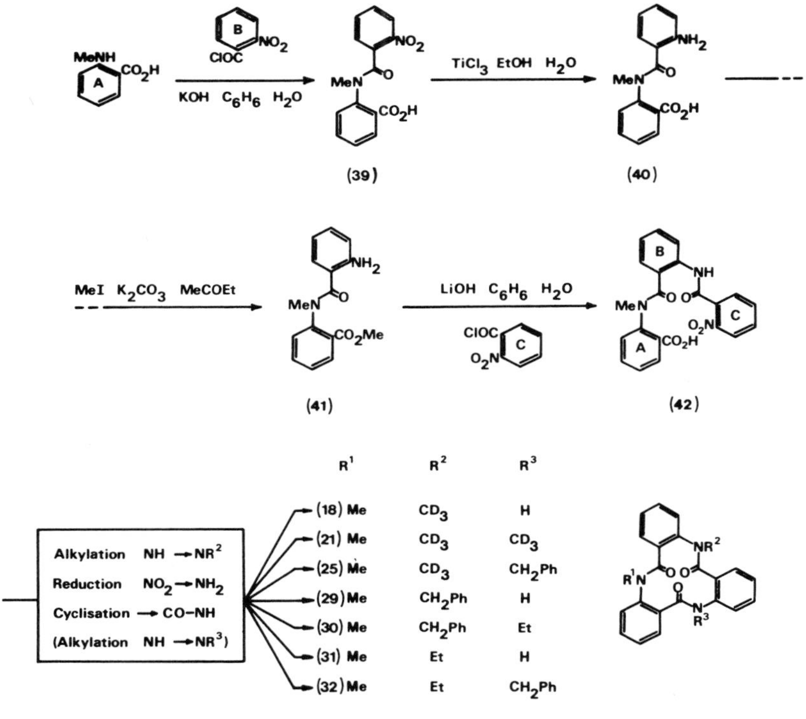

Scheme 3.

sequence of reactions, $39^{40-42} \rightarrow 40 \rightarrow 41^{42} \rightarrow 42$. By ringing the changes on **42** with different alkylating agents (CD_3I, $PhCH_2Br$, and EtI), before reducing the aromatic nitro group to an amino group, and then finally cyclizing in the presence of DCCI, the N,N'-disubstituted trianthranilide derivatives **18**, **29**, and **31** could be obtained. Appropriate further alkylations on these compounds afforded the N,N',N''-trisubstituted derivatives **21**, **25**, **30**, and **32**. The trianthranilides **18**, **20**, **21**, and **25** which contain N-trideuteriomethyl substitutents were prepared[35] in order to aid the investigations (see Section 2.4) of the conformational behaviour of N,N',N''-trimethyltrianthranilide (**19**) and N,N'-dimethyl-N''-benzyltrianthranilide (**24**) in solution by ^1H NMR spectroscopy.

In naming the substituents on the nitrogen atoms of the amide linkages in N,N'-disubstituted and N,N',N''-trisubstituted trianthranilides, it is necessary[35] for ease of constitutional comparison and the presentation of the stereochemical arguments in Section 2.4 for the N,N' and N'' substituents respectively to be named in the order R^1, R^2, and R^3 defined by the constitutional formula rather than to follow the conventional nomenclature practice of naming the substituents in alphabetical order according to the initial letters of their names. In this way, it can be readily appreciated, for example, that N-methyl-N'-benzyl-N''-ethyltrianthranilide (**30**) and N-methyl-N'-ethyl-N''-benzyltrianthranilide (**32**) are constitutional isomers.

Interestingly, it also proved possible[35] to isolate pairs of conformational diastereoisomers of N,N'-dimethyl-N''-benzyl **24**, N,N',N''-tribenzyl **27**, N,N'-dibenzyl-N''-methyl **28** derivatives of trianthranilide by subjecting them to preparative TLC in a cold room ($+5°$ C). In the cases of compounds **27** and **28**, the isomers were well separated after chromatography on silica gel. With N,N',N''-tribenzylanthranilide (**27**), the major component, which corresponded to the faster migrating component on TLC, could also be isolated from the crude reaction mixture as a pure crystalline compound by slow crystallization from light petroleum. In the case of compound **24**, the isomers migrated at almost the same rate on TLC and had to be subjected to subsequent purification by HPLC using ethyl acetate as eluant. The yields, melting points, and diastereoisomeric purities of the crystalline conformational diastereoisomers are listed in Table 1.

2.2. Propeller and helical conformations of constitutionally symmetrical and unsymmetrical trianthranilides

The N,N',N''-trimethyl **19** and $N,N'N''$-tribenzyl **27** derivatives of trianthranilide are both constitutionally symmetrical in so far as $R^1 = R^2 = R^3$. This means that if, by analogy with the known conformational properties

Table 1. *Crystalline conformational diastereoisomers of N,N′,N″-trisubstituted trian-thranilides isolated by preparative TLC on silica gel at +5° C using ethyl acetate–light petroleum (b.p. 60–80° C) (1:1) as eluant*

	Compound						Diastereo-isomeric
Structure	R^1	R^2	R^3	Conformation[a]	Yield (%)	M.p. (° C)	purity (%)[e]
24	Me	Me	CH_2Ph	H-3	21[b],[d]	152–154	>98
				H-1	32[c],[d]	153–154	>98
27	CH_2Ph	CH_2Ph	CH_2Ph	P	59[b]	260–263	>98
				H	19[c]	134–144	93
28	CH_2Ph	CH_2Ph	Me	P	13[b]	120–128	65
				H-2	71[c]	125–140	82

[a] See Section 2.2 for the definitions of these descriptors and Section 2.4 for the conformational assignments.
[b] The faster-moving component.
[c] The slower-moving component.
[d] Purified by HPLC.
[e] Established by ^{1}H NMR spectroscopy.

of trisalicylides[1–28] and trithiosalicylides,[29–31] the three amide linkages are assumed to adopt *trans* geometries, then *two* conformational diastereoisomers of the propeller (P) and helical (H) types can be identified[33–35] (Fig. 3). Since both these conformations are chiral, there also exist enantiomeric propeller (P*) and helical (H*) conformations. Whilst

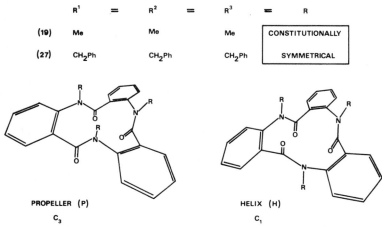

	R^1	=	R^2	=	R^3	=	R	
(19)	Me		Me		Me		CONSTITUTIONALLY	
(27)	CH_2Ph		CH_2Ph		CH_2Ph		SYMMETRICAL	

PROPELLER (P)

C_3

HELIX (H)

C_1

Fig. 3. The two diastereoisomeric conformations for the constitutionally symmetrical N,N′,N″-trisubstituted trianthranilides.

the propeller conformations have averaged C_3 symmetry, the helical conformations are asymmetric. The N,N'-disubstituted trianthranilide derivatives **17**, **26**, **29**, and **31**, and the N,N',N''-trisubstituted derivatives **24**, **28**, **30**, and **32** are constitutionally unsymmetrical because either $R^1 = R^2 \neq R^3$ or $R^1 \neq R^2 \neq R^3$. Assuming that these derivatives will adopt only conformations in which the three amide linkages have *trans*-geometries, then there are *four* diastereoisomeric conformations (Fig. 4), a propeller (P) conformation and three helical conformations, which may be designated[35] H-1, H-2, and H-3 according to whether R^1, R^2, or R^3 are oriented towards the opposite face of the 12-membered ring from R^2 and R^3, R^1 and R^3, and R^1 and R^2, respectively. Since these conformations are all asymmetric, and hence chiral, there are four enantiomeric conformations (P*, H-1*, H-2*, and H-3*) corresponding to each diastereoisomeric conformation. The conformational diastereoisomers in Fig. 4 are represented such that the mean plane of the 12-membered ring lies in the plane of the paper and the substituent groups R^1, R^2, and R^3 on nitrogen atoms are indicated[35] as being oriented above (●) or below (○) this mean plane. The triangular notation,[35] which describes this relative orientation in an abbreviated form, is extremely useful (see Section 2.4) in discussing the conformational behaviour of N,N'-disubstituted and N,N',N''-trisubstituted trianthranilides in solution.

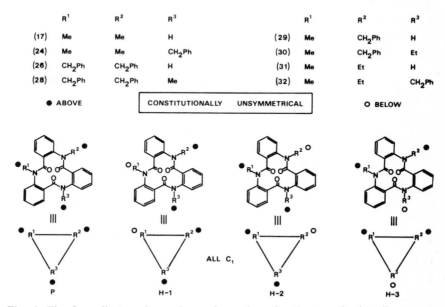

	R^1	R^2	R^3		R^1	R^2	R^3
(17)	Me	Me	H	(29)	Me	CH$_2$Ph	H
(24)	Me	Me	CH$_2$Ph	(30)	Me	CH$_2$Ph	Et
(26)	CH$_2$Ph	CH$_2$Ph	H	(31)	Me	Et	H
(28)	CH$_2$Ph	CH$_2$Ph	Me	(32)	Me	Et	CH$_2$Ph

● ABOVE   CONSTITUTIONALLY   UNSYMMETRICAL   ○ BELOW

ALL C_1

P   H-1   H-2   H-3

Fig. 4. The four diastereoisomeric conformations for the constitutionally unsymmetrical N,N'-di- and N,N',N''-tri-substituted trianthranilides.

2.3. Solid state structures of the N,N′-dimethyl **17** and *N,N′*-dibenzyl **26** derivatives of trianthranilide

N,N′-Dimethyltrianthranilide (**17**) crystallizes[35,43] from both ethanol and *p*-dioxan in a noncentrosymmetric space group (P2₁) which contains molecules of one chirality only; thus this trianthranilide derivative **17** is another example, in addition to tri-*o*-thymotide (**6**)[1–20] and tri-6-methyl-thiosalicylide (**12**),[30,31] of a compound belonging to this tribenzocyclododecatriene category which undergoes spontaneous resolution on crystallization. However, there is no evidence,[35] unlike tri-*o*-thymotide (**6**), that it forms inclusion compounds, i.e. it has solid state characteristics similar to those of tri-6-methylthiosalicylide (**12**), which also undergoes spontaneous resolution but does not form inclusion compounds. Crystal structure analysis (Fig. 5) reveals[43] that *N,N′*-dimethyltrianthranilide (**17**) adopts either a

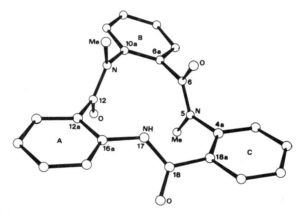

Fig. 5. The solid state structure of *N,N′*-dimethyltrianthranilide (**17**).

distorted H-1 or distorted H-1* conformation in the solid state. Recrystallization from toluene provided[35] sufficiently large dextrorotatory or laevorotatory single crystals for polarimetric examination of selected specimens to be carried out at +2° C in chloroform solution (a dextrorotatory single crystal was estimated to have a specific optical rotation at the sodium D-line of +87° immediately upon dissolution). Under these conditions, the half-life time for racemization was found to be 28.9 min and the $\Delta G^{\ddagger}_{rac}$ value to be 87 kJ mol⁻¹. This value for the free energy of racemization is undoubtedly associated with an H⇌H* ring inversion process, involving either H-1 or H-2 ground state conformations (see Section 2.4).

N,N′-Dibenzyltrianthranilide (**26**) crystallizes[35,43] as a racemate from toluene in a centrosymmetric monoclinic space group (P2₁/a). However, this *N,N'*-disubstituted trianthranilide derivative does form a 1:1 inclusion compound with the solvent (see Section 4). The X-ray crystal structure analysis (Fig. 6) of the *N,N'*-dibenzyl derivative **26** shows[43] that, in common with *N,N'*-dimethyltrianthranilide (**17**), the molecules exist in a distorted H-1 (or H-1*) conformation, but for a different reason.

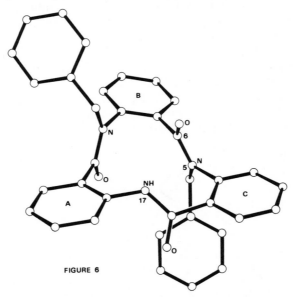

FIGURE 6

Fig. 6. The solid state structure of *N,N'*-dibenzyltrianthranilide (**26**).

In both solid state structures[43] (Figs. 5 and 6), the planes of the *trans*-amide linkages between rings A and B, and between rings B and C, are approximately normal to the planes of the adjacent aromatic rings. However, the plane of the *trans*-amide linkage between rings A and C is inclined at only *c.* 37° and *c.* 43° in the *N,N'*-dimethyl **17** and *N,N'*-dibenzyl **26** derivatives, respectively, i.e. in both cases, the unsubstituted amide bond is tilted by more than 45° compared with the substituted amide bonds. In the *N,N'*-dibenzyl derivative **26**, N(17) has sp^2 geometry with the N–H bond directed towards the amide carbonyl oxygen atom O(6), forming an intramolecular transannular (N–H···O) hydrogen bond (2.94 Å). However, in the *N,N'*-dimethyl derivative **17**, the geometry at N(17) is severely distorted such that it is pyramidal with CNH angles of 108 and 111° and a CNC angle of 125°. The associated OCNH torsional angle is 138° as shown in Fig. 7. The intramolecular transannular hydrogen bond is not to O(6) in this case, but rather to the amide nitrogen atom N(5). The (N–H···N)

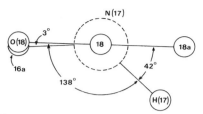

Fig. 7. A projection of the *trans*-amide C(18)–N(17) bond in N,N'-dimethyltrianthranilide (**17**).

hydrogen bond distance is 2.96 Å and the interatomic distance from N(17) to O(6) is increased to 3.26 Å.

Aside from this informative and useful structural detail, the certain knowledge of the conformational types adopted by the N,N'-dimethyl **17** and N,N'-dibenzyl **26** derivatives in the solid state contributes immeasurably to our confidence in interpreting ^{1}H NMR spectra and discussing the conformational behaviour of N,N'-disubstituted and N,N',N''-trisubstituted trianthranilides in solution in the following Section.

2.4. Conformational behaviour of the constitutionally symmetrical and unsymmetrical trianthranilide derivatives in solution[35]

The constitutionally symmetrical N,N',N''-trimethyl **19** and N,N',N''-tribenzyl **27** trianthranilides both exist in solution as an equilibrium mixture of propeller and helical conformations. In the case of the N,N',N''-trimethyl derivative **19**, the predominant diastereoisomer with the helical conformation was isolated by slow recrystallization from chloroform–ether–light petroleum (b.p. 60–80° C). In the case of the N,N',N''-tribenzyl derivative **27**, the propeller and helical conformational diastereoisomers were both characterized as crystalline compounds after preparative TLC at low temperature (+5° C) (see Section 2.1). The free energy barriers to conformational inversion (between enantiomers) and interconversion (between diastereoisomers) processes in solution were obtained for both these derivatives by (i) direct equilibration experiments and by (ii) dynamic ^{1}H NMR spectroscopy. The conformational free energy differences between diastereoisomers were deduced directly from integration of the ^{1}H NMR spectra of fully equilibrated samples.

Immediately after dissolving crystals of N,N',N''-trimethyltrianthranilide (**17**) in CDCl$_3$ at +10° C, *three* singlets were observed at δ 3.13 (C), δ 3.21 (B), and δ 3.33 (A) for the NMe protons (Fig. 8). This indicates that this derivative adopts the asymmetric helical conformation in the solid state. Equilibration

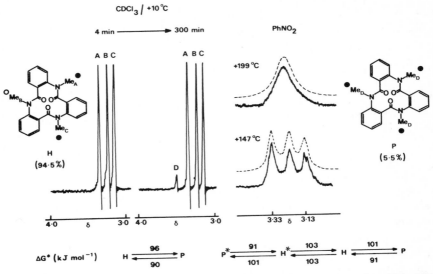

Fig. 8. The ^{1}H NMR spectral behaviour of N,N',N''-trimethyltrianthranilide (**19**) summarizing kinetic experiments in CDCl$_3$ and dynamic experiments in PhNO$_2$.

to 5.5% of a propeller conformation with averaged C$_3$ symmetry, which gives rise to *one* singlet at δ 3.49, is complete inside 5 h. A kinetic treatment of this equilibration process affords $\Delta G^\ddagger$ values for P→H and H→P interconversions which are in good agreement with those obtained (Fig. 8) from variable temperature ^{1}H NMR spectroscopy in nitrobenzene solution between +147° C and +199° C. In addition, the dynamic experiment provides qualitative information on the free energy of activation for the H⇌H* inversion process.

The unambiguous assignment of the NMe groups labelled Me$_A$, Me$_B$, and Me$_C$ in the helical conformation of N,N',N''-trimethyltrianthranilide (**19**) to the singlets appearing at δ 3.33, δ 3.21, and δ 3.17, respectively was made on the basis of kinetically controlled trideuteriomethylations of N,N'-dimethyl **17** and N-methyl-N'-trideuteriomethyl **18** trianthranilides. The absolute identity of these assignments assume importance in the identification of conformational diastereoisomers of constitutionally unsymmetrical N,N',N''-trisubstituted derivatives, such as N,N'-dimethyl-N''-benzyltrianthranilide (**24**) and N,N'-dibenzyl-N''-methyltrianthranilide (**28**). The approach depends upon the assumption that trideuteriomethylation of the CONH function in N,N'-dimethyltrianthranilide (**17**) occurs more rapidly than does conformational reorientation of the two CONMe linkages.

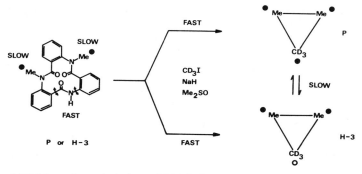

Fig. 9. *N*-Trideuteriomethylation of *N,N'*-dimethyltrianthranilide (**17**) (first case).

Reorientation of the smaller CONH linkage by a pedalling tye of motion, which is faster than reorientation of the larger CONMe group, might be expected to occur at a rate similar to that of its trideuteriomethylation. Following upon these assumptions, two situations are possible: the first is summarized in Figs. 9 and 10, and the second in Figs. 11 and 12. In the first case where the two NMe groups reside on the same side of the 12-membered ring of **17**, as they would in the P and H-3 conformations, trideuteriomethylation can lead (Fig. 9) to both P and H-3 conformations of *N,N'*-dimethyl-*N''*-trideuteriomethyltrianthranilide (**20**). If conformational interconversion between the P and H-3 conformations of *N,N'*-dimethyltrianthranilide (**17**) is occurring more rapidly than its reaction, then the product ratio will be determined by the relative heights of the reaction barriers. There is no problem in assigning the two NMe groups and the NCD$_3$ group in the P conformation of **20** to site D. However, there are six

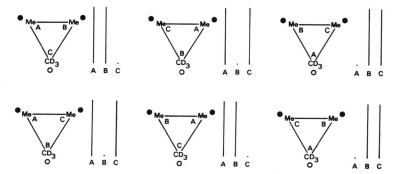

Fig. 10. The six possible ways of assigning NMe groups in the H-3 conformation of *N,N'*-dimethyl-*N''*-trideuteriomethyltrianthranilide (**20**) to sites A, B, and C, and the predicted relative intensities of these sites in the ^{1}H NMR spectrum.

different ways of assigning these same groups in the H-3 conformation to sites A, B, and C (see Fig. 10). Assuming that there is no conformational scrambling of the NCD_3 group with the two NMe groups during the reaction or the isolation of the product, the predicted consequences of the six different assignments upon the partial ^1H NMR spectrum are portrayed as line spectra in Fig. 10. In the second case, where the NMe groups reside on opposite sides of the 12-membered ring of **17**, as they would in the H-1 and H-2* conformations, trideuteriomethylation will then lead (Fig. 11) to both H-1

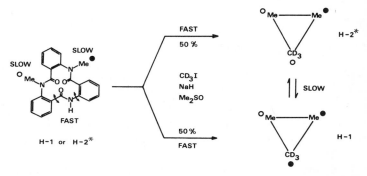

Fig. 11. *N*-Trideuteriomethylation of *N,N'*-dimethyltrianthranilide (**17**) (second case).

and H-2* conformations of *N,N'*-dimethyl-*N"*-trideuteriomethyltrianthranilide (**20**). If we assume for the sake of argument that the two transition states for the reactions are of equal energy, then equal amounts of the two H-1 and H-2* conformations will be formed. Again, there are six different ways in which the two NMe groups and the NCD_3 group in the H-1 and H-2* conformations can be assigned to sites A, B, and C. They are portrayed as pairs in Fig. 12 and the predicted consequences for the partial ^1H NMR spectrum are presented as line spectra where the relative intensities of 2:1:1 assume no conformational scrambling of the NCD_3 group with the two NMe groups. When the ^1H NMR spectra of the *N,N'*-dimethyl-*N"*-trideuteriomethyltrianthranilide (**20**) was recorded immediately after dissolution in $CDCl_3$ at +30° C, three singlets for the NMe groups in the helical conformations were observed (Fig. 13). They had relative intensities of 29:42:29 for $Me_A : Me_B : Me_C$ and underwent equalization of their intensities with time at a rate such that a $\Delta G^{\ddagger}$ value of 95 kJ mol^{-1} was obtained. This value is in good agreement with those obtained by other methods (see Fig. 8) for conformational changes of *N,N',N"*-trimethyltrianthranilide (**19**) in $CDCl_3$ solution. Comparison of the observed spectra (Fig. 13) with predicted ones (Figs. 10 and 12) indicate that only patterns I and VI in Fig. 12 need

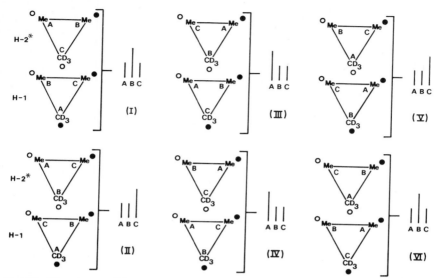

Fig. 12. The six possible ways [(I)–(VI)] of assigning the NMe groups in the H-1 and H-2* conformations of *N,N'*-dimethyl-*N''*-trideuteriomethyltrianthranilide (**20**) to sites A, B, and C, and the predicted relative intensities of these sites in the ¹H NMR spectrum.

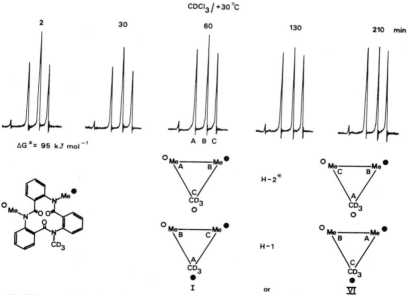

Fig. 13. The equilibration in CDCl₃ at +30° C of the NCD₃ and NMe groups of *N,N'*-dimethyl-*N''*-trideuteriomethyltrianthranilide (**20**) between sites A, B, and C.

to be considered. It proved possible to differentiate between the site assignments I and VI by subjecting N-methyl-N'-trideuteriomethyl-trianthranilide (18) to trideuteriomethylation to yield N-methyl-N',N''-di(trideuteriomethyl)trianthranilide (21). Fig. 14 portrays the predicted consequences of site assignments I and VI upon the relative intensities of sites

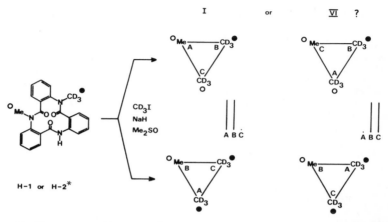

Fig. 14. The two possible ways [(I) and (VI)] of assigning the NMe group in the H-1 and H-2* conformations of N-methyl-N',N''-trideuteriomethyltrianthranilide (21) to sites A, B, and C and the predicted relative intensities of the associated signals in the ^{1}H NMR spectrum.

A, B, and C in the partial ^{1}H NMR spectra of **21** based on the assumptions discussed previously. When the ^{1}H NMR spectrum of N-methyl-N',N''-di(trideuteriomethyl)trianthranilide (**21**) was recorded immediately after dissolution in CDCl$_3$ at +30° C, three singlets for the NMe groups in the helical conformations were observed (Fig. 15). They had relative intensities of 22:39:39 and Me$_A$:Me$_B$:Me$_C$ and underwent equalization of their intensities with time at a rate such that a $\Delta G^{\ddagger}$ value of 95 kJ mol^{-1} was obtained. The observed pattern of relative intensities for Me$_A$, Me$_B$, and Me$_C$ can only arise from site assignment VI and permits the unambiguous designation (see Figs. 8 and 15) of sites A, B, and C to be made to the three diastereotopic NMe groups in the helical conformations of N,N',N''-trimethyltrianthranilide (**19**). On the basis of these experiments, it may be concluded that, in solution N,N'-dimethyltrianthranilide (**17**) populates either H-1 and/or H-2 conformations. Significantly, the distorted H-1 (or H-1*) conformation adopted (Fig. 5) in the solid state is twisted in the direction of the H-2* (or H-2) conformation.

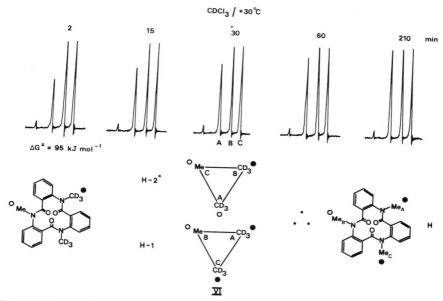

Fig. 15. The equilibration in CDCl$_3$ at +30° C of the NCD$_3$ and NMe groups of *N*-methyl-*N'*,*N''*-ditrideuteriomethyltrianthranilide (**21**) between sites, A, B, and C.

Benzylation of *N,N'*-dibenzyltrianthranilide (**26**) afforded (Fig. 16) two products—one under kinetic control which migrated relatively slowly on TLC, and the other under thermodynamic control which migrated much faster on TLC. These two products could be separted (see Section 2.2) as crystalline compounds by preparative TLC. Equilibration of both pure crystalline compounds in refluxing chloroform solutions at +30° C gave the same equilibrium mixture of 41:59 for the slower:faster migrating components after 30 min. ^{1}H NMR spectroscopy revealed (Fig. 17) that the faster migrating component gave rise to *one* AB system for its benzylic methylene protons whereas the slower moving component gave rise to *three* overlapping AB systems of equal intensities for its benzylic methylene protons. Thus, the product formed under kinetic control from *N,N'*-dibenzyltrianthranilide (**26**) is the helical conformation (H and H*) of the *N,N',N''*-tribenzyl derivative **27**. This proves that the *N,N'*-dibenzyl derivative **26** exists in solution in a helical conformation (e.g. H-1 or H-2*) with its two *N*-benzyl groups residing on opposite sides of the mean plane of the 12-membered ring, leaving the CONH linkage to be *N*-benzylated from either side of this plane and still afford *N,N'N''*-tribenzyltrianthranilide (**27**) in its helical (H or H*) conformations. Conformational interconversions

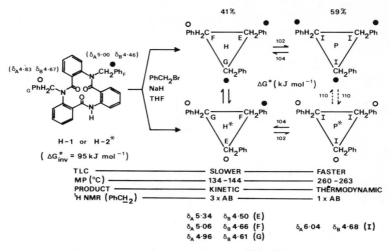

Fig. 16. N-Benzylation of N,N'-benzyltrianthranilide (26).

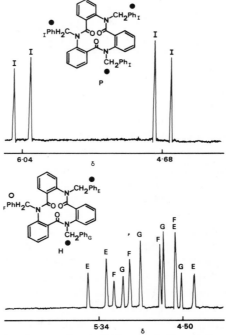

Fig. 17. The partial ¹H NMR spectra recorded in CDCl₃ showing the signals for the benzylic methylene protons in the propeller and helical conformations of N,N',N''-tribenzyltrianthranilide (27).

according to the general scheme P⇌H⇌H*⇌P* led to the thermodynamic product which corresponds to the propeller conformations (P and P*) of the N,N',N''-tribenzyl derivative **27**. The free energies of activation recorded in Fig. 16 for the P⇌H (and P*⇌H*) interconversions were obtained from kinetic ^{1}H NMR spectroscopic experiments involving direct equilibration of both propeller and helical conformational diastereoisomers in CD_3SOCD_3 at +63°C. The value for $\Delta G^{\ddagger}$(P⇌P*) was obtained from dynamic ^{1}H NMR spectroscopy on N,N',N''-tribenzyltrianthranilide (**27**) in $C_6D_5NO_2$ above +175°C. These high free energy barriers to conformational change are completely in accord with the isolation of crystalline conformational diastereoisomers.

We recognized in Section 2.2 that constitutionally unsymmetrical N,N'-di- and N,N',N''-tri-substituted trianthranilides can adopt three helical conformations (H-1, H-2, and H-3) in addition to a propeller conformation. These conformations are related to their enantiomeric counterparts (H-1*, H-2*, H-3*, and P*) by the conformational itinerary represented by the cubic array of interconversions shown in Fig. 18. Both N,N'-dimethyl-N''-benzyltrianthranilide (**24**) and N,N'-dibenzyl-N''-methyltrianthranilide (**28**) populate all four diastereoisomeric conformations at equilibrium in $CDCl_3$ solutions. Correlations between chemical shifts and site assignments (to A, B, C, and D) for the NMe groups in compounds **24** and **28**, based upon direct structural comparisons with the unambiguously designated NMe_A, NMe_B, NMe_C, and NMe_D groups in the helical and propeller conformations of N,N',N''-trimethyltrianthranilide (**19**), allow conformational assignments to be made to these two N,N',N''-trisubstituted trianthranilide derivatives. By way of an example, the site correlation involving NMe groups between

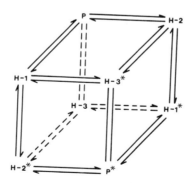

Fig. 18. Cubic array diagram illustrating the conformational itinerary involving P, P*, H-1, H-1*, H-2, H-2*, H-3, and H-3* conformations of unsymmetrically N,N',N''-trisubstituted trianthranilide derivatives.

N,N',N''-trimethyltrianthranilide (**19**) and the N,N'-dibenzyl-N''-methyl derivative **28** is presented in Fig. 19 along with the conformational assignments for this compound. These assignments were confirmed at least partially by the results of a kinetically-controlled methylation of N,N'-dibenzyltrianthranilide (**26**). This experiment allowed a mixture of helical conformations (with H-2/H-2* predominating) to be identified as kinetic products

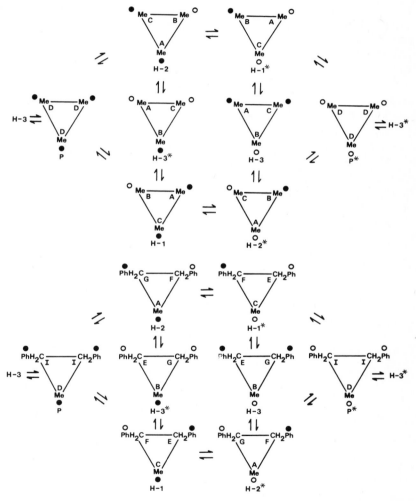

Fig. 19. A comparison of the conformational itineraries and the site exchange schemes for NMe groups in propeller and helical conformation of N,N',N''-trimethyl- **19** and N,N'-dibenzyl-N''-methyl- **18** trianthranilides.

and the thermodynamic product to be assigned to the propeller conformations. This result corresponds with the benzylation of the N,N'-dibenzyl derivative **26** under kinetic control. By careful analysis of the ^{1}H NMR spectra of compounds **24** and **28**, the correlation procedure can be extended to the sites defined (see Figs. 16 and 17) as E, F, G, and I for the stereoheterotopic benzylic methylene groups in the helical and propeller conformations of N,N',N''-tribenzyltrianthranilide (**27**). The correlation of site assignments with chemical shifts in the case of copounds **24**, **27**, and **28** is sufficiently good to put the assignment of stereoheterotopic N-benzylic methylene groups in N,N',N''-tribenzyltrianthranilide (**27**) on an absolute basis. The analysis may be extended in a self-consistent manner to the N-methyl-N'-benzyl **29**, N-methyl-N'-benzyl-N''-ethyl **30**, N-methyl-N'-ethyl **31** and N-methyl-N'-ethyl-N''-benzyl **32** derivatives. For example, Fig. 20 shows the partial ^{1}H NMR spectrum of equilibrated N-methyl-N'-ethyl-N''-benzyltrianthranilide (**32**) in $CDCl_3$ solution with conformational assignments indicated beside the signals for the N-methyl and N-benzylic methylene protons. Tables 2 and 3 present correlation charts relating conformational assignments (i) to sites (A, B, C, and D) associated with N-methyl protons in compounds **17**, **19**, **24**, and **28–31**, and (ii) to sites (E, F, G, and

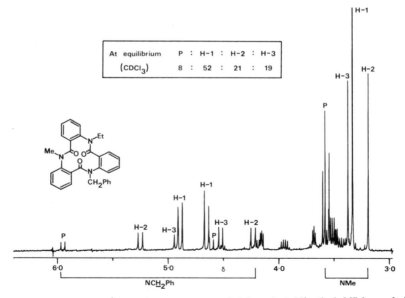

Fig. 20. The partial ^{1}H NMR spectrum of N-methyl-N'-ethyl-N''-benzyltrianthranilide (**32**).

Table 2. *The correlation between conformational assignments to N,N'-di- and N,N',N''-tri-substituted trianthranilide derivatives and the ^1H NMR sites associated with the singlets for their N-methyl protons based upon the known site assignments to the N-methyl groups in N,N',N''-trimethyltrianthranilide* (19)

	Compound[a],[b]				Chemical shifts (δ) for N-Me groups[c]		
Structure	R^1	R^2	R^3	Conformation[a]	R^1	R^2	R^3
19	Me$_B$	Me$_A$	Me$_C$	H/H*	3.21(B)	3.33(A)	3.13(C)
24	Me$_B$	Me$_A$	CH$_2$Ph	H-1/H-1*	3.18(B)	3.42(A)	—
28	CH$_2$Ph	CH$_2$Ph	Me$_C$	H-1/H-1*	—	—	3.20(C)
32	Me$_B$	Et	CH$_2$Ph	H-1/H-1*	3.33(B)	—	—
17	Me$_C$	Me$_B$	H	H-2/H-2*	3.17(C)	3.27(B)	—
24	Me$_C$	Me$_B$	CH$_2$Ph	H-2/H-2*	3.35(C)[d]	3.25(B)	—
28	CH$_2$Ph	CH$_2$Ph	Me$_A$	H-2/H-2*	—	—	3.41(A)
29	Me$_c$	CH$_2$Ph	H	H-2/H-2*	3.24(C)	—	—
31	Me$_c$	Et	H	H-2/H-2*	3.19(C)	—	—
32	Me$_c$	Et	CH$_2$Ph	H-2/H-2*	3.18(C)	—	—
24	Me$_A$	Me$_C$	CH$_2$Ph	H-3/H-3*	3.32(A)	3.14(C)	—
28	CH$_2$Ph	CH$_2$Ph	Me$_B$	H-3/H-3*	—	—	3.37(B)
30	Me$_A$	CH$_2$Ph	Et	H-3/H-3*	3.40(A)	—	—
32	Me$_A$	Et	CH$_2$Ph	H-3/H-3*	3.36(A)	—	—
19	Me$_D$	Me$_D$	Me$_D$	P/P*	3.49(D)	3.49(D)	3.49(D)
24	Me$_D$	Me$_D$	CH$_2$Ph	P/P*	3.54(D)	3.54(D)	—
28	CH$_2$Ph	CH$_2$Ph	Me$_D$	P/P*	—	—	3.60(D)
32	Me$_D$	Et	CH$_2$Ph	P/P*	3.57(D)	—	—

[a] For the designation of conformation, see Figs. 3 and 4.
[b] Site assignments to N-methyl groups are shown by means of a subscript.
[c] The site associated with each signal is shown in parentheses after each δ value.
[d] The chemical shift for Me$_C$ appears at lower field than is expected. Examination of molecular models of the H-2 and H-2* conformations of 24 reveals that the N-methyl group at R^1 could experience deshielding by the phenyl ring of the N-benzyl group at R^3 on the same side of the 12-membered ring.

I) associated with N-benzylic methylene protons in compounds 24, 26–30, and 32. A diagrammatic representation of this correlation between conformational and site assignments is given in Fig. 21. It reveals that, for most sites, the chemical shift variations between sites in different conformations of different compounds is small. The ground state conformer populations of the constitutionally unsymmetrical N,N'-di- and N,N',N''-tri-substituted trianthranilides in CDCl$_3$ are summarized in Table 4. (The N,N'-dimethyl-N''-acetyl 22 and N,N'-dimethyl-N''-benzoyl 23 trianthranilides both exist in solution in only *one* of the three possible diastereoisomeric helical conformations. However, it is not possible in the case of either derivative to make a conformational assignment.) The data in the Table show that five compounds adopt only one of the three helical conformations whereas three

Table 3. *The correlation between conformational assignments to N,N'-di- and N,N',N''-tri-substituted trianthranilide derivatives and the ¹H NMR sites associated with the AB systems for the N-benzylic methylene protons deduced from the known site assignments to the N-methyl groups in N,N',N''-trimethyltrianthranilide (19)*

Structure	Compound[a],[b] R¹	R²	R³	Conformation[a]	Chemical shifts (δ_A and δ_B) for N-benzylic methylene groups[c] R¹ δ_A	R¹ δ_B	R² δ_A	R² δ_B	R³ δ_A	R³ δ_B
27	CH_2Ph_F	CH_2Ph_E	CH_2Ph_G	H/H*	5.07(F)	4.66(F)	5.33(E)	4.50(E)	4.96(G)	4.61(G)
24	Me	Me	CH_2Ph_G	H-1/H-1*	—	—	—	—	4.74(G)	4.58(G)
32	Me	Et	CH_2Ph_G	H-1/H-1*	—	—	—	—	4.90(G)	4.65(G)
24	Me	Me	CH_2Ph_E	H-2/H-2*	—	—	—	—	—	4.24(E)
26	CH_2Ph_G	CH_2Ph_F	H	H-2/H-2*	4.83(G)	4.67(G)	5.00(F)	4.46(F)	—	—
28	CH_2Ph_G	CH_2Ph_F	Me	H-2/H-2*	4.78(G)	4.57(G)	4.91(F)	4.70(F)	—	—
29	Me	CH_2Ph_F	H	H-2/H-2*	—	—	5.02(F)	4.45(F)	—	—
32	Me	Et	CH_2Ph_E	H-2/H-2*	—	—	—	—	5.26(E)	4.24(E)
24	Me	Me	CH_2Ph_F	H-3/H-3*	—	—	—	—	4.98(F)	4.67(E)
30	Me	CH_2Ph_G	Et	H-3/H-3*	—	—	4.72(G)	4.56(G)	—	—
32	Me	Et	CH_2Ph_F	H-3/H-3*	—	—	—	—	4.94(F)	4.53(F)
24	Me	Me	CH_2Ph_I	P/P*	—	—	—	—	5.94(I)	—
27	CH_2Ph_I	CH_2Ph_I	CH_2Ph_I	P/P*	6.04(I)	4.68(I)	6.04(I)	4.68(I)	6.04(I)	4.68(I)
28	CH_2Ph_I	CH_2Ph_I	Me	P/P*	5.99(I)	4.64(I)	5.99(I)	4.64(I)	—	—
32	Me	Et	CH_2Ph_I	P/P*	—	—	—	—	5.96(I)	4.61(I)

(a) For conformational assignments, see Figs. 3 and 4.

(b) Site assignments to N-benzylic methylene groups are shown by means of a subscript.

(c) The site associated with each signal is shown in parentheses after each δ value.

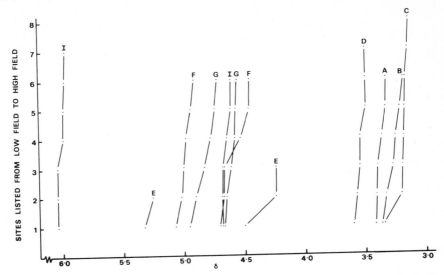

Fig. 21. A diagrammatic representation of the correlation between conformational and site assignments (see Tables 2 and 3) in N,N'-di- and N,N',N''-tri-substituted trianthranilides.

compounds exist as equilibrium mixtures of the four conformational diastereoisomers. This situation is to be contrasted with those previously observed (i) for trisalicylide derivatives[23-28] where the propeller conformation is generally the more stable, and (ii) for trithiosalicylide derivatives[30,31] where only helical conformations are populated.

The free energies of activation for the conformational inversion and interconversion processes in the various N,N'-di- and N,N',N''-tri-substituted trianthranilides are listed in Table 5. It has been noted previously in this Section that the values for $\Delta G^{\ddagger}$ lie in a range which permits the isolation of conformational diastereoisomers (see Section 2.2). Examination of space-filling molecular models demonstrates that operation of a pedalling mechanism during the reorientation of *trans*-amide linkages is restricted to *trans*-CONH linkages in trianthranilide derivatives. Since neither carbonyl oxygen atoms or bulky R substituents can pass through the middle of the 12-membered ring, reorientation of CONR linkages must involve *cis*-intermediates. This proposal seems entirely reasonable in the knowledge of the solid state structures (Fig. 22) of N,N',N''-trimethyltetra-anthranilide (**43**)[44,45] and 1,9,17-trimethyl-1,9,17-triaza[2.2.2]metacyclophane-2,10,18-trione (**44**):[46,47] structure **43** contains one *cis*-CONMe linkage, while structure **44** has all three CONMe linkages with *cis*-geometry, albeit with significant departures from the ideal synperiplanar arrangement for the C–CO–NMe–C torsional angles in all three cases.

Table 4. *Ground state conformer populations (%) in CDCl₃ at room temperature for N,N'-di- and N,N',N''-tri-substituted trianthranilides*

	Compound				Conformations			
Structure	R¹	R²	R³	P/P*	H-1/H-1*	H-2/H-2*	H-3/H-3*	
17	Me	Me	H	0	←———— 100$^{(a)}$ ————→		0	
24	Me	Me	CH₂Ph	19	41	7	33	
26	CH₂Ph	CH₂Ph	H	0	←———— 100$^{(a)}$ ————→		0	
28	CH₂Ph	CH₂Ph	Me	36	7	48	9	
29	Me	CH₂Ph	H	0	←———— 100$^{(a)}$ ————→		0	
30	Me	CH₂Ph	Et	0	0	0	100	
31	Me	Et	H	0	←———— 100$^{(a)}$ ————→		0	
32	Me	Et	CH₂Ph	8	52	21	19	

[a] Although the chemical shift data (see Tables 2 and 3) suggest that the *N,N'*-disubstituted derivatives **17, 26, 29,** and **31** prefer to adopt H-2/H-2* conformations, the fact that distorted H-1 and/or H-1* conformations, stabilized by intramolecular hydrogen bonding involving the unsubstituted amide linkages, are observed (see Figs. 5 and 6, respectively) for *N,N'*-dimethyl- **17** and *N,N'*-dibenzyl- **26** trianthranilides in the solid state persuades us to be cautious about the assignment of ground state conformations to **17, 26, 29,** and **31** in CDCl₃.

Table 5. *The free energies of activation (ΔG‡) for conformational changes in N,N'-di and N,N',N''-trisubstituted trianthranilides*

	Compound					ΔG‡
Structure	R¹	R²	R³	Solvent	Process	(kJ mol⁻¹)
17	Me	Me	H	CHCl₃	H⇌H*	87 (+2° C)$^{(a)}$
19	Me	Me	Me	PhNO₂	P→H	91$^{(b)}$
					H→P	101$^{(b)}$
					H⇌H*	103$^{(b)}$
				CDCl₃	P→H	90 (+10° C)$^{(c)}$
24	Me	Me	CH₂Ph	PhNO₂	H→P	96 (+10° C)$^{(c)}$
					P→H	100$^{(b)}$
					H→P	105$^{(b)}$
26	CH₂Ph	CH₂Ph	H	CD₃SOCD₃	H⇌H*	110$^{(b)}$
27	CH₂Ph	CH₂Ph	CH₂Ph	CD₃SOCD₃	H⇌H*	95$^{(b)}$
					P→H	104 (+63° C)$^{(c)}$
					H→P	102 (+63° C)$^{(c)}$
28	CH₂Ph	CH₂Ph	Me	PhNO₂	P⇌P*	110$^{(b)}$
					P→H	106$^{(b)}$
					H→P	107$^{(b)}$
					H⇌H*	108$^{(b)}$
29	Me	CH₂Ph	H	CD₃SOCD₃	H⇌H*	90 (+160° C)$^{(d)}$
30	Me	CH₂Ph	Et	CD₃SOCD₃	P⇌H	>113
32	Me	Et	CH₂Ph	CD₃SOCD₃	P⇌H	>113

[a] Determined by polarimetry.
[b] Determined by line shape analysis.
[c] Determined by equilibration.
[d] Calculated at the coalescence temperature.

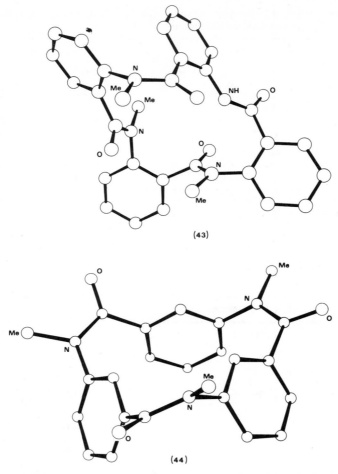

Fig. 22. The solid state structures of N,N',N''-trimethyltetra-anthranilide (**43**) and 1,9,17-trimethyl-1,9,17-triaza[2.2.2]metacyclophane-2,10,18-trione (**44**).

3. Triarylmethyltrianthranilides

In view of the demonstrated influence that the introduction of alkyl sub-stituents has upon the solid state properties and the conformational behaviour in solution of trisalicylides **1–8** and trithiosalicylides **9–13**, it was of considerable interest to us to examine[48,49] the consequences of introducing alkyl substituents into either position−3 or position−6 of the aromatic rings of the trianthranilide nucleus. While we could anticipate a further raising of the barriers to ring inversions and interconversions in these trimethylaryl derivatives relative to those observed (see Section 2.4) in the parent trianthranilides 17–32, we could not predict how the positions of the conformational equilibria in solution would be influenced or what the consequences would be for spontaneous resolution and/or inclusion compound formation on crystallization.

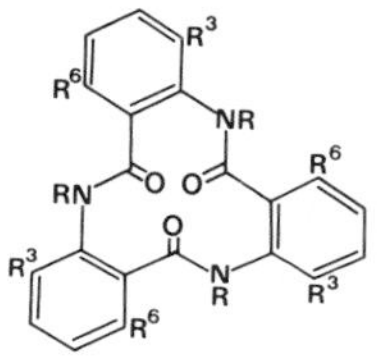

	R^3	R^6
(46)	Me	H
(47)	H	Me

All attempts so far to prepare tri-6-methyltrianthranilide derivatives 47 have been unsuccessful. When 2-amino-*o*-toluic acid (48)[50] was treated with nitro-*o*-toluoyl chloride (49), according to the conditions cited in Scheme

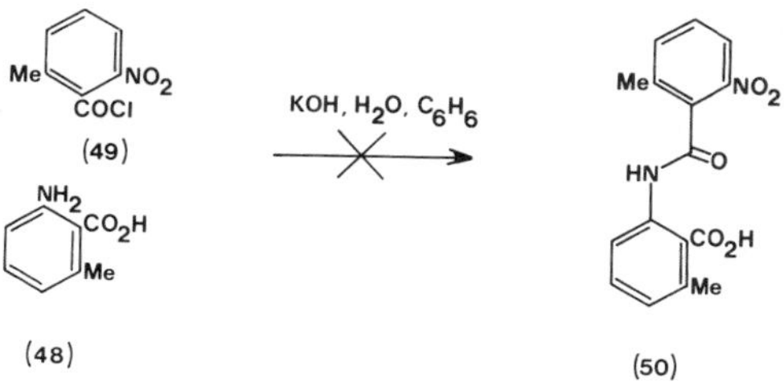

Scheme 4.

4, there was no evidence for the presence of the expected amide 50. Presumably, attack by the nucleophilic amino group at the carbonyl carbon atom of the acid chloride is hindered by the 2-nitro- and 6-methyl-substituents. By contrast, however, it was possible to synthesize a range of tri-3-methyltrianthranilide derivatives 46 by a stepwise approach analogous to that which proved so successful in the preparation of the parent compounds. The N,N'-di- and N,N',N''-tri-substituted derivatives 53–59 listed in Scheme 5 were synthesized by stepwise procedures from 2-amino-*m*-toluic acid (51)[51] and 2-nitro-*m*-toluoyl chloride (52).[52] The N,N'-dibenzyl derivative 58 was isolated in very low yield after HPLC as a mixture of conformational diastereoisomers, one (H-1/H-1*) crystalline and the other (P/P*) as an oil. N,N',N''-Tribenzyltri-3-methyltrianthranilide (59) could be

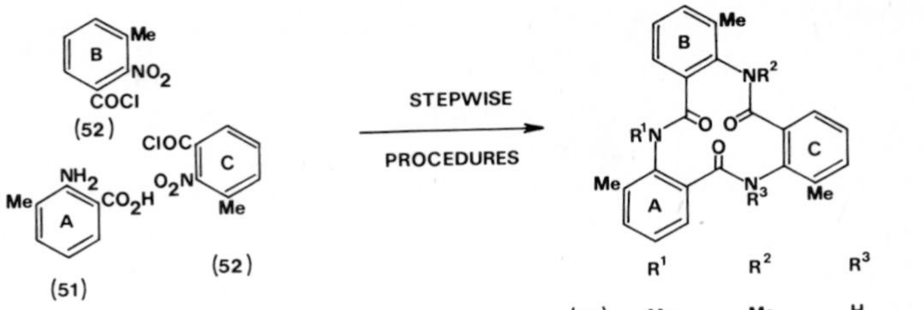

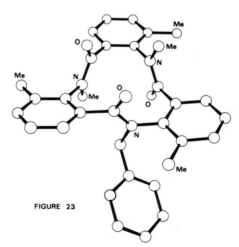

Scheme 5.

separated by HPLC into propeller and helical conformational diastereo-isomers, both of which were crystalline.

The key compound which allowed conformational assignments to be made to other constitutionally unsymmetrical derivatives was *N,N'*-dimethyl-*N''*-benzyltri-3-methyltrianthranilide (**55**). It afforded good single crystals from toluene which, on X-ray crystal structure analysis (see Fig. 23) revealed that *either* the H-2 or H-2* conformation is adopted in the

FIGURE 23

Fig. 23. The solid state structure of *N,N'*-dimethyl-*N''*-benzyltri-3-methyltrian-thranilide (**55**).

solid state. Moreover, ^{1}H NMR spectroscopy shows that this is also the conformation the molecule adopts in solution. Conformational assignments can be made (see Fig. 24) to all these derivatives on the basis of the excellent correlation between chemical shifts exhibited by protons in three constitutionally different ^{1}H NMR probes, namely N-benzylic methylene protons, N-methyl protons, and aryl-methyl protons. While N-methyl-N′,N″-dibenzyltri-3-methyltrianthranilide (57) adopts H-2/H-2* conformations, the N,N′-dimethyl 53, N-methyl-N′-benzyl 56, and N,N′-dibenzyl 58

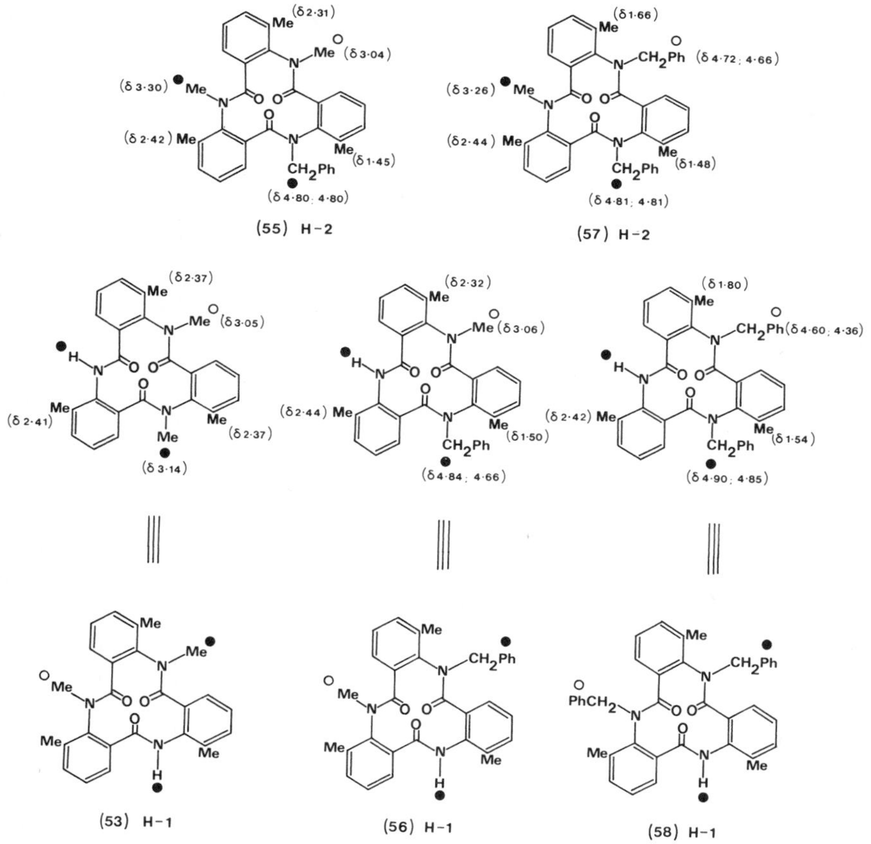

Fig. 24. Correlations between conformational assignments to compounds 53 and 55–58 and the chemical shifts of (i) benzylic methylene protons, (ii) N-methyl protons, and (iii) aryl-methyl protons in CDCl$_3$. Note that the H-1 conformations for compounds 53, 56, and 58 have been drawn in such a way as to emphasize their relationship with the H-2 conformations for compounds 55 and 57.

derivatives all adopt H-1/H-1* conformations. Confidence in the diagnostic nature of the chemical shift correlations leading to these conformational assignments emanates principally from a dramatic shielding of c. 0.7–0.9 p.p.m. experienced by aryl-methyl protons when they are in the vicinity of an N-benzyl group. The X-ray crystal structure (Fig. 23) of the N,N'-dimethyl-N''-benzyl derivative 55 shows that it is possible for an aryl-methyl group to find itself in the shielding zone of the phenyl ring of a neighbouring N-benzyl group. N,N',N''-Trimethyltri-3-methyltrianthranilide (54) exists in solution solely as its helical conformation with a free energy barrier of 112 kJ mol⁻¹ to H⇌H* inversion. In the case of the other N,N',N''-trisubstituted derivatives, the barriers to conformational change in solution are in excess of 113 kJ mol⁻¹.

4. Inclusion compounds

In Section 2.3, it was noted that N,N'-dibenzyltrianthranilide (26) forms an inclusion compound with toluene when it is recrystallized from this solvent. X-Ray crystallography and ¹H NMR spectroscopy have established that it has 1:1 stoichiometry. The entrapped toluene molecules can be removed under vacuum. It was discovered quite accidentally that other related 12-membered ring derivatives—namely the helical conformational isomer of N,N',N''-tribenzyltrianthranilide (27), the hexadeuteriated cyclic triamine 60, derived from 27 on reduction of the amide linkages with LiAlD₄, and the N,N'-dibenzyl derivative 63 of the cyclic bislactam 61 obtained[53] on condensation of o-phenylenediamine with stilbene-2,2'-dicarboxylic acid—all form 1:1 inclusion compounds with ethanol when these compounds are dissolved in chloroform containing 2% of ethanol (present as a stabilizer) and the solvent distilled off under vacuum at room temperature. In all cases, the ethanol can be removed under reduced pressure (<1 Torr)

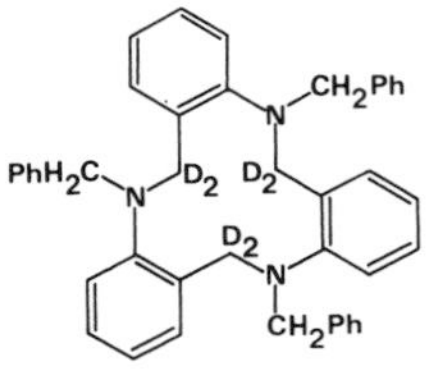

(60)

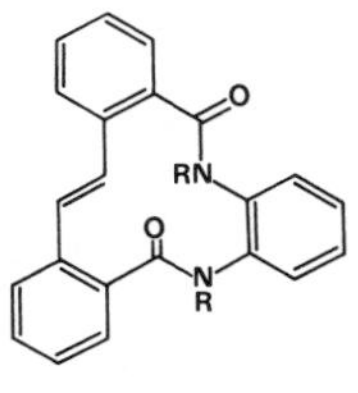

(61)   R = H
(62)   R = Me
(63)   R = CH₂Ph

at +70 to +90° C in a few hours. Quite remarkably, the 1:1 inclusion compound formed between the N,N'-dibenzyl derivative **63** and ethanol has the ability to survive recrystallization from aprotic solvents, e.g. ether–light petroleum (b.p. 60–80° C). Other compounds which have recently been reported[56,57] to form inclusion compounds (host:guest, 1:2) with alcohols include cyclotriveratrylene (propan-1-ol) (see Chapter 4, Volume 2) and tetraphenylsquaramidine (methanol). Also, certain pyridino crown compounds form[58] inclusion compounds (host: guest, 2:1) selectively with methanol, ethanol, and the homologous series of unbranched alcohols from propan-1-ol to hexan-1-ol.

Regioselectivity has been found[54,55] to accompany the crystallization of the N,N'-dimethyl derivative **62** of the cyclic bislactam **61** from a mixture of constitutional isomers of xylene. X-Ray crystallography has revealed that **62** forms a 1:1 inclusion compound with *o*-xylene.

In none of the examples discussed so far in this Section is the solid state relationship between the host and guest known. Fortunately, the situation is better defined for the 1:1 inclusion compound formed between N,N'-dimethyl-N''-benzyltri-3-methyltrianthranilide (**55**) and toluene. An X-ray crystallographic examination (see Fig. 22) showed that in this situation the host adopts a helical conformation (*either* H-2 *or* H-2*) in the solid state and also undergoes spontaneous resolution on crystallization. Plotting the van der Waals radii of all the host atoms in the unit cell discloses (see Fig. 25) large continuous channels in the crystal running parallel to the cell translation *c*. These channels have an average minimum dimension of approximately 550 pm and an average maximum dimension of approximately 900 pm. Significant portions of the channels, which are occupied by disordered guest molecules, are bound by the aryl-methyl groups and the phenyl rings of the benzyl groups. Obviously, since these channels are chiral, they should have the potential to effect chiral recognition of racemic guest species, cf. tri-*o*-thymotide (**6**). (See Chapter 9, Volume 3).

Interestingly, N-methyl-N'-benzyltri-3-methyltrianthranilide (**56**) also crystallizes from toluene in a helical conformation (H-1 and/or H-1*) as a 1:1 inclusion compound with the solvent.[55] The toluene can be removed under vacuum at +60° C in a few hours indicating that it may be entrapped in channels in the crystals in much the same way as toluene is included in crystalline N,N'-dimethyl-N''-benzyltri-3-methyltrianthranilide (**55**). It is not known, however, if the N-methyl-N''-benzyl derivative **56** undergoes spontaneous resolution on crystallization.

Finally, the 30-membered hexalactam **64**, obtained on cyclization of terephthaloyl chloride and N,N'-dibenzylethylene diamine has very recently been reported[59] to form a 1:1 inclusion compound with chloroform on crystallization from that solvent. X-Ray crystallography of the inclusion

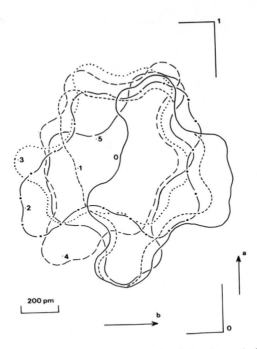

Fig. 25. A contour map representing the bounds of the channels formed in crystals of **55**. The contours are spaced at equal intervals of 0.1 (approximately 110 pm) of a unit cell translation in the *c* direction. Only half of the channel is illustrated, the remainder being generated by the crystallographic two-fold screw axis.

complex reveals that the chloroform molecule is embedded in the *molecular cavity* of the 30-membered hexalactam **64** in such a way that the hydrogen atom of the chloroform projects into the centre of the macro ring. The inclusion compound remains stable after several hours of drying at 50° C and 12 Torr, and after recrystallization from ethyl acetate.

(64)   R = CH$_2$Ph

5. Conclusion

The trianthranilides do comprise a new class of organic hosts. Their potential is epitomized in the discovery of *N,N'*-dimethyl-*N"*-benzyl-tri-3-methyl-trianthranilide (**55**), which shares with tri-*o*-thymotide (**6**) the ability to undergo spontaneous resolution and also form inclusion compounds. However, some important distinctions emerge in comparing the properties and assessing the potential of these two 12-membered ring compounds. Firstly, tri-*o*-thymotide crystallizes in chiral *propeller* conformations whereas the tri-3-methyltrianthranilide derivative **55** crystallizes in chiral *helical* conformations. Secondly, the opportunity to modify the constitution of a trianthranilide derivative by changing the nature of the substituents on the nitrogen atoms during stepwise syntheses is much greater than could ever be feasible with tri-*o*-thymotide (**6**). This is an obvious advantage of all trianthranilide hosts when the necessity to introduce functionality into the host as a prelude to the investigation of reactions in the solid state on entrapped guest substrates is considered.

References

1. W. Baker, W. D. Ollis and T. S. Zealley, *Nature (London)*, 1949, **164**, 1049.
2. W. Baker, W. D. Ollis and T. S. Zealley, *J. Chem. Soc.*, 1951, 201.
3. W. Baker, B. Gilbert, W. D. Ollis and T. S. Zealley, *Chem. Ind. (London)*, 1950, 333.
4. W. Baker, B. Gilbert, W. D. Ollis and T. S. Zealley, *J. Chem. Soc.*, 1951, 209.
5. W. Baker, B. Gilbert and W. D. Ollis, *J. Chem. Soc.*, 1952, 1443.
6. W. Baker, J. B. Harborne, A. J. Price and A. Rutt, *J. Chem. Soc.*, 1954, 2042.
7. H. M. Powell, *Nature (London)*, 1952, **170**, 155.
8. H. M. Powell, *Endeavour*, 1956, **15**, 20.
9. A. C. D. Newman and H. M. Powell, *J. Chem. Soc.*, 1952, 3747.
10. D. Lawton and H. M. Powell, *J. Chem. Soc.*, 1958, 2339.
11. D. J. Williams and D. Lawton, *Tetrahedron Lett.*, 1975, 111.
12. S. Brunie, A. Navaza and G. Tsoucaris, *Acta Crystallogr.*, 1975, **A31**, S127.
13. S. Brunie, A. Navaza, G. Tsoucaris, J. P. Declercq and G. Germain, *Acta Crystallogr.*, 1977, **B33**, 2645.
14. R. Arad-Yellin, S. Brunie, B. S. Green, M. Knossow and G. Tsoucaris, *J. Am. Chem. Soc.*, 1979, **101**, 7529.
15. R. Arad-Yellin, B. S. Green, M. Knossow and G. Tsoucaris, *Tetrahedron Lett.*, 1980, **21**, 387.
16. R. Arad-Yellin, B. S. Green and M. Knossow, *J. Am. Chem. Soc.*, 1980, **102**, 1157.
17. R. Gerdil and J. Allemand, *Tetrahedron Lett.*, 1979, 3499.
18. R. Gerdil and J. Allemand, *Helv. Chim. Acta*, 1980, **63**, 1750.

19. J. Allemand and R. Gerdil, *Cryst. Struct. Commun.*, 1981, **10**, 33.
20. J. Allemand and R. Gerdil, *Acta Crystallogr.*, 1982, **B38**, 1473.
21. E. Gil, A. Quick and D. J. Williams, *Tetrahedron Lett.*, 1980, **21**, 4207.
22. S. Brunie and G. Tsoucaris, *Cryst. Struct. Commun.*, 1974, **3**, 481.
23. W. D. Ollis and I. O. Sutherland, *Chem. Commun.*, 1966, 402.
24. A. P. Downing, W. D. Ollis and I. O. Sutherland, *Chem. Commun.*, 1967, 171.
25. A. P. Downing, W. D. Ollis, I. O. Sutherland, J. Mason and S. F. Mason, *Chem. Commun.*, 1968, 329.
26. A. P. Downing, W. D. Ollis and I. O. Sutherland, *J. Chem. Soc. B*, 1970, 24.
27. W. D. Ollis, J. F. Stoddart and I. O. Sutherland, *Tetrahedron*, 1974, **30**, 1903.
28. W. D. Ollis, J. S. Stephanatou and J. F. Stoddart, *J. Chem. Soc., Perkin Trans. 1*, 1982, 1629.
29. W. Baker, A. S. El-Nawawy and W. D. Ollis, *J. Chem. Soc.*, 1952, 3163.
30. G. B. Guise, W. D. Ollis, J. A. Peacock, J. S. Stephanatou and J. F. Stoddart, *Tetrahedron Lett.*, 1980, **21**, 4203.
31. G. B. Guise, W. D. Ollis, J. A. Peacock, J. S. Stephanatou and J. F. Stoddart, *J. Chem. Soc., Perkin Trans. 1*, 1982, 1637.
32. H. Meyer, *Justus Liebigs Ann. Chem.*, 1907, **351**, 267.
33. W. D. Ollis, J. A. Price, J. S. Stephanatou and J. F. Stoddart, *Angew. Chem., Int. Ed. Engl.*, 1975, **14**, 169.
34. W. D. Ollis, J. S. Stephanatou, J. F. Stoddart and A. G. Ferrige, *Angew. Chem., Int. Ed. Engl.*, 1976, **15**, 223.
35. A. Hoorfar, W. D. Ollis, J. A. Price, J. S. Stephanatou and J. F. Stoddart, *J. Chem. Soc., Perkin Trans. 1*, 1982, 1649.
36. R. P. Staiger and E. B. Miller, *J. Org. Chem.*, 1959, **24**, 1214.
37. N. P. Peet and S. Sunder, *J. Org. Chem.*, 1974, **39**, 1931.
38. E. Gil and D. J. Williams, unpublished results.
39. P. A. Petyunin, V. S. Shklyaev and M. E. Konshin, *Zh. Prikl. Khim.*, 1960. **33**, 1428 (*Chem. Abs.*, 1960, **54**, 22426i).
40. G. Schroeter and O. Eisleb, *Justus Liebigs Ann. Chem.*, 1909, **367**, 101.
41. G. Schroeter, *Chem. Ber.*, 1919, **52**, 2224.
42. D. H. Hey and D. G. Turpin, *J. Chem. Soc.*, 1954, 2471.
43. D. J. Williams, *J. Chem. Soc., Chem. Commun.*, 1977, 170.
44. A. Hoorfar, W. D. Ollis, J. F. Stoddart and D. J. Williams, *Tetrahedron. Lett.*, 1980, **21**, 4211.
45. A. Hoorfar, W. D. Ollis and J. F. Stoddart, *J. Chem. Soc., Perkin Trans. 1*, 1982, 1721.
46. F. E. Elhadi, W. D. Ollis, J. F. Stoddart, D. J. Williams and K. A. Woode, *Tetrahedron Lett.*, 1980, **21**, 4215.
47. F. E. Elhadi, W. D. Ollis and J. F. Stoddart, *J. Chem. Soc., Perkin Trans. 1*, 1982, 1727.
48. S. J. Edge, W. D. Ollis, J. S. Stephanatou, J. F. Stoddart, D. J. Williams and K. A. Woode, *Tetrahedron Lett.*, 1981, **22**, 2229.
49. S. J. Edge, W. D. Ollis, J. S. Stephanatou and J. F. Stoddart, *J. Chem. Soc., Perkin Trans. 1*, 1982, 1701.
50. S. Gronowitz and G. Hansen, *Ark. Kemi*, 1967, **27**, 145.
51. E. D. Bergmann and Z. Pelchowicz, *J. Am. Chem. Soc.*, 1953, **75**, 2663.
52. E. Müller, *Chem. Ber.*, 1909, **42**, 423.
53. W. D. Ollis, J. S. Stephanatou, J. F. Stoddart, A. Quick, D. Rogers and D. J. Williams, *Angew. Chem., Int. Ed. Engl.*, 1976, **15**, 757.

54. W. D. Ollis, J. S. Stephanatou, J. F. Stoddart, G. G. Unal and D. J. Williams, *Tetrahedron Lett.*, 1981, **22**, 2225.
55. W. D. Ollis, J. S. Stephanatou and J. F. Stoddart, *J. Chem. Soc., Perkin Trans. 1*, 1982, 1715.
56. J. A. Hyatt, E. N. Duesler, D. Y. Curtin and I. C. Paul, *J. Org. Chem.*, 1980, **45**, 5074.
57. H. J. Bestmann, E. Withelm and G. Schmid, *Angew. Chem., Int. Ed. Engl.*, 1980, **19**, 1012.
58. E. Weber and F. Vögtle, *Angew. Chem., Int. Ed. Engl.*, 1980, **19**, 1030.
59. F. Vögtle, H. Puff, E. Friedrichs and W. M. Müller, *J. Chem. Soc., Chem. Commun.*, 1982, 1398.

7 · INCLUSION COMPOUNDS OF DEOXYCHOLIC ACID

E. GIGLIO

Universita di Roma, Rome, Italy

1. Deoxycholic acid

$3\alpha,12\alpha$-Dihydroxy-5β-cholan-24-oic acid (deoxycholic acid, hereafter referred to as DCA) is shown in Fig. 1 together with its numbering system. DCA has a molecular weight of 392.6, corresponding to the formula $C_{24}H_{40}O_4$. This typical bile acid of chiral character, isolated from the biles of some animals by saponification and subsequent processing, has a per-hydro-1,2-cyclopentenophenanthrene ring system.[1] Two sides can be recognized in the DCA molecule: one polar, characterized by two secondary hydroxyl groups at C(3) and C(12), and the other apolar, characterized by two angular methyl groups at C(10) and C(13). Therefore, the molecule is two-faced, showing both one hydrophilic and one hydrophobic side, and this feature is mainly responsible for its remarkable chemical behaviour. A short aliphatic chain, with another methyl group at C(20), sticks out from the cyclopentane ring at C(17) and terminates in a carboxyl group.

The *arched* shape of the DCA molecule is due to the *cis* fusion of the A and B rings and to the side-chain conformation. The A, B and C rings can be considered rigid whereas ring D is relatively flexible and its conformation

INCLUSION COMPOUNDS 2
ISBN 0-12-067101-8

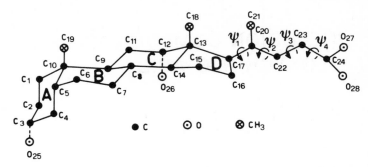

Fig. 1. Atomic numbering of DCA without hydrogen atoms. The torsion angles of the side-chain are shown.

can be related to that of the side-chain. In order to establish the permissible conformations of the side-chain, its intramolecular van der Waals energy was computed[2] as a function of the four torsion angles $C(13)C(17)C(20)C(21)$, $C(17)C(20)C(22)C(23)$, $C(20)C(22)C(23)C(24)$ and $C(22)C(23)C(24)C(28)$, indicated as ψ_1, ψ_2, ψ_3 and ψ_4 respectively and shown in Fig. 1. Semi-empirical atom–atom potentials, previously verified in known and unknown crystal structures,[3,4] were employed in the generalized form:

$$V(r) = a\,\exp(-br)/r^d - cr^{-6}$$

where r is the interatomic distance and a, b, c and d are coefficients listed

Table 1. *The coefficients of the van der Waals potentials. The energy is in Kcal per atom pair with the interatomic distance in Å*

Interaction	$a \times 10^{-3}$	b	c	d
H–H	6.6	4.080	49.2	0
H–C	44.8	2.040	125.0	6
H–N	52.1	2.040	132.0	6
H–O	42.0	2.040	132.7	6
H–CH$_3$	49.1	3.705	380.5	0
C–C	301.2	0.000	327.2	12
C–N	340.0	0.000	340.0	12
C–O	278.7	0.000	342.3	12
C–CH$_3$	291.1	1.665	981.1	6
N–N	387.0	0.000	354.0	12
N–O	316.2	0.000	356.0	12
N–CH$_3$	325.9	1.665	1020.5	6
O–O	259.0	0.000	358.0	12
O–CH$_3$	272.7	1.665	1026.3	6
CH$_3$–CH$_3$	273.9	3.329	2942.0	0

in Table 1. The methyl group is treated as one atom. The coefficients of the potentials concerning the nitrogen atom, to which it will be referred later, are also included in Table 1. The values of the dihedral angles hereafter are evaluated according to the convention of Klyne and Prelog.[5]

The calculations pointed out that ψ_1 is always confined within a very narrow range around $-60°$, in agreement with the experimental data, since the C(21) methyl group and the hydrogen atoms bonded to C(20) and C(22) are hindered from rotating around C(17)–C(20), owing to short contacts with the C(18) methyl group and with two hydrogen atoms linked to C(12) and C(16). Moreover, the energy minima observed for ψ_4 do not correspond to the arrangement of the carboxyl group, found later in the DCA crystal structures, which mainly depends on the ability of this group to form hydrogen bonds rather than on the intramolecular van der Waals energy. Values of about 60 and 190° on one hand and of about 50, 165 and 265° on the other are allowed for ψ_2 and ψ_3 respectively, so that five minimum regions are observed (Fig. 2). Three of them give rise to a "folded back" carboxyl group, which can achieve contacts energetically favoured with the remaining part of the side-chain, but cannot form a good hydrogen-bonding scheme in a crystal. The other two, with $\psi_3 \sim 165°$, are populated by the DCA molecules found in the crystal structures solved so far and give rise

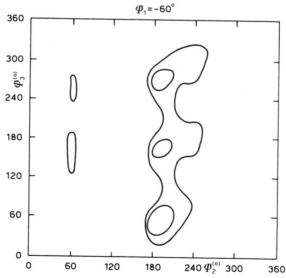

Fig. 2. Section $\psi_1 = -60°$ of the van der Waals energy of the DCA side-chain. Contours are drawn at arbitrary intervals.

to an efficient hydrogen-bonding scheme (see later). The corresponding conformations can be indicated as "*trans*" ($\psi_2 \sim 190°$) and *gauche* ($\psi_2 \sim 60°$) if the arrangement of C(17), C(20), C(22) and C(23) is taken into account.

The cyclopentane ring (D) assumes many intermediate conformations between the half-chair and the β-envelope[6] in the known crystal structures of DCA. Owing to the interactions involving the hydrogen atoms of C(16), C(17), C(22) and C(23), the *trans* and *gauche* states are coupled with a conformation approaching the β-envelope or the half-chair symmetry, respectively.[2]

2. Choleic acids

DCA forms stable inclusion compounds of the "channel" type with a wide variety of molecules: for example, aliphatic, aromatic and alicyclic hydrocarbons, alcohols, ketones, fatty acids, esters, ethers, phenols, azo dyes, nitriles, peroxides and amines. The guest components are generally imprisoned in channels running though a host lattice composed of DCA molecules. The crystalline inclusion compounds, called choleic acids, are often obtained by slow evaporation from a solution prepared by dissolving DCA and the guest component in ethanol, or DCA in a liquid guest component, sometimes by means of a small amount of methanol, and present a well-defined stoichiometry. Several review articles[1,7-11] deal more or less in detail with the history and the structure of the choleic acids.

One class of the choleic acids widely investigated is that formed between DCA and fatty acids. A mixture of deoxycholic, palmitic and stearic acids was isolated from ox bile by Latschinoff,[12,13] who believed it to be DCA, which had been previously characterized by Mylius.[14] Later Wieland and Sorge[15] recognized that the mixture was composed of a combination of two pairs of unreactive components (DCA-palmitic acid and DCA-stearic acid). Several investigations have contributed to clarify the stoichiometry and some physicochemical properties of the fatty acid choleic acids. Formic acid fails to form a choleic acid.[16] Acetic and propionic acids give with DCA a host/guest ratio of $1:1$[15] and $3:1$[15,17] or $2:1$[18] respectively. Ratios of $4:1$, $6:1$ and $8:1$ are found for the C_4-C_8, C_9-C_{14} and $C_{15}-C_{29}$ saturated fatty acids respectively.[15,17,19] Butyric acid, alone in the C_4-C_8 series, also gives a $2:1$ ratio.[18] Dicarboxylic acids, as well as branched and unsaturated acids also form choleic acids.[18,20,21]

Normal and branched alkanes from C_5 to C_{43} are included by DCA[22,23] together with a certain number of esters.[17,18,24-27] In general, the host/guest ratio increases with the length of the guest molecule. However, the ratio of

8:1 reported in the literature for very long guest molecules is puzzling in view of the choleic acids crystal structures known at present (see Section 3).

Aromatic hydrocarbons form very stable choleic acids. DCA combines with bromobenzene,[28] *p*-diiodobenezene,[28] *p*-xylene,[15,28] naphthalene,[15,28] acenaphthene[15] and styrene[29] in the ratio of 2:1, with phenanthrene[20,28,29] and 1,2-benzanthracene[28,30] in the ratio of 3:1 and with anthracene,[20] 1,2,5,6-dibenzanthracene,[30] methylcholanthrene[30] and hexahydromethyl-cholanthrene[30] in the ratio of 4:1.

A few choleic acids of azo dyes were prepared by Cilento[31-33] with *p*-aminoazobenzene, *p*-dimethylaminoazobenzene, *p*-diethylaminoazobenzene, *o*-aminoazotoluene and *m*'-methyl-*p*-dimethylaminoazobenzene and were claimed to have a 4:1 ratio.[33] Further studies on compounds of this class, such as methyl orange, were reported later.[34,35]

Finally, it should be mentioned that inclusion complexes formed between DCA and alcohols,[15,18,19,24,36] ketones[20,26] and ethers,[15,37] have been reported so that DCA can be considered to be a very versatile host lattice.

3. Structural investigations of choleic acids

3.1. Introduction

Several pioneering X-ray crystallographic studies were carried out in the past,[27,38-49] especially by Giacomello, Kratky and coworkers, on some choleic acids. However, owing to the remarkable difficulty encountered in the solution of the crystal structures, which were too complex for that time, many aspects of the choleic acids' basic characteristics remained obscure and many misconceptions arose. Only recently a number of choleic acid crystal structures have been solved and, hence, the description of these molecular compounds can now be considered to be satisfactory.[50-62] Table 2 reports the abbreviations used for indicating the choleic acids, the guest molecules, the host/guest ratios, the unit cell constants, the space groups and the references. The most frequently used and useful methods of checking the composition of the choleic acids are gas-chromatography, X-ray and density measurements, and NMR.

By inspection of Table 2 it is seen that the choleic acids listed there can be grouped into three crystal systems, orthorhombic, tetragonal and hexagonal. The first system is the most commonly observed and represents the best including ability. The main structural features of these three forms will be described briefly.

Table 2. Crystal data and composition of the choleic acids whose X-ray crystal structures are known

Abbreviation	Guest	DCA/guest	a(Å)	b(Å)	c(Å)	Space group	Ref.
DCAACA	acetic acid	1:1	25.55	13.81	7.11	$P2_12_12_1$	50
DCAACE	acetone	5:3	25.81	13.61	7.23	$P2_12_12_1$	53, 56, 60
DCAEMK	ethyl methyl ketone	2:1	25.81	13.59	7.23	$P2_12_12_1$	56
DCADEK	diethyl ketone	2:1	25.83	13.56	7.24	$P2_12_12_1$	56
DCACLK	chloroacetone	2:1	25.83	13.63	7.19	$P2_12_12_1$	56
DCAPAL	palmitic acid, ethanol[a]	8:1:1	26.02	13.54	7.27	$P2_12_12_1$	59
DCAAPH	acetophenone	5:2	25.59	13.71	7.25	$P2_12_12_1$	56, 60
DCADAB	p-dimethylaminoazobenzene	4:1	25.68	13.73	7.16	$P2_12_12_1$	73
DCAPIB	p-diiodobenzene	2:1	26.59	13.58	7.17	$P2_12_12_1$	51
DCAPHE	phenanthrene	3:1	26.81	13.60	21.66[b]	$P2_12_12_1$	51
DCACHX	cyclohexanone	2:1	26.99	13.35	14.16	$P2_12_12_1$	56
DCARMC	(R)3-methyl cyclohexanone	2:1	26.98	13.51	14.18	$P2_12_12_1$	56
DCASMC	(S)3-methyl cyclohexanone	2:1	26.90	13.52	14.16	$P2_12_12_1$	56
DCANBD	norbornadiene	2:1	27.13	13.46	14.21	$P2_12_12_1$	61
DCADTB	di-t-butyl diperoxycarbonate	4:1	27.16	13.48	14.17	$P2_12_12_1$	52, 56
DCADCP	d-camphor	2:1	27.35	13.81	7.23	$P2_12_12$	62
DCAWAT	water	2:3	14.00	14.00	48.90	$P4_12_12$	56
DCAETW	ethanol, water	2:1:1	14.07	14.07	49.33	$P4_12_12$	58
DCAETH	ethanol, water	3:2:1	15.12	15.12	18.68	$P6_5$	54
DCADMS	dimethyl sulphoxide, water	2:1:1	15.12	15.12	18.70	$P6_5$	55

(a) Ethanol can be replaced by acetone if the choleic acid crystallizes from acetone.
(b) The translation period of the DCA molecules along c is 7.22 Å.

3.2. Orthorhombic crystals

Orthorhombic choleic acids, containing as guest components all the compounds previously cited together with others, are generally grown, by slow evaporation from ethanolic solutions, as prismatic needles, elongated along *c*, although various solvents, such as dioxan, acetone and methanol, can also be used. If DCA is directly soluble in a liquid guest component the crystallization can occur without addition of solvent.

A typical crystal packing is shown in Fig. 3, the guest component being phenanthrene. The DCA side-chain conformation is always *gauche* and the D ring approaches the half-chair symmetry. The characteristic structural unit is the bilayer, visible in the projection in the central zone of Fig. 3,

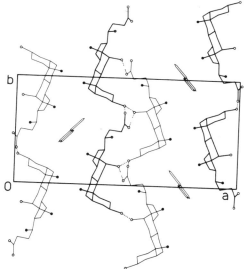

Fig. 3. Molecular packing of the DCAPHE crystal viewed along *c*. Black and open circles represent methyl groups and oxygen atoms respectively. The dashed lines indicate hydrogen bonding.

which extends into planes parallel to *bc*. Rows of DCA molecules, developed along *b* and linked in each row by head-to-tail hydrogen bonds involving O(25) and O(27) of two adjacent molecules, are connected by hydrogen bonding, thus forming the bilayer (see Figure 4). This is composed of two halves, hereafter called monolayers, shown with thick and thin lines in Fig. 3 and 4 and shifted from each other by about 3.6 Å along *c*, and is stabilized mainly by an efficient network of hydrogen bonds involving hydroxyl and carboxyl groups. The repetitive unit along *c* consists of three hydrogen

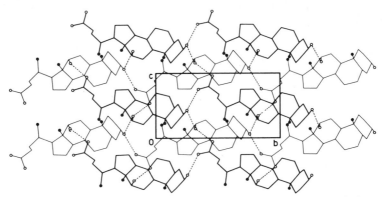

Fig. 4. Molecular packing of a bilayer in the DCAPHE crystal viewed along *a*. The symbols have the same meaning as in Fig. 3.

bonds, about 2.7 Å long, between $O(25)$ and $O(27)^1$, $O(28)^1$ and $O(26)^2$, $O(26)^2$ and $O(25)^3$, where the 1, 2 and 3 superscripts refer to the DCA molecules at $(x, 1+y, z)$, $(-x, 1/2+y, 3/2-z)$ and $(x, y, 1+z)$ respectively (see Fig. 4).

Two adjacent bilayers are antiparallel along *b*, being related by a 2_1 axis, and give rise, because of their wavy shape, to empty spaces, called channels, centered on crystallographic two-fold screw axes (for example at $a/4$, $b/2$) and running along *c* (see Fig. 3). The outer surfaces of a bilayer are hydrophobic, so that the guest molecules included in the channels are surrounded by non-polar groups. The channels have a variable size and shape depending on the mutual positions along *b* of two adjacent bilayers and on their mean distance, corresponding to $a/2$. This accounts for the ability of the DCA host lattice to accommodate guest molecules of very different dimensions. Schematic drawings of three crystal packings of choleic acids and of the "without channels" packing, which will be discussed later in this Section, are shown in Fig. 5 in order to illustrate the flexible size of the channel section, which is perpendicular to *c* and of approximately rectangular shape. It is clear that the area of the section increases and the orientation of the rectangle changes by shifting the central bilayer towards more positive values of *b*, passing from DCAACE to DCANBD and to DCAPHE. The DCAACE packing is shared with slight modifications by the choleic acids from DCAACA to DCADAB (see Table 2), that of DCAPHE by DCAPIB and that of DCANBD by the choleic acids from DCACHX to DCADTB. The walls of the channels are covered by rings A and D, C(6) methylenic groups and side-chains, with the exception of the carboxyl and C(23) methylenic groups, more distant from the centre of the canal, and, generally, the strongest interactions with the guest molecules are given by the DCA atoms belonging to the longer edges of the rectangle.

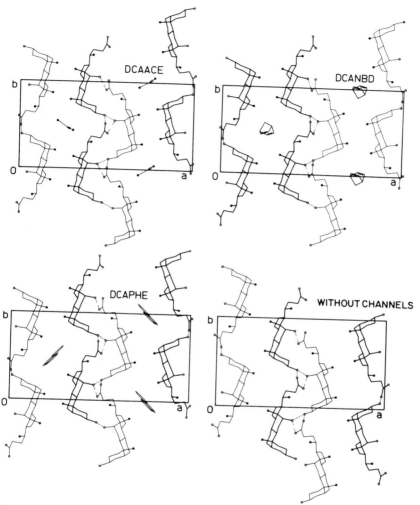

Fig. 5. DCAACE, DCAPHE, DCANBD and "without channels" crystal packings viewed along *c*. The symbols have the same meaning as in Fig. 3.

Two groups of orthorhombic choleic acids, hereafter referred to as α and β, can be recognized in Table 2.[11,57] The α and β structures, which crystallize in the space group $P2_12_12_1$, are distinguished by a translation along c of the DCA molecules of about 7.2 and 14.2 Å respectively. Their bilayer structures are practically the same (see Fig. 6), although in the β group monolayers, the intermediate row, almost half way between the rows at

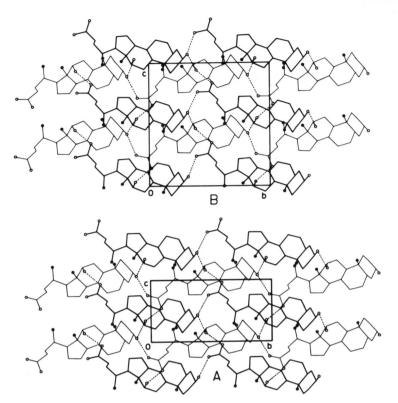

Fig. 6. Structure of DCAACE (A) and DCANBD (B) bilayers viewed along *a*. The symbols have the same meaning as in Fig. 3.

$c \sim 0$ and $c \sim 14.2$ Å, cannot be reproduced by an exact translation of about 7.1 Å, as in the α group. However, the bilayer packing differs in the α and β group. In fact, two facing monolayers, belonging to two adjacent bilayers, are shifted along c by about 3.6 Å in the α group, while they are at nearly the same height on c in the β group.

The values of the unit cell parameters given in Table 2 show that a, influenced by the bilayer separation, tends to increase with the bulkiness of the guest molecules, which are thread-like or flat in the α group and are, or have substituents, of approximately spherical shape in the β group. Moreover, the parameters b and c, connected with the bilayer structure, undergo small changes both in the α and β group, which is to be expected owing to the close similarity between their bilayers. Therefore, van der Waals energy calculations were accomplished for the choleic acids in the space group $P2_12_12_1$ in order to establish if it were possible to predict the packing of the bilayers as a function of a and of the bilayer shift along b, neglecting the contribution of the guest molecules.[57] Similar calculations were carried out independently by Tang.[56]

The b and c parameters and the atomic coordinates of DCAACE and DCANBD corresponding to a good R factor were assumed as representative of the α and β structures respectively. The hydrogen atoms, except those of the carboxyl, hydroxyl and methyl groups, were generated by putting C–H = 1.08 Å and H–Ĉ–C = 109.5°. The hydrogen atoms of the tertiary carbon atoms give rise to two H–Ĉ–C of 109.5°. The methyl group was considered as one atom. The coefficients of the potentials are listed in Table 1. All the interactions between an asymmetric unit and the sixteen nearest DCA molecules of a monolayer facing that of the asymmetric unit were computed by assuming a cut-off distance of 7.5 Å. The energy curves of the α and β structures are reported for different values of a in Figs. 7 and 8 respectively. The energy values of Fig. 8 must be halved when compared with those of Fig. 7, since the DCANBD asymmetric unit is composed of two molecules and that of DCAACE of one. The y value represents the increment given to the y atomic coordinates of DCAACE and DCANBD, so that these choleic acids are placed at $y = 0$. Three minima were found both for the α group (A, B, C) and for the β group (A', B', C').

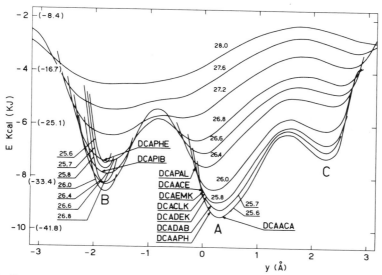

Fig. 7. Curves of the van der Waals energy at various values of a for the α structures. The arrows show the positions of the choleic acids' crystal structures.

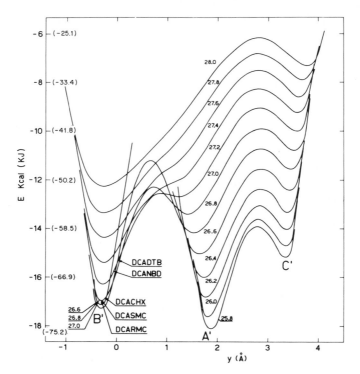

Fig. 8. Curves of the van der Waals energy at various values of a for the β structures. The arrows show the positions of the choleic acids' crystal structures.

The minimum-point energy values of the curves of Figs. 7 and 8, taken in the regions A, B, C, A′, B′ and C′, are given as a function of a in Fig. 9.

The deepest minimum of the α structures, A, corresponds to the DCAACE packing of Fig. 5 and is populated by DCA channel complexes including mostly thread-like guest molecules. The approximate minimum sizes of the channel section in DCAACA are about 2.6×6.0 Å, although aromatic molecules such as acetophenone and p-dimethylaminoazobenzene can also be accommodated. Generally, the molecular plane of the guest, passing through the non-hydrogen atoms, is sandwiched between rings A and the C(6) methylenic groups, lying on the longer edges of the channel rectangular section. The second minimum, B, corresponding to the DCAPHE packing of Fig. 5, begins to be the lowest one for a values greater than 26 Å (see Fig. 9). It is populated by choleic acids containing as guests flat molecules with a lateral dimension too large, or with substituents too bulky, for the A cavity, so that the approximate minimum sizes of the channel section in

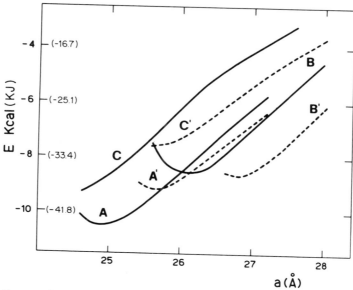

Fig. 9. Curves of the lowest values of the van der Waals energy for the A, B, C, A′, B′ and C′ minimum regions vs *a*.

DCAPHE increase to about 5.0×7.1 Å. The guest molecular plane is sandwiched between rings D and side-chains, constituting the longer edges of the channel cross-section. The third minimum, C, represented by the "without channels" packing of Fig. 5, is always energetically less stable, for a given *a* value, than at least one of the other two (see Fig. 9). The C packing, which does not correspond to an inclusion compound, has not been observed so far.

A similar situation has been found for the β structures of choleic acids. However, the deepest minimum, A′, which corresponds to a host lattice similar to that of A, is not populated because of its higher energy than that of A for *a* values less than 26 Å (see Fig. 9). On the other hand, the B packing is favoured for *a* values greater than 26 Å and less than 27 Å, while the B′ structure becomes the most stable from about 27 Å upward. The B′ packing is represented by that of DCANBD of Fig. 5 and is preferred by guest molecules which have larger cross-sections of roughly circular shape causing the lengthening of *a*. The approximate minimum sizes of the channel section in DCANBD, more square than those of DCAACE and DCAPHE, are about 4.9×5.4 Å, with a further narrowing (about 4.4 Å) along the line joining the two C(18) methyl groups. The same considerations mentioned above for the minimum C hold for C′.

These results illustrate that the packing of the bilayers and the shape of the cavity can be approximately inferred for an orthorhombic choleic acid if its unit-cell constants are known. Moreover, the size of the guest component controls the shape of the cavity and, hence, it is possible to establish which molecules can be included.

It is clear from the energy calculations that the analysis of the minimum residual[63] allows the DCA molecules of the orthorhombic choleic acids to be easily located. In fact, the best R factor, obtained by translating along a, b and c the DCA asymmetric unit of which the geometry and conformation are known, corresponds to the actual packing. This method was satisfactorily applied in the solution of the crystal structures of DCAPAL,[59] DCANBD[61] and DCADAB.[73] However, there are often insurmountable difficulties in the location of the guest molecules by means of standard X-ray methods. In these cases van der Waals energy calculations, performed using the potentials of Table 1, allowed the determination of the positions of the included molecules. This procedure was satisfactorily adopted for DCAPAL, DCANBD and DCADAB.

The DCAPAL crystal structure has shown that the DCA/guest ratios of 4 and 6 can be easily explained for the fatty acids series C_4–C_8 and C_9–C_{14}. In fact, the longest members of these two series, C_8 and C_{14}, can be accommodated in channels about 14.4 and 21.6 Å long, corresponding to the ratios of 4 and 6 respectively, while the other members give rise to channels with empty spaces between guest molecules, very probably occupied by molecules of crystallization (very often ethanol), as in DCAPAL. On the other hand, Giacomello and Bianchi[46] demonstrated the formation of choleic acids containing esters (and ethanol) with a DCA/ester ratio of 10 and 12 in contrast to the work of Rheinboldt and coworkers,[17,19] who found a maximum value of 8.

A last point deserves some attention. The DCA–guest interactions are usually normal van der Waals interactions when the guest component is non-aromatic. If the included molecule is aromatic, short contacts and, hence, strong interactions with the π charge cloud occur, so that choleic acids such as DCAPHE, DCAPIB, DCADAB and, very likely, DCAAPH are the most stable among the orthorhombic ones of Table 2. The atoms of DCA mainly engaged in these interactions are the hydrogens of C(5), C(6), C(20) and C(22) and the methyl groups C(18) and C(21).

3.3. Tetragonal crystals

Tetragonal crystals of DCA were identified by Bonamico and Giacomello[49] many years ago. More recently the crystal structures of two tetragonal forms, grown by slow evaporation from an acetone/ethanol mixture[58] and from methanol[56] as colourless tetragonal-bipyramidal crystals elongated along *c* and indicated as DCAETW and DCAWAT respectively (see Table 2), were solved. DCAETW can be also obtained by crystallization from aqueous ethanol. A lowering of the temperature generally favours the formation of the tetragonal phase (DCAETW) with respect to the hexagonal one (DCAETH, see Table 2 and later). The side-chain and ring D torsion angles of DCA are similar to those observed in the orthorhombic crystals, the *gauche* side-chain conformation being coupled with the nearly half-chair form. The crystal packing (see Fig. 10) can be described as an assembly of wavy bilayers related to each other by a 4_1 axis, instead of a 2_1 axis as in

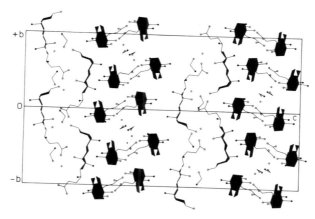

Fig. 10. Crystal packing of DCAETW viewed along *a*. The overall shape of the DCA molecules is drawn. Methyl groups and oxygen atoms are represented by filled and open circles respectively. Hydrogen bonds are omitted.

the orthorhombic phases. The bilayer extends into the *ab* plane and is composed by two monolayers, each one formed by rows in which DCA molecules are linked by head-to-tail hydrogen bonds involving O(25) and O(27) of two consecutive molecules (see Fig. 11). The rows of the two monolayers face each other and are at about the same *a* height, at variance with the orthorhombic bilayer where they are shifted by about 3.6 Å. The driving force in the formation of the bilayer is an efficient scheme of hydrogen bonds in which ethanol and water molecules also participate. These molecules are located in channels of small section, approximately rectangular in shape (see Fig. 10), running inside the bilayer along *a* or *b*.

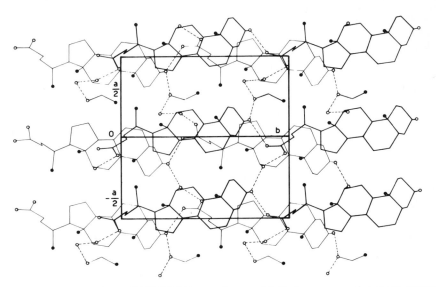

Fig. 11. A view of a DCAETW bilayer along *c*. Methyl groups and oxygen atoms are represented by filled and open circles respectively. Broken lines indicate hydrogen bonding.

The outer surfaces of a bilayer are hydrophobic with the methyl groups protruding outward, whereas the interior surfaces of the channels are covered mainly by hydrophilic groups. Adjacent bilayers, held together by weak van der Waals interactions, are closer than the orthorhombic ones, so that no empty space occurs between them. Thus, the density of the tetragonal crystals ($\sim$1.16 g cm^{-3}) is slightly higher than those of the ortho-rhombic phases (average value $\sim$1.15 g cm^{-3}, excluding DCAPIB), although these generally receive a more massive contribution from the guest molecules. The unit cell constants can easily be related to the orthorhombic ones: a_{tetr} (or b_{tetr}) $\sim b_{orth}$; b_{tetr} (or a_{tetr}) $\sim 2\,c_{orth}$; $c_{tetr} \sim 2\,a_{orth}$. In fact, the DCA molecule, $\sim$14 Å long, lies parallel to b_{orth}, a_{tetr} and b_{tetr} and determines the length of these axes ($\sim$14 Å); the separation between two rows of the same monolayer is about 7 Å, which corresponds to $\sim c_{orth}$ and to about half of a_{tetr} and b_{tetr}; c_{tetr} is about four times 12.5 Å, the approximate distance between two bilayers in both the tetragonal and orthorhombic crystals.

Other tetragonal crystals have been obtained, but not structurally charac-terized, in the laboratory of the author by crystallizing DCA from small polar molecules. Only in the case of propargyl alcohol, however, it is sure, by means of I.R. spectra, that the solvent molecules have been included.

3.4. Hexagonal crystals

The DCAETH and DCADMS crystals (see Table 2) are grown by slow evaporation respectively from ethanol (or acetone/ethanol mixture) and dimethyl sulphoxide as colourless hexagonal prisms elongated along *c*. Their crystal structures[54,55] show that DCAETH and DCADMS are isomorphous with the host molecules occupying the same positions. The side-chain conformation of DCA, very similar to that of cholic acid,[64] is *trans*, namely nearly extended, and populates the central minimum of Fig. 2. Ring D approaches a β-envelope symmetry, giving rise to satisfactory intramolecular van der Waals interactions with the side-chain. The orientation of the carboxyl group differs from those found in the orthorhombic and tetragonal crystals, thus supporting the ability of DCA to form various hydrogen-bonding schemes, which seem to be the driving force in the aggregation of the DCA molecules. The energy gain, due to the formation of the hydrogen bonds, largely counterbalances the high van der Waals energy which arises from the unfavourable contacts existing between the carboxyl group and some atoms of the side-chain[2] in all the crystal systems.

The DCAETH and DCADMS crystal structures are characterized by the packing of helices of DCA molecules, generated by 6_5 axes, held together by van der Waals forces, since the outer surfaces of the helices, which come into contact, are apolar (see Fig. 12 for DCADMS). The cavity of the helix, magnified in Fig. 13, has a section of approximately circular shape with a diameter of about 4 Å, which allows the inclusion only of small size or thread-like molecules. The interior surface of the helix is covered mainly by polar groups which give rise to a spiral of hydrogen bonds. O(25) is engaged with O(27) of the asymmetric unit at $(y - x, -x, z - 2/3)$ and with O(28) of that at $(y, y - x, z - 5/6)$; O(26) with two oxygen atoms belonging to dimethyl sulphoxide and water. The guest molecules fill the helical channel, which extends along the *c* axis. A similar situation is observed in DCAETH.

Standard X-ray methods failed to locate the methyl groups of ethanol and dimethyl sulphoxide. Their positions were established by computing the van der Waals energy of the guest molecule, interacting with the three nearest DCA molecules, as a function of the rotation ϕ along the C–O or S–O bond (see Figure 14 and, for the definition of $\phi = 0°$.[54,55] The nonequivalent crystallographic positions occupied by the methyl groups of two adjacent ethanol molecules correspond to the two minima of curve A of Fig. 14. The directions of the two S–C bonds of dimethyl sulphoxide, corresponding to the minimum of curve B of Fig. 14, nearly coincide with those permitted for the two C–C bonds of the adjacent ethanol molecules.

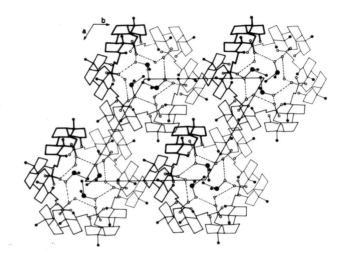

Fig. 12. DCADMS crystal packing viewed along *c.* Oxygen and sulphur atoms and methyl groups are represented by open circles and large and small black circles. The hydrogen bonds among DCA and dimethyl sulphoxide or water molecules are indicated by broken lines, whereas those among DCA molecules are omitted for the sake of clarity.

Potential energy calculations also allowed the determination of both the ratio between host and guest components, confirmed by gas-chromatographic measurements, and that ordered sequences (ethanol-ethanol-water)$_n$ and (dimethyl sulphoxide-water)$_n$ are present in the channels of DCAETW and DCADMS respectively.

In conclusion, the hexagonal host lattice can be utilized to receive polar guest molecules at variance with the orthorhombic ones which include, preferably, apolar guest molecules.

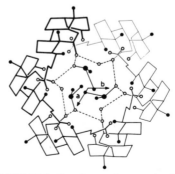

Fig. 13. A view of a DCADMS helix along the *c* axis. The legend is as that of Fig. 12.

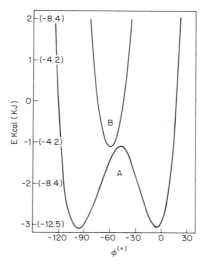

Fig. 14. Van der Waals energy of the ethanol (curve A) and dimethyl sulphoxide (curve B) molecules vs ϕ in the DCAETH and DCADMS crystals respectively.

4. Stability of choleic acids by vapour pressure measurements

The orthorhombic choleic acids show a remarkable inclusion ability which is far better than that of the tetragonal and hexagonal crystals. Therefore, it is desirable to predict which molecules can be preferably included in the orthorhombic phases in order to establish an affinity scale between DCA and guest molecules, owing to the many potential applications of the DCA channel complexes in the fields of polymerization[65] (see Chapter 10, Volume 3) and photochemical[60] reactions, the transport of drugs,[66–68] the stabilization of autoxidizable materials,[69] chromatographic techniques and other separation methods.[20,36] Only a rough qualitative estimate of the host–guest attractive energy can be provided by energy calculations because of the semi-empirical character of the potentials employed. Thus, the nature and the strength of the host–guest interactions were investigated by vapour pressure measurements accomplished by means of a torsion–effusion method[70,71] on the styrene[29] (DCASTY), naphthalene[29] (DCANAF), phenanthrene, 1,2-benzanthracene (DCABAN) and 11,12-benzofluoranthene (DCABKF) choleic acids (unpublished results of D. Ferro, P. Imperatori and C. Quagliata). The vapour pressure of the gaseous guest in

equilibrium with its choleic acid as a function of temperature was determined, having ascertained that the contribution of the vapour of DCA to
the total pressure is negligible and below the instrument sensitivity. The
vapour pressure–temperature dependence of the above-mentioned choleic
acids in the first step of vaporization is shown in Fig. 15. From the
pressure–temperature equation, the enthalpy change associated with the
release of the guest molecule from the crystal of the choleic acid (ΔH_r) and
their heat of formation (ΔH_f) at 298 K were derived by knowing or measuring

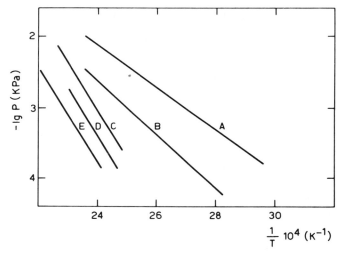

Fig. 15. The first step of vaporization of DCASTY (A), DCANAF (B), DCAPHE
(C), DCABAN (D) and DCABKF (E).

the guest vaporization enthalpies. These results are reported in Table 3.
All the choleic acids have unit cell constants which allow these to be assigned
to the B minimum region of the α group, so that the main interactions
involve the C(18) methyl groups of two DCA molecules, related by a 2_1 axis,
and the π charge cloud of the aromatic molecule (see Fig. 3). The methyl
groups point approximately towards the centres of aromatic rings, giving
rise to "polarization bonding", and behave as hooks to which the guest
molecules are hung. Therefore, the host–guest interactions become stronger
the more polarizable is the guest molecule and the greater is the number
of hooks for each guest molecule. By inspection of Fig. 15 and Table 3 it
is clear that the expectations are upheld in the case of DCASTY (one hook),
DCANAF (one hook) and DCAPHE (two hooks), whereas for DCABAN

Table 3. ΔT (K, experimental temperature range), ΔH_r and ΔH_f (KJ) of some orthorhombic choleic acids. The errors, estimated by taking into account the uncertainties in the temperature measurements, are in parentheses

	ΔT	ΔH_r	ΔH_f
DCASTY	353–414	56.9 (5.4)	15.1 (5.4)
DCANAF	361–419	72.0 (5.4)	23.8 (5.4)
DCAPHE	403–441	128.9 (5.9)	50.2 (7.1)
DCABAN	405–429	126.4 (7.5)	21.8 (9.2)
DCABKF	414–454	124.3 (3.8)	4.6 (5.0)

and DCABKF the ΔH_r values are nearly equal to that of DCAPHE, although the number of the condensed rings increases. However, it can be assumed that the number of hooks for 1,2-benzanthracene and 11,12-benzofluoranthene is always two, as for phenanthrene, since the nearest C(18) in the channel is about 7 Å from those engaged with phenanthrene and, hence, the guest molecule must increase considerably its length to bind another methyl group.

The vapour pressure measurements can provide useful information for the replacement by new molecules of the guest molecules released from the channels of the choleic acids by heating.[72] Of course, the DCA-guest interaction energy, together with the size and the shape of the channels, control the mechanism of release and replacement, so that the release temperature, which influences the replacement procedure, may be related empirically to the interaction energy.

References

1. L. F. Fieser and M. Fieser, *Steroids*, Reinhold Publishing Corp., New York, 1959, Chapter 3, pp. 53 and 56.
2. E. Giglio and C. Quagliata, *Acta Crystallogr.*, 1975, **B31**, 743.
3. N. V. Pavel, C. Quagliata and N. Scarcelli, *Z. Kristallogr.*, 1976, **144**, 64.
4. E. Gavuzzo, F. Mazza and E. Giglio, *Acta Crystallogr.*, 1974, **B30**, 1351.
5. W. Klyne and V. Prelog, *Experientia*, 1960, **16**, 521.
6. C. Altona, H. J. Geise and C. Romers, *Tetrahedron*, 1968, **24**, 13.
7. H. Sobotka, *Chem. Rev.*, 1934, **15**, 311.
8. M. Baron, *Phys. Methods in Chem. Anal.*, 1961, **4**, 223.
9. C. Asselineau and J. Asselineau, *Ann. Chim.*, 1964, **9**, 461.
10. W. C. Herndon, *J. Chem. Educ.*, 1967, **44**, 724.
11. E. Giglio, *J. Mol. Struct.*, 1981, **75**, 39.
12. P. Latschinoff, *Chem. Ber.*, 1885, **18**, 3039.

13. P. Latschinoff, *Chem. Ber.*, 1887, **20**, 1043.
14. F. Mylius, *Chem. Ber.*, 1886, **19**, 369.
15. H. Wieland and H. Sorge, *Z. Physiol. Chem.*, 1916, **97**, 1.
16. H. Rheinboldt, *Justus Liebigs Ann. Chem.*, 1926, **451**, 256.
17. H. Rheinboldt, H. Pieper and P. Zervas, *Justus Liebigs Ann. Chem.*, 1927, **451**, 256.
18. H. Sobotka and A. Goldberg, *Biochem. J.*, 1932, **26**, 555.
19. H. Rheinboldt, E. Flume and O. König, *Z. Physiol. Chem.*, 1929, **180**, 180.
20. W. Marx and H. Sobotka, *J. Org. Chem.*, 1936, **1**, 275.
21. E. Chargaff and G. Abel, *Biochem. J.*, 1934, **28**, 1901.
22. E. H. Huntress and R. F. Phillips, *J. Am. Chem. Soc.*, 1949, **71**, 458.
23. H. Rheinboldt, P. Braun, E. Flume, O. König and A. Lauber, *J. Prakt. Chem.*, 1939, **153**, 313.
24. H. Rheinboldt, O. König and P. Otten, *Justus Liebigs Ann. Chem.*, 1929, **473**, 249.
25. Ng. Buu-Hoi, *Z. Physiol. Chem.*, 1943, **278**, 230.
26. H. Sobotka and J. Kahn, *Biochem. J.*, 1932, **26**, 898.
27. G. Giacomello and E. Bianchi, *Ric. Sci.*, 1942, **12**, 345.
28. A. Damiani, E. Giglio, N. Morosoff, R. Puliti and I. Rosen, *Ric. Sci.*, 1967, **37**, 42.
29. D. Ferro, C. Quagliata, E. Giglio and V. Piacente, *J. Chem. Eng. Data*, 1981, **26**, 192.
30. L. F. Fieser and M. S. Newman, *J. Am. Chem. Soc.*, 1935, **57**, 1602.
31. G. Cilento, *J. Am. Chem. Soc.*, 1950, **72**, 4272.
32. G. Cilento, *J. Am. Chem. Soc.*, 1951, **73**, 1355.
33. G. Cilento, *J. Am. Chem. Soc.*, 1952, **74**, 968.
34. W. Lautsch, W. Bandel and W. Broser, *Z. Naturforsch.*, 1956, **11B**, 282.
35. E. Angelescu and G. Nicolau, *Rev. Roum. Chim.*, 1965, **10**, 355.
36. H. Sobotka and A. Goldberg, *Biochem. J.*, 1932, **26**, 905.
37. L. F. Fieser and S. Rajagopalan, *J. Am. Chem. Soc.*, 1949, **71**, 3935.
38. R. O. Herzog, O. Kratky and S. Kuriyama, *Naturwiss.*, 1931, **19**, 524.
39. Y. Go and O. Kratky, *Z. Phys. Chem.*, 1934, **B26**, 439.
40. Y. Go and O. Kratky, *Z. Kristallogr.*, 1935, **A92**, 310.
41. G. Giacomello and O. Kratky, *Z. Kristallogr.*, 1936, **A95**, 459.
42. O. Kratky and G. Giacomello, *Monatsh. Chem.*, 1936, **69**, 427.
43. G. Giacomello, *Atti Accad. Nazl. Lincei Rend. Classe Sci. Fis. Mat. Nat.*, 1938, **27**, 101.
44. V. Caglioti and G. Giacomello, *Gazz. Chim. Ital.*, 1939, **69**, 245.
45. G. Giacomello, *Gazz. Chim. Ital.*, 1939, **69**, 790.
46. G. Giacomello and E. Bianchi, *Gazz. Chim. Ital.*, 1943, **73**, 3.
47. G. Giacomello and M. Romeo, *Gazz. Chim. Ital.*, 1943, **73**, 285.
48. H. Fischmeister, *Monatsh. Chem.*, 1954, **85**, 182.
49. M. Bonamico and G. Giacomello, *Gazz. Chim. Ital.*, 1962, **92**, 647.
50. B. M. Craven and G. T. DeTitta, *J. Chem. Soc. Chem. Commun.*, 1972, 530.
51. S. Candeloro De Sanctis, E. Giglio, V. Pavel and C. Quagliata, *Acta Crystallogr.*, 1972, **B28**, 3656.
52. N. Friedman, M. Lahav, L. Leiserowitz, R. Popovitz-Biro, C. P. Tang and Z. Zaretzkii, *J. Chem. Soc. Chem. Commun.*, 1975, 864.
53. M. Lahav, L. Leiserowitz, R. Popovitz-Biro and C. P. Tang, *J. Am. Chem. Soc.*, 1978, **100**, 2542.
54. S. Candeloro De Sanctis, V. M. Coiro, E. Giglio, S. Pagliuca, N. V. Pavel and C. Quagliata, *Acta Crystallogr.*, 1978, **B34**, 1928.

55. S. Candeloro De Sanctis, E. Giglio, F. Petri and C. Quagliata, *Acta Crystallogr.*, 1979, **B35**, 226.
56. C. P. Tang, Ph. D. Thesis, The Feinberg Graduate School, The Weizmann Institute of Science, Rehovot, Israel, 1979.
57. S. Candeloro De Sanctis and E. Giglio, *Acta Crystallogr.*, 1979, **B35**, 2650.
58. V. M. Coiro, A. D'Andrea and E. Giglio, *Acta Crystallogr.*, 1979, **B35**, 2941.
59. V. M. Coiro, A. D'Andrea and E. Giglio, *Acta Crystallogr.*, 1980, **B36**, 848.
60. R. Popovitz-Biro, H. C. Chang, C. P. Tang, N. R. Shochet, M. Lahav and L. Leiserowitz, *Pure Appl. Chem.*, 1980, **52**, 2693.
61. A. D'Andrea, W. Fedeli, E. Giglio, F. Mazza and N. V. Pavel, *Acta Crystallogr.*, 1981, **B37**, 368.
62. J. G. Jones, S. Schwarzbaum and L. Lessinger, Abstract of the Twelfth Congress of the International Union of Crystallography, Ottawa, August 1981, *Acta Crystallogr.*, 1981, **A37 suppl.**, C 76.
63. A. Damiani, E. Giglio, A. M. Liquori and A. Ripamonti, *Acta Crystallogr.*, 1967, **23**, 681.
64. P. L. Johnson and J. P. Schaefer, *Acta Crystallogr.*, 1972, **B28**, 3083.
65. K. Takemoto and M. Miyata, *J. Macromol. Sci. Rev. Macromol. Chem.*, 1980, **C18**, 83.
66. J. L. Lach and W. A. Pauli, *J. Pharm. Sci.*, 1966, **55**, 32.
67. M. H. Malone, H. I. Hochman and K. A. Nieforth, *J. Pharm. Sci.*, 1966, **55**, 972.
68. K. H. Frömming and R. Sandmann, *Arch. Pharm.* (Weinheim), 1970, 371.
69. H. Schlenk, D. M. Sand and J. A. Tillotson, *J. Am. Chem. Soc.*, 1955, **77**, 3587.
70. R. D. Freeman, in *The Characterization of High Temperature Vapours*, (ed. J. L. Margrave) Wiley, New York, 1967.
71. V. Piacente and G. De Maria, *Ric. Sci.*, 1960, **39**, 545.
72. M. Miyata and K. Takemoto, *Makromol. Chem.*, 1978, **179**, 1167.
73. V. M. Coiro, E. Giglio, F. Mazza, N. V. Pavel and G. Pochetti, *Acta Crystallogr.*, 1982, **B38**, 2615.

8 · STRUCTURAL ASPECTS OF CYCLODEXTRINS AND THEIR INCLUSION COMPLEXES

W. SAENGER

Institut für Kristallographie, Freie Universität Berlin, Berlin, FRG

1. Introduction: general description of cyclodextrins

Cyclodextrins comprise a family of cyclic oligosaccharides obtained from starch by enzymatic degradation. In the course of this process, catalysed by a group of amylases called glycosyltransferases, one turn of the starch helix is hydrolysed off and its ends are joined together.[1,2] Since, however,

INCLUSION COMPOUNDS 2
ISBN 0-12-067101-8

these enzymes are not very specific as to the site of hydrolysis, rings with different numbers of glucose units are produced, with the six units of cyclohexaamylose (α-cyclodextrin, α-CD) representing the smallest number. Higher analogues consisting of up to 12 glucose units can also be identified.[3] They have not all been subjected to detailed analysis, however, mainly because they are more difficult to obtain in suitable quantities except for the second and third smallest members, cyclohepta- and octaamylose (β- and γ-cyclodextrin or β-, γ-CD) which are comprised of seven and eight units respectively. Cycles composed of less than six members are not possible due to excessive steric strain[4] and for reasons inherent in the sixfold character of the starch helix.[5]

In Figs. 1 and 2, the chemical structures and overall molecular shapes of α-, β- and γ-cyclodextrin are presented. As their appearance immediately suggests, these molecules, if dissolved in water, are able to accommodate smaller guest molecules within the cavities formed by the annular structures. This particular feature distinguishes cyclodextrins (CD's) from most other host molecules which *per se* are unable to form inclusion complexes and require crystallization into a lattice in order to provide a matrix with suitable cavities.

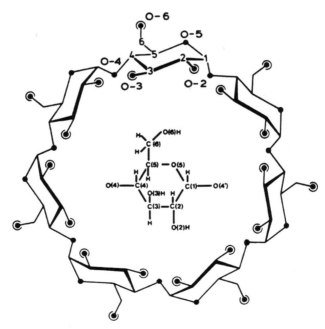

Fig. 1. Chemical structure of β-cyclodextrin, with oxygen ● and hydroxyl groups ⊙ marked. All glucoses are in C$_1$ chair form.[8]

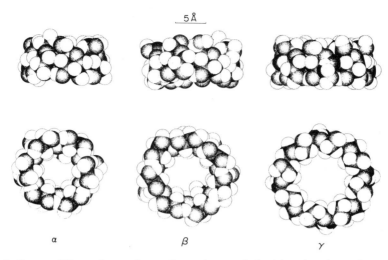

Fig. 2. Space filling plots of α-, β- and γ-cyclodextrins in views from the O(2)H/O(3)H and O(6)H rims and from the side. Drawn with coordinates given in,[11,18,41] using the program SCHAKAL.[43]

As in starch, the glucose units in CD's are linked by $\alpha(1-4)$ bonds. The chair forms of the six-membered glucose rings are all aligned in register, i.e. the O(2)H/O(3)H hydroxyls point to one side and the O(6)H hydroxyls to the other. This geometry gives the CD's the overall shape of a truncated cone with the wider side formed by the secondary O(2)H/O(3)H and the narrower side by the primary O(6)H hydroxyls.

One of the most important features of CD's concerns the distribution of hydrophilic and hydrophobic groups. Thus, hydroxyls occupy both rims of the cone and render the CD's soluble in aqueous solution. On the other hand, the inside of the cavity is hydrophobic in character because it is lined by C(3)–H and C(5)–H hydrogens and by the ether-like oxygens O(4). In solution, therefore, these cavities provide a hydrophobic matrix in hydrophilic surroundings, described by the term "microheterogeneous environment".[6]

The unique properties of the CD cavity explain some of the unusual features of these molecules. Thus, they form inclusion complexes rather unspecifically with a wide variety of guest molecules, even in solution; the only obvious requirement being that the guest must fit into the cavity—even if only partially. On this basis, it is not surprising to find that noble gases, paraffins, alcohols, carboxylic acids, aromatic dyes, benzene derivatives, salts etc. are included, just to name a few of a long list of potential guest substances.[7,8,9]

Table 1. A list of α-, β- and γ-cyclodextrin complexes for which crystal structure analyses have been carried out. Shown are the chemical formulae of complexes and their cavity types: herringbone cage (C,H); brick cage (C,B); channel with head-to-head arranged cyclodextrins (Ch, HH); channel head-to-tail (Ch, HT), and pertinent references

Cyclodextrin	Formula of guest	Cavity type	Reference
α	$6H_2O$ (Form I)	C, H	11
α	$6H_2O$ (Form II)	C, H	12
α	$7.57H_2O$ (Form III)	C, H	13
α	methanol.$5H_2O$	C, H	14
α	n-propanol.$4.8H_2O$	C, H	15
α	iodine.$4H_2O$	C, H	16
α	0.7 krypton.$5H_2O$	C, H	17
β	$12H_2O$	C, H	18
β	$2HI.8H_2O$	C, H	19
β	methanol.$6H_2O$	C, H	19
β	ethanol.$8H_2O$	C, H	20
β	nicotinamide.$6H_2O$	C, H	63
γ	$17H_2O$	C, H	21
α	dimethylsulfoxide.methanol .H_2O(α-CD:1:2)	C, B	22
α	p-nitrophenol.$3H_2O$	C, B	23
α	p-hydroxybenzoic acid.$3H_2O$	C, B	23
α	p-iodoaniline.$3H_2O$	C, B	24, 25
α	p-toluidine.$3H_2O$	C, B	26
α	p-iodophenol.$3H_2O$	C, B	27
α	2-pyrrolidone.$5H_2O$	C, B	28
α	N,N-dimethylformamide.$5H_2O$	C, B	28
α	2(m-nitrophenol).$6H_2O$	Ch, HT	29
α	1-propanesulfonate-Na.$9H_2O$	Ch, HT	30
α	benzenesulfonate-Na.$10H_2O$	Ch, HT	31
$2α^{(a)}$	methylorange-Na and -K.$9.75H_2O$	Ch, HT	32
$2α^{(a)}$	3.08 K-acetate.$19.4H_2O$	Ch, HT	33
α	benzaldehyde.$6H_2O$	Ch, HT	34
α	γ-aminobutyric acid-K.$10H_2O$	Ch, HT	35
$2α^{(a)}$	$LiI_3.I_2.8H_2O$	Ch, HH	36
$2α^{(a)}$	$Cd_{0.5}I_5.27H_2O$	Ch, HH	36
β	2,5-diiodobenzoic acid.$10H_2O$	Ch, HH	37
β	p-nitroacetanilide.$7H_2O$	Ch, HH	38
β	p-iodophenol	Ch, HH	39
β	n-propanol	Ch, HH	39
$2β^{(a),(b)}$	$NaI_3.2I_2.XH_2O$	Ch, HH	40
$γ^{(b)}$	X(n-propanol).YH_2O	Ch, HH	41
2,6-Tetra-deca-O-methyl-β-cyclodextrin	adamantol.$12H_2O$		42

(a) In these cases, formulae require two cyclodextrins, e.g. (α-CD)₂ methylorange-Na.$9.75H_2O$.
(b) X, Y mean that exact formula is unknown due to disorder within the crystal structure.

General studies on CD's in aqueous solution and their behaviour when guest molecules are included can be performed using spectroscopic methods such as circular dichroism, UV absorption and nuclear magnetic resonance.[8,9] These methods allow investigation of changes that occur upon inclusion of guest molecules into the host cavity and are therefore valuable tools for the determination of thermodynamic and kinetic parameters of the inclusion process. If, however, precise details, surpassing the information obtained from spectroscopic data, are required, recourse has to be taken to X-ray or even neutron diffraction analysis of crystallized complexes. These latter techniques not only provide a picture of the geometry of host and guest molecules, but also of host–guest interactions. In addition, packing of complexes within the crystal lattice indicates intermolecular aggregation.

A number of crystal structure analyses on α-, β- and γ-CD have been performed and are summarized in Table 1. They cannot all be treated in detail within the framework of this chapter and therefore only the main characteristics are outlined.[8,10]

2. General structural features of cyclodextrin molecules

2.1. The glucose is a "rigid" unit

By and large, the overall shapes and conformations of individual glucose units in all CD's so far investigated are comparable. In no case has a glucose conformation other than the classical C1 chair been observed, no matter what guest is included within the cavity. Moreover, the endocyclic torsion angles in the glucose rings are rather rigidly confined to + or − *gauche* values within a narrow range, suggesting that the glucose can be considered to be a fairly "rigid" building block (Table 2).

In contrast, the primary O(6)H hydroxyl group can rotate about the C(5)–C(6) bond. In principle, three staggered orientations + *gauche*, − *gauche* and *trans* can be adopted by the O(5)–C(5)–C(6)–O(6) torsion angle, yet only the first two have actually been observed; *trans* has not been found thus far. A reason for this behaviour could be that, in the *trans* orientation, adverse steric interactions might occur between O(6)H and atoms of the adjacent glucose unit. In addition, the *gauche*-effect,[44] operative in O–C–C–O systems, should also destabilize the *trans* arrangement.

The two *gauche* conformers are not equally common, the −*gauche* form, with O(6)H pointing away from the centre of the cavity (Fig. 3) being largely

Table 2. Some averaged geometrical data for α-, β- and γ-cyclodextrins. The γ-cyclodextrins are not sufficiently refined yet to allow for proper comparison and therefore some data are omitted. Data from[10,18,41]

Torsion angles (°)	α-CD	β-CD	γ-CD
C(1)–C(2)–C(3)–C(4)	−54.4	−54.7	
C(2)–C(3)–C(4)–C(5)	52.6	55.4	
C(3)–C(4)–C(5)–O(5)	−53.4	−56.5	
C(4)–C(5)–O(5)–C(1)	59.8	60.0	
C(5)–O(5)–C(1)–C(2)	−62.5	−60.4	
O(5)–C(1)–C(2)–C(3)	58.3	56.5	
O(5)–C(1)–O(4′)–C(4′)	107.7	109.9	
C(2)–C(1)–O(4′)–C(4′)	−132.7	−129.3	
C(1)–O(4′)–C(4′)–C(3)	130.6	128.3	
C(1)–O(4′)–C(4′)–C(5′)	−110.3	−112.2	
Distances in macrocycle (Å)			
O(4)···O(4′)	4.23	4.36	4.48
O(2)···O(3′)	3.00	2.86	2.81
Angles in macrocycle (°)			
$\phi \equiv$ O(4)···C(1)–O(4′)–C(4′)	166	169	165
$\psi \equiv$ C(1)–O(4′)–C(4′)···O(4″)	−169	−172	−169
Distances of atoms from planes defined by C(2), C(3), C(5), O(5) (Å)			
C(1)	0.675	0.661	
C(4)	−0.629	0.675	
Angle at glycosyl link (°)			
C(1′)–O(4)–C(4)	119.0	117.7	112.6

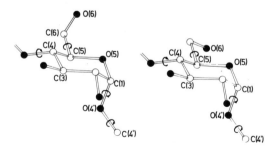

Fig. 3. Glucose chair conformation with O(6)H in − *gauche* (left) and + *gauche* (right) orientation. Bonds about which rotations are likely to occur in cyclodextrins are indicated by arrows.[10]

preferred. The +*gauche* orientation with O(6)H "towards" the cavity is only found in crystal structures if certain packing requirements are met or if a hydrogen bond is formed with an included guest.

2.2. The macrocyclic geometry is rather well defined

Some remarkable features are observed concerning the overall shape of the CD's. First, the macrocyclic hexa-, hepta- and octagons defined by O(4) atoms of α-, β- and γ-CD are virtually planar, with less than *c.* 0.25 Å deviation from the common mean plane. In all cases, the O(4)⋯O(4') distances (primed atom belongs to the next glucose) forming the edges of the macrocycle are more or less constant within each member of the CD family (See Table 2). Individual differences between α-, β- and γ-CD arise because the glucose unit has to adjust to the respective radius of the CD, as illustrated in Fig. 4.

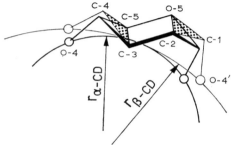

Fig. 4. Mechanical model describing how glucose in the chair form has to change shape in order to adapt to the different macrocyclic radii in α- and β-CD. When going from α-CD to β-CD, the O(4)⋯O(4'), distance is enlarged and C(4) moves away from the plane defined by atoms C(2), C(3), C(5), O(5) whereas C(1) approaches this plane (pointed areas).[18]

In line with these observations are distributions of O(4)⋯O(4')⋯O(4'') angles, being around 120° in α-CD, 128° in β-CD and 135° in γ-CD, with variations within ±5°. Exceptions from these rather constant macrocyclic data are only observed in the α-CD hexahydrate (Forms I and II) and in α-CD complexes with *para*-disubstituted benzene derivatives, the distortions of which will be discussed further in Section 5. Otherwise, the data given in Table 2 for the O(4) macrocycles and in Fig. 5 for the averaged geometry of the glucose unit are more or less constant and not influenced by inclusion of guest molecules. This means that upon complex formation the guest has little or no influence on the overall shape of the enclosing CD.

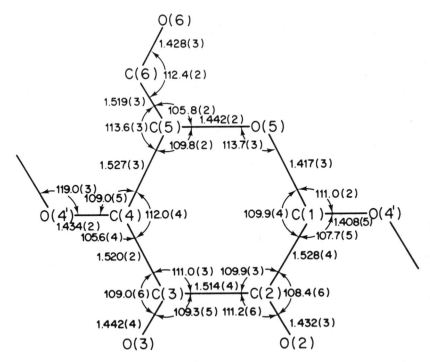

Fig. 5. Average geometry of the glucose unit, taken from the data for α-cyclodextrin methanol.$5H_2O$.[14] Standard deviations in parentheses were obtained according to $\sigma_i = [\Sigma(X_{ij} - \bar{X}_i)^2/5]^{1/2}$ with X_{ij} individual observation, $\bar{X}_i$ = mean of the 6 observations.

2.3. The torsion angle index

French and Murphy[45] noted a correlation between the O(4)$\cdots$O(4'), distance and a certain combination of endocyclic torsion angles in several pyranoses.

This interrelation can also be studied for α-cyclodextrin complexes in different crystal forms, the diagram shown in Fig. 6 being obtained. There is an obvious and clear correlation observed for brick-type cage structures formed by *para*-disubstituted benzene derivatives (*vide infra*), indicating that the distortions of the α-CD as expressed in O(4)$\cdots$O(4'), variations of ~0.5 Å are also reflected in changes of endocyclic torsion angles.

2.4. A ring of intramolecular hydrogen bonds

The question arises as to why the CD macrocycles are conformationally so stable. The remarkable structural rigidity of the CD ring appears to be due,

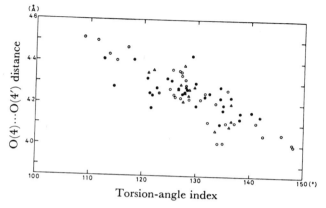

Fig. 6. Correlation of O(4)···O(4′) distance with torsion angle index in α-cyclodextrin complexes with *para*-disubstituted benzenes (○, brick-type cage), complexes with herringbone-type cages (●) and with channel-type structures (△). Torsion angle index defined after French and Murphy[45] is $|\phi C(1)-C(2)| + |\phi C(2)-C(3)| + |\phi C(5)-O(5)| + |\phi O(5)-C(1)| - |\phi C(3)-C(4)| - |\phi C(4)-C(5)|$, with $\phi C(1)-C(2)$ describing torsion angle O(5)–C(1)–C(2)–C(3) etc.[23]

primarily, to the formation of a ring of hydrogen bonds which, in every CD crystal structure so far investigated, is formed intramolecularly between O(2)H and O(3)H hydroxyl groups of adjacent glucose units. The O(2)···O(3′) distance is not constant for all CD's but decreases from α-CD (3.01 Å), through β-CD (2.86 Å), to γ-CD (2.81 Å), if average values are taken, indicating that hydrogen bonds become increasingly stronger going from α- to γ-CD, (Table 2). This statement is corroborated by measurements of deuterium/hydrogen exchange in aqueous solution (rather slow *per se* in CD's, indicative of strong hydrogen bonding) which have established that even slower exchange occurs for β-CD than for α-CD.[46,47]

2.5. Glycosyl torsion angles ϕ, ψ are rather constant

The intramolecular O(2)H···O(3)H hydrogen bonds stabilize the macrocyclic conformation and also limit torsion angles ϕ, ψ describing rotations about glycosyl C(1)–O(4′), and O(4′)–C(4′), bonds. In this nomenclature, primed and unprimed atoms belong to different glucoses and ϕ, ψ are defined as ϕ, O(4)···C(1)–O(4′)–C(4′) and ψ, C(1)–O(4′)–C(4′)···O(4″). The average ϕ, ψ torsion angles are virtually identical for all three kinds of CD's and, in general, do not deviate markedly from the values given in Table 2 except for the exceptional α-CD hexahydrate, *vide infra*.

3. Two types of crystal structures: channels and cages

Cyclodextrins are easily dissolved in water (Table 3) and can be crystallized as "empty" species with their cavities not really empty but filled by water molecules. On the other hand, if a molar excess of guest substance is added, the inclusion complex can be isolated in crystalline form. For α-, β-, and γ-CD, packing of the CD molecules within the crystal lattice occurs in one of two modes, described as cage and channel structures according to the overall appearance of the cavities formed; see Fig. 7.

Table 3. Some physical properties of cyclodextrins[8]

Cyclo-dextrin	Molecular weight	Solubility in water (g 100 ml^{-1} solution)	$[\alpha]_D^{25°}$	Diameter of cavity[a] (Å)	Diameter of outer periphery[a] (Å)
α	972	14.5	150.5 ± 0.5	$4.7 - 5.2$	14.6 ± 0.4
β	1135	1.85	162.5 ± 0.5	$6.0 - 6.4$	15.4 ± 0.4
γ	1297	23.2	177.4 ± 0.5	$7.5 - 8.3$	17.5 ± 0.4

[a] Measured with CPK-models. The height of all cyclodextrins is 7.9–8.0 Å.

3.1. Channel-type structures

In channel-type complexes, CD molecules are stacked on top of each other like coins in a roll, the now linearly aligned cavities producing "endless" channels in which guest molecules are embedded. The stacks of CD's are stabilized by hydrogen bonds either between O(2)H/O(3)H and O(6)H sides producing head-to-tail pattern or between O(2)H/O(3)H and O(2)H/O(3)H on one side and between O(6)H and O(6)H on the other, leading to head-to-head arrangement. In general, in one particular crystal structure only one type of stack is formed, but in γ-CD.X(n-propanol).YH$_2$O, a special situation is encountered because head-to-tail and head-to-head orientations alternate within the same stack;[41] see Fig. 8.

3.2. Cage-type structures in herringbone and brick arrangement

In crystal structures belonging to the cage type, the cavity of one CD molecule is blocked off on both sides by adjacent CD's, thereby leading to isolated cavities.

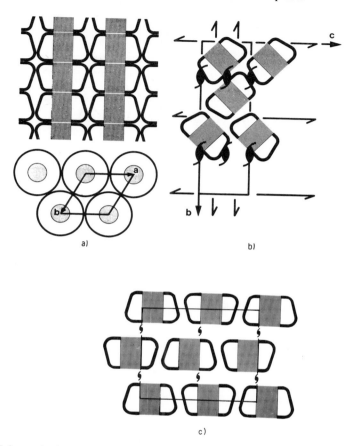

a) b)

c)

Fig. 7. Schematic description of (a) channel type, (b) cage *herringbone* type, (c) brick type crystal structures formed by crystalline cyclodextrin inclusion complexes.

In this type of arrangement, two categories are encountered, depending on the packing of the CD molecules. In one, CD's are packed crosswise in *herringbone* fashion, a pattern most common in α-, β- and γ-CD. In the other, thus far observed only for α-CD complexed with some *para*-disubstituted benzene derivatives, or with a dimethylsulfoxide/methanol mixture (Table 1), a motif reminiscent of *bricks* in a wall is found. In it, the α-CD's are arranged in sheets with all O(2)H/O(3)H hydroxyls on one side and all O(6)H's on the other side. Adjacent sheets face each other with lateral displacements so that cavities of α-CD's are closed on both sides by adjacent molecules (Fig. 7(c)).

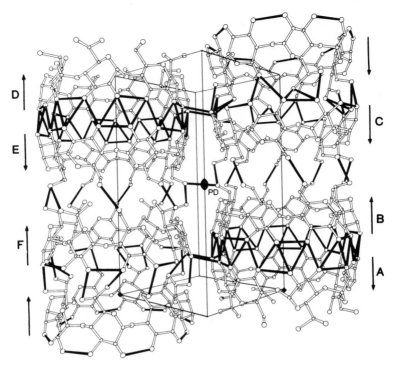

Fig. 8. The channel-type structure of γ-cyclodextrin.X(n-propanol).YH$_2$O. Water molecules located between the γ-CD stacks omitted for clarity, n-propanol molecules within the cavity could not be positioned due to disorder. Channel axes coincide with fourfold rotation axes (space group P4), i.e. each γ-CD has fourfold symmetry and the crystal structure is composed of six quarter molecules labelled A, B, C, D, E, F. Arrows indicate the orientations of γ-CDs, parallel orientation being associated with head-to-tail, and antiparallel with head-to-head interactions.[41] Hydrogen bonds are drawn solid.

3.3. What are the conditions for channel or cage type formation?

The answer is that we know with certainty only for α-CD. As discussed in an early survey of crystalline α-CD inclusion complexes,[48] small molecular guests form cages whereas long and ionic guests prefer channels. This behaviour is especially obvious with carboxylic acids, α-CD complexes with acetic, propionic and butyric acid crystallize in cages whereas valeric acid and higher analogues form channels—a clear and well-defined size-selectivity. As far as ionic compounds are concerned, molecular acetic acid crystallizes in a cage structure contrasting with the inclusion complex of the salt, potassium acetate, which produces a channel structure. In the latter, acetate ions and water molecules are located within the channel (and are disordered) and potassium ions hydrated by water are located in external interstices.

The situation for β- and γ-CD appears to be different and clear separation into channel and cage forming guests is, thus far, not possible.[49] In any case, hydrates of "empty" β- and γ-CD as well as methanol, ethanol, and hydroiodide complexes of β-CD crystallize to give the cage type of complex, whereas *n*-propanol, which easily fits into the cavity, induces a change to the channel type for β- and γ-CD, in contrast to α-CD which crystallizes in the cage type with *n*-propanol. It appears, therefore, that generalizations can be made only for one kind of CD and extrapolations to the other CD's are not possible.

Is there any rationale for the preferred channel complex formation of β- and γ-CD? As can be seen from Table 1, it becomes clear that the channels formed by α-CD are mostly of the head-to-tail (HT) form and only in one case, the polyiodides, head-to-head (HH). In contrast, in all β-CD channel structures the β-CD's are arranged HH and in the γ-CD.X(*n*-propanol).Y H$_2$O complex, an alternating HH/HT distribution is found. It appears likely that, as proposed for the α-CD.polyiodides,[36] dimer formation with β-CD's occurs already in aqueous solution with appropriate guests, the β-CD's being hydrogen bonded head-to-head (HH) *via* O(2)H/O(3)H hydroxyl groups. These "baskets" contain the guest and crystallize in linear aggregates to produce channels. Arguments for the "basket" hypothesis are the many channel-type structures exhibiting different molecular packing as expressed by space group symmetries. Against, one could argue that these dimers should also crystallize in other than the channel form, but this has not yet been observed.

In the following, let us look more closely at some channel and cage type crystal structures and discuss the most salient features of a few characteristic examples.

4. The "empty" cages formed by hydrated cyclodextrins and the "induced-fit" type complex formation of α-cyclodextrin

If α-, β- or γ-CD are crystallized from water, hydrates are obtained with overall formulae given in Table 1. In these, cavities and interstices between molecules are filled by water molecules which are statistically disordered in the "round" α-CD.7.57H$_2$O (Form III) and in β- and γ-CD, whereas they are in fixed positions in the "collapsed" α-CD.6H$_2$O (Forms I and II).

As thermodynamic studies have indicated, α-CD is able to form four different hydrates,[50] three of which have been analyzed crystallographically.[11,12,13] In one of them,[13] the α-CD macrocycle adopts a "round" shape with all O(2)H$\cdots$O(3)H hydrogen bonds formed and 2.57 water molecules enclosed and statistically distributed over 4 disordered sites. In the other two complexes, all water molecules are fully ordered, yet the α-CD torus is partly collapsed by rotation of one glucose unit out of alignment with the other five. This movement disrupts two of the six O(2)H$\cdots$O(3)H hydrogen bonds, brings the O(6)H "inside" to hydrogen bond to one of the included water molecules and, as a consequence, reduces the volume of the cavity (Fig. 9).

The water molecule bound to O(6)H is in hydrogen bonding contact with another water close to the O(2)H/O(3)H rim in α-CD.6H$_2$O (Form I); in

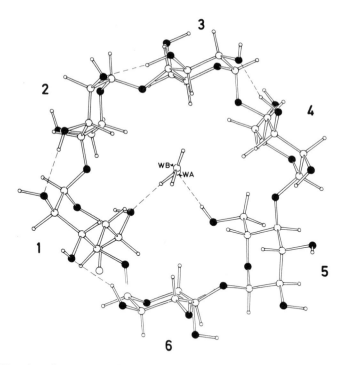

Fig. 9. Results of a neutron diffraction study of α-cyclodextrin.6H$_2$O (Form I). All hydrogen atoms are drawn and hydrogen bonding interactions are indicated by dashed lines. WA and WB mark enclosed water molecules, WA being hydrogen bonded to two O(6)H groups whereas WB is bonded to WA and to an adjacent cyclodextrin molecule, not to the enclosing one.[11]

Form II, this second water is replaced by the O(6)H of an adjacent α-CD, thus forming a self-complex.

Based on these crystal structures, a mechanism for inclusion compound formation of α-CD, reminiscent of the induced-fit in enzyme-substrate interaction, has been proposed.[51] If a substrate occupies the void, the ring of six O(2)H···O(3)H hydrogen bonds is formed and α-CD adopts a "round" structure corresponding to minimum potential energy, the "relaxed" state (Fig. 10). In the "empty" α-CD, with water enclosed in the cavity, the torus is collapsed and the ring of hydrogen bonds is disrupted; the torsion angles about glycosyl C(4)–O(4) and O(4)–C(1'), bonds indicate steric strain[11] and therefore this state is called "tense". The inclusion of a substrate S can

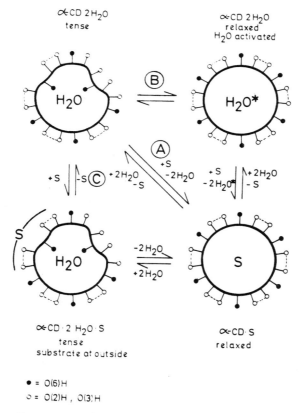

$\bullet$ = O(6)H

$\circ$ = O(2)H , O(3)H

Fig. 10. Schematic representation of the "induced fit"-type complex formation of α-CD. S = substrate, H_2O^* = "activated water". Hydrogen bonds marked by broken lines, $\circ$ = O(2)H/O(3)H hydroxyls, $\bullet$ = O(6)H hydroxyls.[51]

proceed *via* routes A, B or C (Figure 10). In A, cavity water is directly replaced by the substrate whereas B involves first the formation of a "relaxed", round α-CD with enclosed "activated" water, which is then expelled by the substrate. The term "activated water" was chosen because disordered water cannot satisfy its hydrogen bonding potential and therefore exists in a higher energy state compared with bulk water. (This kind of α-CD hydrate was directly observed in Form III crystals). In route C, the substrate first binds "outside" and in a second step enters the cavity.

This kind of mechanism is also supported by circular dichroism data showing that α-CD undergoes conformational change when forming a complex.[52] For β-CD, such a change was not detected, in agreement with crystallographic analysis which revealed "round" β-CD with 6.5 water molecules inside the cavity disordered over 9 sites—the activated water (Fig. 11). It is therefore assumed that β-CD does not change conformation as drastically as α-CD when inclusion occurs and for γ-CD the same appears to hold. This "rigid" behaviour of β- and γ-CD, as outlined above, is probably caused by the ring of rather stable O(2)H$\cdots$O(3)H hydrogen bonds, as already discussed in Section 2.

5. Order and disorder in the cage

What happens if guest molecules of different size and shape are enclosed? Starting with guests which fit snugly into the cavity, *e.g.* benzene rings in α-CD, we find that they are, in general, well ordered and even lead to elliptical distortions of the host, with the O(4) hexagon elongated by ~ 0.8 Å. This is especially well documented for the channel-type methyl orange complex and the brick-type cages formed with *p*-disubstituted benzene derivatives. In the latter, the host–guest fit is so tight that the intermolecular distances $H_{aromatic}\cdots H_{\alpha\text{-CD}}$, around 2.2 Å, are 0.2 Å shorter than expected from the sum of the respective van der Waals radii (Fig. 12).

Of particular interest are crystal structure analyses of *p*- and *m*-nitrophenol (Fig. 13). The former crystallizes to give a brick-type cage structure and the latter in a channel array with one *m*-nitrophenol molecule enclosed, while another is located "outside" and between α-CD molecules. The penetration in both cases is with the nitro group head-on into the cavity from the O(2)H/O(3)H side, with the OH group sticking out. Insertion is deeper for the *para*- than for the *meta*-isomer because the hydroxyl group in the latter interferes sterically with the α-CD hydroxyls O(2)H/O(3)H. This finding lends support to the interpretation of kinetic data for CD-catalysed hydrolysis of *para* and *meta* substituted phenylesters.[9]

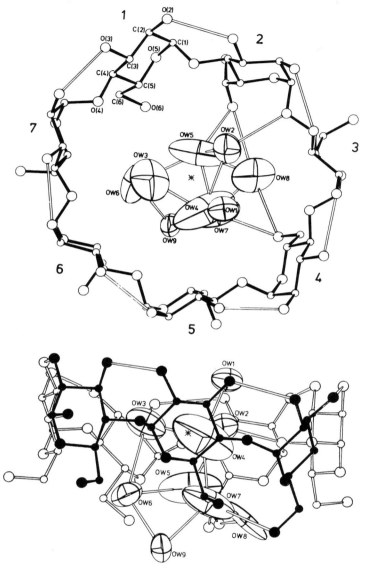

Fig. 11. The "round" structure of the "empty" β-cyclodextrin.12H$_2$O. Within the cavity, 6.5 water molecules are distributed statistically over 9 sites. They are drawn as thermal ellipsoids (50% probability) in order to show that they display excessive motion and are not well defined. The β-cyclodextrin torus exhibits a "round" structure with all O(2)H···O(3)H hydrogen bonds formed, contrasting with α-cyclodextrin.6H$_2$O (Fig. 9).[18]

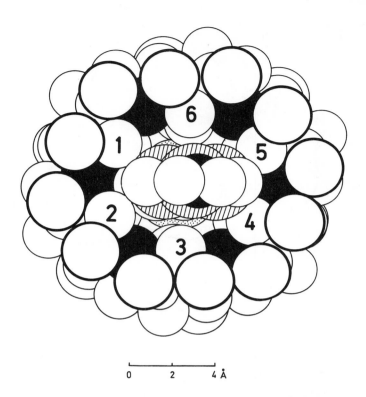

0   2   4 Å

Fig. 12. Structure of the α-cyclodextrin.*p*-iodoaniline complex, showing elliptical distortion of the macrocycle.[25]

For β- and γ-CD, no obvious distortions of the O(4) hepta- and octagons have been observed so far, probably because benzene rings just fit into these larger cavities and other, more bulky guests have not yet been studied.

If guests are smaller than the diameter of the cavity, they tend to be disordered; this is found for all three CD's. In α-CD, the cage-type structures of small guests, like methanol or *n*-propanol, show disorder over two statistically half occupied sites, with alcohol hydroxyls hydrogen-bonded to O(6)H groups which are then rotated "inside" and in the + *gauche* orientation. With the noble gas Kr, an atom slightly larger than the water molecule, α-CD does not adopt the collapsed form, with a diminished cavity, so characteristic for α-CD.6H$_2$O (Forms I and II). Rather it is "round" and the Kr atom of 4 Å diameter is statistically disordered over 6 different sites filling the 5 Å wide cavity.[17]

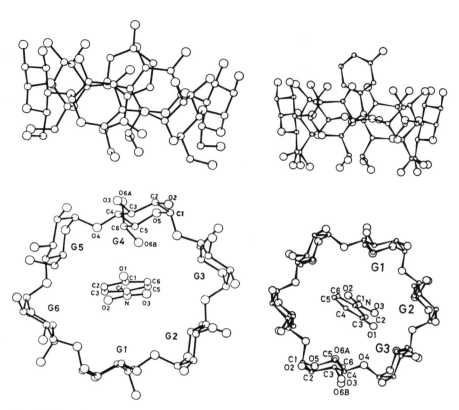

Fig. 13. Results of crystal structure analyses of p- and m-nitrophenol complexed by α-cyclodextrin. In both cases, the nitro group enters from the O(2)H/O(3)H side, yet the guests penetrate differently due to steric hindrance in case of the *meta*-isomer.[23,29]

6. Channel-type complexes are mostly disordered

In general, the channel-type CD complexes so far investigated display disorder. However, some cases exist, such as the β-CD.p-nitroacetanilide complex, where the guest is well ordered within the channel matrix. The reason might be that in this particular case, the guest molecule is asymmetric and wedges into the channel such that adjacent guests are in close contact, with nitro...nitro and acetyl...acetyl distances around 3.4 Å (Fig. 14).

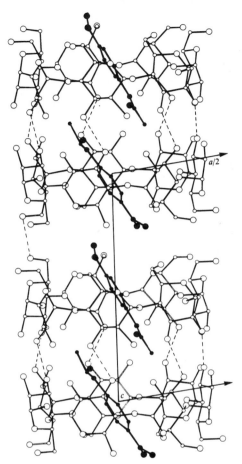

Fig. 14. Channel-type complex formed by β-cyclodextrin. *p*-nitroacetanilide. Note the short distances between adjacent nitro- and acetyl groups ($\sim$3.4 Å).[38]

In other channel-type complexes, more or less defined disorder exists, be it with guests as large as methyl orange, γ-aminobutyric acid (GABA) and *m*-nitrophenol in α-CD, or as small as acetate in α-CD or *n*-propanol in γ-CD. The latter represents an especially "pathological" case because only the host matrix and water molecules outside the channel could be located whereas the contents of the channel remained obscure due to excessive disorder (Fig. 8).

It should be noted that α and γ-CD possess internal rotational symmetry which, in principle, could coincide with space group symmetries of respective crystal structures. This, however, is only rarely observed, for instance in α-CD potassium acetate, in $(\alpha\text{-CD})_2.\text{Cd}_{0.5}\,\text{I}_5.26\text{H}_2\text{O}$, and in α-CD.methyl orange which display twofold symmetry. In γ-CD.X(n-propanol).YH$_2$O, even a fourfold axis aligns with the γ-CD torus axis. The three- or sixfold symmetry for α-CD has not been observed except for $(\alpha\text{-CD})_2\,\text{BaI}_2.\text{I}_2.12\text{H}_2\text{O}$ of which only the packing scheme is known, suggesting that α-CD displays truly hexagonal symmetry and BaI$_2$.I$_2$ is aligned along the sixfold axis.[36]

7. Linear polyiodide in α-CD channels as a model for the blue starch-iodine complex

The polyiodide complex of α-CD has been of interest for some time because it serves as a model compound for blue starch iodine discovered in 1814.[53] The unusual colour properties of the latter are of practical as well as of theoretical interest and there have been numerous attempts to try to explain their appearance.[54] The polymer cannot be crystallized for structural studies, which should give a first indication as to possible starch···iodine interactions, but can only be obtained in the form of *quasi*-crystalline fibres. These yield sparse X-ray diffraction patterns providing only overall, averaged structural data.[55,56] It could, however, be established that along the fibre axis, polyiodide is aligned, with a mean I···I distance of 3.1 Å, and embedded into a matrix consisting of a helical starch molecule with pitch height of 7.9 Å. This helix was later defined more closely as left-handed, with six glucoses per turn, *i.e.* the overall dimensions of one turn starch helix and of α-CD are comparable.[5]

If α-CD is cocrystallized with molecular iodine from aqueous solution, brownish, needle-shaped crystals are obtained with I$_2$ included in a herringbone cage-type matrix. However, if iodide is added to this solution, near-black crystals with a blue, green or golden metallic lustre are formed. A systematic survey of such crystals grown with different cations indicated that, depending on the type of cation, triclinic, tetragonal, and hexagonal space groups are formed. They all contain channel-type α-CD.polyiodide complexes which are aggregated in a hexagonal close packing arrangement, or in a square array, or in layers of parallel channels, the latter being stacked and rotated by 120° with respect to the next layer.[57]

In the two crystal structures investigated in greatest detail, Li$^+$ and Cd^{2+} served as counterions (Fig. 15). In the Li$^+$ complex, α-CD's are arranged head-to-head with molecules tilted ~8° with respect to the channel axis, whereas they are exactly perpendicular to the axis when Cd^{2+} is the counterion because the channel axis coincides with a crystallographic 4$_2$ axis. Iodine

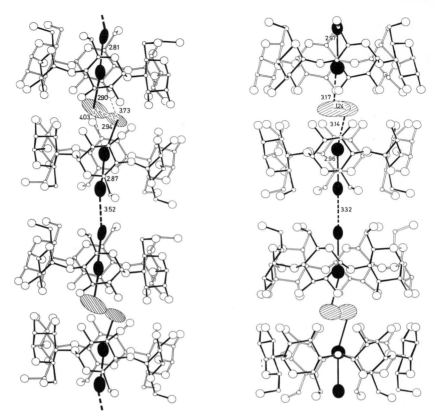

Fig. 15. Simplified view of the channel structure formed by lithium (left) and cadmium (right) polyiodide complexes with α-cyclodextrin. For formulae see Table 1 $\circ$ = C, $\bigcirc$ = O, $\bullet$ = I, $\oslash$ = disordered iodine close to the wide O(2)H/O(3)H side of the α-cyclodextrin torus. Iodine atoms drawn with their thermal ellipsoids. I$\cdots$I distances indicated in Å units. In the Li$^+$ complex, polyiodide is best described as alternating I$_3^-$ I$_2$ I$_3^-$ I$_2$... whereas in the Cd^{2+} complex, it is rather I$_5^-$ I$_5^-$ ... Taken from ref. 36.

atoms are located within the channel and form infinite I$_2$ I$_3^-$ I$_2$ I$_3^-$ chains in the Li$^+$ complex whereas I$_5^-$ I$_5^-$ predominates in that of Cd^{2+}, and these units also appear to prevail in the starch iodine complex, with the I$_5^-$-chain preferred as demonstrated by spectroscopic methods.[58,36] In both cases, the polyiodide chains display some disorder, with atoms close to the wider O(2)H/O(3)H rim of the α-CD torus distributed over two sites ~1.2 Å apart whereas all the other iodines within the α-CD cavities are fully ordered. Cations are located outside the α-CD cylinder, and are coordinated with

O(2)H, O(3)H and water in the case of Li^+ and only with water for the Cd^{2+} case. This difference in cation coordination leads to and explains the observed variety of α-CD.polyiodide crystal structures.

The deep colour of these complexes must be attributed to charge transfer *along* the polyiodide chain, facilitated by $I\cdots I$ distances in the range 2.9 Å to 3.5 Å (average 3.1 Å as in starch iodine). These distances are much shorter than expected from the sum of the van der Waals radii, 4.3 Å. There is no obvious interaction between iodine and the surrounding α-CD that could cause the colour (studied spectroscopically[54]), because all $I\cdots O$ distances exceed van der Waals contacts. The only short interactions are between iodine and C–H hydrogens lining the α-CD cavity which, however, cannot produce the colour.

The situation is different if β-CD.polyiodide is considered.[58,59] Here different cations do not produce the variety of packing patterns observed for α-CD. Only one type of complex is found, with β-CD's in a head-to-head arrangement, the hosts being not directly stacked on top of each other but slightly displaced laterally to yield an oblique channel. The cavities are too wide to force the polyiodide into exact alignment. Therefore, a zigzag chain of alternating I_3^- and I_2 units is formed which does not give rise to the dark colour of the linear polyiodides but rather produces a brownish appearance.

8. Hydrated cyclodextrins display circular and flip-flop hydrogen bonds

Neutron diffraction studies of α-CD.6H$_2$O[11] and of β-CD.12H$_2$O[60] allowed the location of all the hydrogen atoms and thus gave detailed hydrogen bonding schemes for these complicated systems (with 30 hydroxyls in α-CD.6H$_2$O and 45 hydroxyls in β-CD.12H$_2$O). As an extra bonus, in α-CD.6H$_2$O a general scheme of hydrogen bonds organized into circular patterns was discovered which could later also be verified in other α-CD crystal structures.[61,12,13] In β-CD.12 H$_2$O, 25 out of the 45 hydroxyl groups are statistically disordered and lead to systems O–H$\cdots$H–O called flip-flop hydrogen bonds because of their bi-directionality discussed below.[62]

The term "circular hydrogen bonds" means that chains of O–H$\cdots$O–H$\cdots$O–H hydrogen bonds are arranged circularly to yield 4, 5, 6 and higher membered rings (Fig. 16). There are three kinds of circular hydrogen bonds, *homodromic* if all hydrogen bonds run in the same direction, *antidromic* if two chains emanate from one H$_2$O and collide at one oxygen and *hetero-dromic* for randomly oriented hydrogen bonds. The *homodromic* circles constitute a special case of the frequently observed infinite chains of

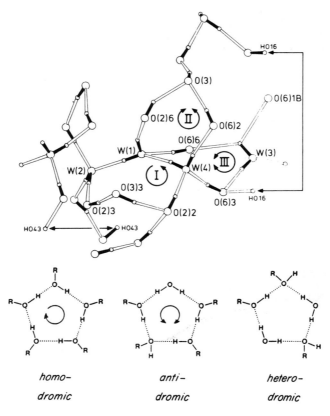

Fig. 16. Circular hydrogen bonds (left) observed in the crystal structure of α-cyclodextrin.6H₂O. Only a section of the asymmetric unit is shown, with chain-like hydrogen bonds O–H···O–H···O–H indicated by arrows, and circular structures by circular arrows. Circle I is *homodromic* and II, III are *antidromic*. ∘ = H, ○ = O, covalent O–H bonds drawn solid, O···H hydrogen bonds as open lines. Taken from ref. 12. Nomenclature of circular hydrogen bonds (right) taken from ref. 60.

unidirectional hydrogen bonds and both are energetically favoured. This is due to the cooperative effect which leads to increased hydrogen bonding activity of an OH group if it is already accepting or donating a hydrogen bond.

Flip-flop hydrogen bonds were found in β-CD.12H₂O, where 19 out of 58 hydrogen bonds are not simply O–H···O but rather O–H···H–O. Oxygens are at the usual hydrogen bonding distance between 2.8 and 3.0 Å, yet the two hydrogen atom positions are only ~1 Å apart, too close to be simultaneously present. The occupancy factors, ~0.5, also indicate that what is observed, is in fact, a dynamic system O–H···O⇌O···H–O. Since several of these O–H···H–O hydrogen bonds are linked, a change from O–H···O to the other side, O···H–O, requires cooperative, concerted change of *all* involved hydrogen bonds (Fig. 17), hence the name "flip-flop".

Combination of circular and flip-flop hydrogen bonds leads to oscillating systems which, if interconnected and fluctuating, can serve as models for hydration of macromolecules and for "flickering clusters" in bulk water. It should be stressed that the flip-flop circles are energetically favourable not only because of cooperative effects but also due to entropic contributions from the two structurally and energetically equivalent flip-flop states.

9. Increasing the extent of the hydrophobic cavity by methylation: 2,6-tetradeca-*O*-methyl-β-cyclodextrin

Thus far, only native, unmodified cyclodextrins have been discussed. The crystal structure analysis of 2,6-tetradeca-*O*-methyl-β-cyclodextrin complexed with adamantol provides the first picture of a modified species which, because of attachment of hydrophobic methyl groups at O(6)H and O(2)H hydroxyls, exhibits a markedly extended hydrophobic cavity (Figure 18). In contrast to the unmodified CD's, torsion angles O(5)–C(5)–C(6)–O(6) correspond to the + *gauche* conformation so that the O(6) atoms point towards the interior of the cavity. The attached methyl groups are directed approximately along the cavity, with torsion angles C(5)–C(6)–O(6)–Me in the *trans* conformational range. All methyl groups at O(2) point "away" from the cavity, probably because they would otherwise interfere with the guest.

It should be noted that methylated CD's display unusual solubility properties in water. They crystallize out when the solution is heated up and dissolve again at lower temperatures. This behaviour suggests hydrophobic interactions probably correlated with hydration around the methylated CD's.

10. Why study cyclodextrins?

Cyclodextrins are of more general significance for several reasons.[8] They serve as tools in the study of weak intermolecular interactions which are of importance in biochemistry but which are difficult to investigate due to the size of the molecules involved. This model character of the CD's is further extended to enzymatic processes because CD's, as will be outlined in other chapters, possess broad-spectrum catalytic activity. Going far

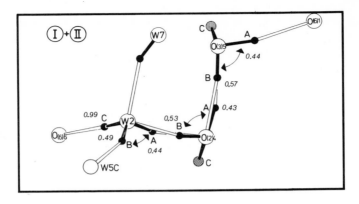

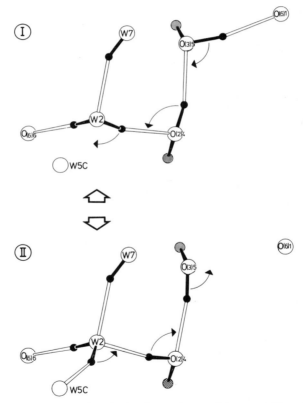

Fig. 17. Flip-flop hydrogen bonds (top) as observed in crystalline β-cyclodextrin.12H$_2$O. O–H$\cdots$H–O represent an average over O–H$\cdots$O and O$\cdots$H–O and the system is broken down into different states I and II. Arrows indicate changes of hydrogen bonds which occur in a concerted, cooperative fashion.

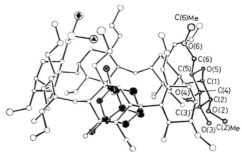

Fig. 18. Result of crystal structure analysis of 2,6-tetradeca-*O*-methyl-β-cyclodextrin.adamantol.12H$_2$O. Note the elongated hydrophobic cavity lined at both ends by methyl groups.[42]

beyond the model aspects, CD's have found application in industry because a substance, if included within the CD cavity, displays in some cases favourable properties not achieved otherwise. This is of importance in the pharmaceutical industry and a search is in progress for further uses of these interesting compounds. Chapters 11, 12, 13 and 14 in Volume 3 discuss these uses of cyclodextrins.

References

1. S. Chiba, S. Okada, S. Kitahata and T. Shimomura, *Agric. Biol. Chem.*, 1975, **39**, 2353.
2. H. Bender, *Carbohydr. Res.*, 1978, **65**, 85.
3. A. O. Pulley and D. French, *Biochem. Biophys. Res. Commun.*, 1961, **5**, 11.
4. P. R. Sundarajan and V. S. R. Rao, *Carbohydr. Res.*, 1970, **13**, 351.
5. V. G. Murphy, B. Zaslow and A. D. French, *Biopolymers*, 1975, **14**, 1487.
6. F. Cramer, *Einschlußverbindungen*, Springer, Heidelberg, 1954.
7. F. Cramer and F. M. Henglein, *Chem. Ber.*, 1957, **90**, 2561.
8. W. Saenger, *Angew. Chem., Int. Ed. Engl.*, 1980, **19**, 344.
9. M. L. Bender and M. Komiyama, *Cyclodextrin Chemistry*, Springer, Berlin, 1978.
10. W. Saenger, in *Environmental Effects on Molecular Structure and Properties*, (ed. B. Pullman) D. Reidel, Dordrecht-Holland, 1976, p. 265.
11. (a) P. C. Manor and W. Saenger, *J. Am. Chem. Soc.*, 1974, **96**, 3630. (b) For neutron diffraction study, see B. Klar, B. Hingerty and W. Saenger, *Acta Crystallogr.*, 1980, **B36**, 1154.
12. W. Saenger and K. Lindner, *Angew. Chem., Int. Ed. Engl.*, 1980, **19**, 398.
13. K. K. Chacko and W. Saenger, *J. Am. Chem. Soc.*, 1981, **103**, 1708.
14. B. Hingerty and W. Saenger, *J. Am. Chem. Soc.*, 1976, **98**, 3357.
15. W. Saenger, R. K. McMullan, J. Fayos and D. Mootz, *Acta Crystallogr.*, 1974, **B30**, 2019.

16. R. K. McMullan, W. Saenger, J. Fayos and D. Mootz, *Carbohydr. Res.*, 1973, **31**, 211.
17. W. Saenger and M. Noltemeyer, *Chem. Ber.*, 1976, **109**, 503.
18. K. Lindner and W. Saenger, *Carbohydr. Res.*, 1982, **99**, 103.
19. K. Lindner and W. Saenger, *Carbohydr. Res.*, 1982, **107**, 7.
20. R. Tokuoka, M. Abe, T. Fujiwara, K.-I. Tomita and W. Saenger, *Chem. Lett.*, 1980, 491.
21. J. M. Maclennan and J. J. Stezowski, *Biochem. Biophys. Res. Commun.*, 1980, **92**, 926.
22. K. Harata, *Bull. Chem. Soc. Jpn.*, 1978, **51**, 1644.
23. K. Harata, *Bull. Chem. Soc. Jpn.*, 1977, **50**, 1416.
24. K. Harata, *Bull. Chem. Soc. Jpn.*, 1975, **48**, 2409.
25. W. Saenger, K. Beyer and P. C. Manor, *Acta Crystallogr.*, 1976, **B32**, 120.
26. W. Saenger and H. Brand, unpublished results.
27. K. Harata, *Carbohydr. Res.*, 1976, **48**, 265.
28. K. Harata, *Bull. Chem. Soc. Jpn.*, 1979, **52**, 2451.
29. K. Harata, H. Uedaira and J. Tanaka, *Bull. Chem. Soc. Jpn.*, 1978, **51**, 1627.
30. K. Harata, *Bull. Chem. Soc. Jpn.*, 1977, **50**, 1259.
31. K. Harata, *Bull. Chem. Soc. Jpn.*, 1976, **49**, 2066.
32. K. Harata, *Bull. Chem. Soc. Jpn.*, 1976, **49**, 1493.
33. A. Hybl, R. E. Rundle and D. E. Williams, *J. Am. Chem. Soc.*, 1965, **87**, 2779.
34. K. Harata, K. Uekama, M. Otagiri, F. Hirayama and H. Ogino, *Bull. Chem. Soc. Jpn.*, 1981, **54**, 1954.
35. R. Tokuoka, M. Abe, K. Matsumoto, K. Shirakawa, T. Fujiwara and K.-I. Tomita, *Acta Crystallogr.*, 1981, **B37**, 445.
36. M. Noltemeyer and W. Saenger, *J. Am. Chem. Soc.*, 1980, **102**, 2710.
37. J. A. Hamilton, M. N. Sabesan, L. K. Steinrauf and A. Geddes, *Biochem. Biophys. Res. Commun.*, 1976, **73**, 659.
38. M. M. Harding, J. M. Maclennan and R. M. Paton, *Nature (London)*, 1978, **274**, 621.
39. J. J. Stezowski, K. H. Jogun, E. Eckle and K. Bartels, *Nature (London)*, 1978, **274**, 617.
40. Ch. Betzel, M. Noltemeyer, G. Weber and W. Saenger, to be published.
41. K. Lindner and W. Saenger, *Biochem. Biophys. Res. Commun.*, 1980, **92**, 933.
42. M. Czugler, E. Eckle and J. J. Stezowski, *J. Chem. Soc., Chem. Commun.*, 1981, 1291.
43. Computer Plot Program SCHAKAL, E. Keller, Institut für Anorganische Chemie der Universität Freiburg, Freiburg, FRG.
44. S. Wolfe, *Acc. Chem. Res.*, 1972, **5**, 102.
45. A. D. French and V. G. Murphy, *Carbohydr. Res.*, 1973, **27**, 391. (b) D. A. Pensak and A. D. French, *Carbohydr. Res.*, 1980, **87**, 1.
46. B. Casu, M. Reggiani, G. G. Gallo and A. Vigevani, *Chem. Soc., Spec. Publ.*, 1968, **23**, 217.
47. R. Bergeron and M. A. Channing, *Bioorg. Chem.*, 1976, **5**, 437.
48. R. K. McMullan, W. Saenger, J. Fayos and D. Mootz, *Carbohydr. Res.*, 1973, **31**, 37.
49. J. A. Hamilton, L. K. Steinrauf and R. L. vanEtten, *Acta Crystallogr.*, 1968, **B24**, 1560.
50. N. Wiedenhof and J. N. J. J. Lammers, *Carbohydr. Res.*, 1968, **7**, 1.

51. W. Saenger, M. Noltemeyer, P. C. Manor, B. Hingerty and B. Klar, *Bioorg. Chem.*, 1976, **5**, 187.
52. D. A. Rees, *J. Chem. Soc., B*, 1970, 877.
53. C. de Claubry, *Ann. Chim. (Paris)*, 1814, **90**, 87.
54. W. Banks and C. T. Greenwood, eds. Starch and its Components, Edinburgh University Press, 1975.
55. R. E. Rundle and D. French, *J. Am. Chem. Soc.*, 1943, **65**, 558 and 1707.
56. R. E. Rundle and F. C. Edwards, *J. Am. Chem. Soc.*, 1943, **65**, 2200.
57. F. Cramer, U. Bergmann, P. C. Manor, M. Noltemeyer and W. Saenger, *Justus Liebigs Ann. Chem.*, 1976, 1169.
58. R. C. Teitelbaum, S. L. Ruby and T. J. Marks, *J. Am. Chem. Soc.*, 1978, **100**, 3215.
59. Ch. Betzel, M. Noltemeyer, G. Weber and W. Saenger, *J. Incl. Phenom.*, 1983, **1**, 181.
60. Ch. Betzel, B. Hingerty and W. Saenger, to be published (1984).
61. W. Saenger, *Nature (London)*, 1979, **279**, 343.
62. W. Saenger, Ch. Betzel, B. Hingerty and G. M. Brown, *Nature (London)*, 1982, **296**, 581.
63. K. Harata, K. Kawano, K. Fukunaga and Y. Ohtani, *Chem. Pharm. Bull.*, 1983, **31**, 1428.

9 · COMPLEXES OF CROWN ETHERS WITH MOLECULAR GUESTS

I. GOLDBERG

Tel-Aviv University, Ramat Aviv, Israel

1. Introduction

The discovery of the remarkable complexing properties of macrocyclic polyethers (termed "crown" ethers because of the appearance of their molecular models) toward metal cations, primary alkylammonium salts and other neutral as well as charged potential substrates has provided new challenges in chemistry during the past fifteen years. One of the primary goals was to develop methods for new design and synthesis of macrocyclic host molecules in order to better understand the process of structured molecular complexation which is central in enzyme chemistry. A major effort has been concentrated on the exploration of possible applications of crown-ether macrocycles as model compounds in reactions of stereoselective complexation, ion-transport and molecular catalysis. Indeed, hundreds or even thousands of original papers relating to various aspects of crown

INCLUSION COMPOUNDS 2
ISBN 0-12-067101-8

chemistry have been published over recent years and considerable progress
has been achieved.

Within the broad field of inclusion phenomena this survey intends to give
a description of the stereochemical features of host–guest molecular systems
formed by monocyclic crown ethers and the closely related acyclic polyether
species. Macropolycyclic systems and inclusion compounds of other natural
and synthetic ionophores are discussed in other chapters of this volume
and will not be treated here. (See Chapters 8 and 10, Volume 2.) Even within
such limitations the amount of relevant data available in the literature is
so large that any attempt to present a comprehensive and encyclopaedic
survey of all structural properties within a single chapter would obviously
be unsuccessful. In fact, while extensive reviews (some gathered in
specialists' volumes) have already been written concerning the structures
of crown compounds, it seems that every new issue of the journals contains
some novel determinations which aim at defining even more effective applications for the macrocyclic ligands. Extensive discussions of structural
studies of crown ethers and their complexes with metal cations are given
in a number of reviews.[1-3] Complexation of uncharged molecules and anions
by crown-type host molecules has recently been reviewed[4] and the general
concepts of binding metal and ammonium substrates to macrocyclic polyether ligands have been dealt with.[5] Several other review-type accounts,
some of a more concise nature, have also appeared, including one by the
present author.[6]

We have chosen therefore to include in this compilation only a selected
group of representative polyether systems with an emphasis on reports that
have appeared in recent years. In order to provide as general a view as
possible within the limited space of this chapter several different types of
ligands and complexes will be considered, most attention being drawn to
details which are relevant to the subject of inclusion phenomena. X-ray
diffraction crystallography is at present still the most convenient method
for the investigation of moderately complex molecules. It has been widely
used in accurate determinations of host-to-guest binding geometries, as such
features cannot be anticipated reliably from molecular models. We shall
consider, therefore, mainly results obtained from these studies; the discussions in other chapters of this series will undoubtedly provide complementary data from different areas of chemistry. Inclusion compounds
of the polyether ligands with uncharged molecules, organic and metal ions
or metal-to-ligand coordinated entities can be considered as host–guest
complexes. The usually smaller guest is bound to a cavity-like framework
formed by the larger polyether host in a structure reflecting a complementary
stereoelectronic arrangement of the binding sites. The experimental elucidation of detailed molecular structures seems therefore essential for the charac-

terization and understanding of the variety of binding forces by which host and guest constituents are held together in a structured way.

The first section of this chapter will illustrate the various types of complexes that form between crown ether ligands and neutral and ionic substrates; the 18-crown-6 host provided an outstanding model for a description of the most fundamental features of binding in crown ether adducts. The second section refers, with some more detail, to complexes of various crown ethers acting as host and alkylammonium ions acting as guest moieties. It will emphasize to some extent the molecular model development process that has been necessary to increase the ligand capacity for the recognition of enantiomers of a potential chiral substrate. Next, we shall deal with compounds containing either water or hydronium species as exclusive guests within the crown cavities, as well as with substituted monocyclic ligands exhibiting phenomena of intramolecular complexation. In general, we shall refer throughout this chapter to macrocyclic ligands which in their size are equal to or larger than the 18-crown-6 model. Smaller crown ethers are mostly effective in direct association with various metal cations. However, since the subject of metal ion complexation with crown ethers has already been extensively reviewed elsewhere (see above), only a small number of relevant examples will be discussed here. The last two sections of the present review are devoted to two other groups of synthetic polyether ligands which, although differing greatly in some basic characteristics of their molecular structure, also provide excellent hosts for various (mostly) cationic guests. Thus, open chain polyethers are of considerable interest because of their high conformational flexibility, a property which is inherent to natural receptors. On the other hand, hemispherands and particularly spherands have enforced conformational organization, which is not allowed to change much upon complexation, and often exhibit a significant metal ion selectivity.

For easier recognition of the cyclic ligand type, trivial names, derived from those proposed by Pedersen in his pioneering work,[7] will be used as much as possible instead of the cumbersome IUPAC nomenclature. The "crown" terminology (e.g., 18-crown-6) defining clearly the size of the macroring and the number of heteroatoms in it, although not unequivocal, is now commonly accepted in chemistry. Complex ligands will also be elucidated by drawings of their structural formulae. For simplicity, the characteristic conformational features of polyether macrocycles will, wherever possible, be described only in general terms rather than by the actual torsion angle values. Symbols such as g for *gauche* and a for *anti* will often be used to specify overall conformations about C–C or C–O bonds; g^+, g^- and a representing torsion angles from 40° to 80°, from −80° to −40° and from ±140° to 180° respectively.

2. Host–guest complexes of the macrocyclic 18-crown-6 system

The unsubstituted 1,4,7,10,13,16-hexaoxacyclooctadecane (18-crown-6) ligand (1) is one of the most extensively studied crown ethers. It reacts readily with a large variety of substrates, providing suitable structural models for a detailed analysis of types and geometries of the interaction between polyether crown hosts and various molecular guests. In many of these

(1)

complexes the 18-crown-6 molecule adopts a nearly ideal "crown" conformation with the C and O atoms lying alternately about 0.2–0.3 Å above and below the mean plane of the ring. The six oxygens are turned toward the centre of the molecule, forming a hexagonal cavity with sides of approximately 2.8 Å. The molecular conformation is characterized by D_{3d} symmetry, with all torsion angles about the C–O bonds close to 180° and about the C–C bonds close to 65–70°; correspondingly, there are six equivalent CH_2–O–CH_2–CH_2–O–CH_2 units with *anti-gauche-anti* (*aga*) conformations.

2.1. Association with metal cations

Crystal structure analyses of 18-crown-6 complexes with alkali thiocyanates have already been reviewed in great detail,[8] and we shall refer to them only briefly. In the potassium thiocyanate complex, the K^+ ion is located at the centre of the hexagon of ligating oxygen atoms (*nesting* arrangement). The Rb^+ and Cs^+ ions are too large to fit into the cavity of 18-crown-6; they adopt, therefore, a *perching* arrangement being displaced by 1.19 Å and 1.44 Å from the mean oxygen plane, in direct correspondence with their relative size. Cations smaller than K^+ (e.g., Na^+) are also expected to be entirely encircled by the macrocycle, although a completely different arrangement of binding sites may be required to optimize dipole–dipole interactions. The coordination sphere around the alkali cations is completed by the thiocyanate counter ions. The K^+ complex was found to be the most stable one, possibly because of the complementary relationships between host and guest. The ideal D_{3d} crown conformation of the ligand is also preserved, however, in complexes with the larger cations.

Interestingly, a very similar series of metal–cation complexes with 1,4,7,10,13-pentaoxa-16-thiacyclooctadecane, a closely related species to 18-crown-6 with one S heteroatom instead of O, has recently been reported.[9] In the complexes of this ligand with NaSCN, KSCN, RbSCN and $AgNO_3$, the small Na^+, K^+ and Ag^+ cations (of ionic radii 0.95, 1.33 and 1.26 Å, respectively) are accommodated within the centre of the cavity, while the larger Rb^+ ion (ionic radius 1.48 Å) is located about 1 Å above the mean plane of the ligating atoms. The various cations coordinate to all five ether oxygens of the ligand. Their interaction with sulphur depends, however, on the cation. Thus, the S-atom is involved in relatively strong and even partially covalent interaction with the "soft" Ag^+, it interacts only weakly with K^+ and Rb^+, but there is no coordination at all with the "hard" Na^+. Correspondingly, the S-atom is turned towards the centre of the ligand, more in the silver ion complex than in the potassium or rubidium complexes. On the other hand, in the sodium compound the sulphur heteroatom is directed away from the cavity. Moreover, the distributions of intramolecular distances involving S and their relation to relevant sums of the van der Waals radii indicate that in the K^+, Rb^+ and Ag^+ complexes the outer electrons on the S atom are slightly polarized toward the cation. As in the previous series of complexes with 18-crown-6, the alkali cations coordinate in the crystal to the counter ions as well.

From among the four adducts of the thia-18-crown-6 ligand, the Ag^+ complex appears to be the most stable one, probably because of the strong interaction of the soft Ag^+ cation with the soft S atom. A very similar reasoning can be applied to explain the high stability constant found for the Pb^{2+} complex with 1,10-dithia-18-crown-6.[10] Effects of other heteroatoms on the binding properties of crown ethers will be discussed below.

2.2. Guests with –CH_3 and $\geq CH_2$ coordinating entities

Crystalline complexes of 18-crown-6 with a variety of neutral organic guests are known. The first crystal structure analysis reported in the literature was that of the 18-crown-6 adduct with dimethyl acetylenedicarboxylate[11] (Fig. 1). The overall stoichiometric ratio between host and guest in this complex is 1:1. However, the formal definition of the stoichiometry is in a sense misleading, because the centrosymmetric bifunctional guest species extends between and interacts simultaneously with two 18-crown-6 rings. All six oxygen atoms of each ligand seem to participate in binding the terminal methyl groups (the coordinating entities) of

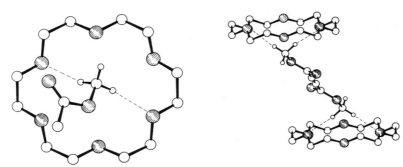

Fig. 1. Illustration of the complex between 18-crown-6 and dimethyl acetylenedicarboxylate; in the crystal, both molecules are located on inversion centres. View down the normal to the best plane of the ligand (left) shows only one half of the substrate. The shaded atoms represent an oxygen, as in most of the figures throughout this chapter.

dimethyl acetylenedicarboxylate molecules which approach opposite sides of the crown in a symmetric manner. We note that these methyl groups are in a perching position with respect to the crown ring, their C atoms lying 1.89 Å above and below the mean plane of the six ligating oxygens; the observed C(guest)$\cdots$O(host) distances vary between 3.09 and 3.37 Å. Two of the hydrogens on each methyl group are directed towards the ring, forming relatively short O$\cdots$H contacts (2.32 and 2.45 Å) and a nearly linear C–H$\cdots$O arrangement. The third methyl hydrogen is not involved in direct interaction and points away from the crown. Since the methyl groups are certainly activated because they are bound to electron-withdrawing acetylenecarboxylate fragments, their interactions could perhaps be regarded as weak hydrogen bonds.

As shown in Fig. 1, the stoichiometry of the host–guest interaction in this compound is best described as involving one molecule of the polyether and two coordinating units of the guest species. It will be shown in the following examples that many crystalline complexes of 18-crown-6 with uncharged and partially charged guests are characterized by the 1:2 stoichiometry of interaction, where two substrate units converge on a single hexaether ring from opposite sides. Guest moiety located above the mean plane of the ring usually interacts with the upper triangle of the crown-ether oxygens, while that below the ring attracts lone-pairs from the lower triangle of the oxygen nucleophiles.

The structure of the 1:2 18-crown-6 adduct with dimethylsulphone provides another example of a host–guest complex in which methyl groups of the substrate interact directly with the crown.[12] Figure 2 illustrates the geometry of the interaction between two molecules of dimethylsulphone

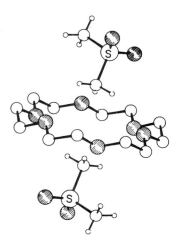

Fig. 2.

and one molecule of 18-crown-6. The methyl carbons are 1.63 Å distant from the mean plane of the hexaether cavity, all three methyl hydrogens being apparently involved in C–H⋯O interactions with alternate O-atoms of the crown. The observed H⋯O distances are within 2.5–2.6 Å, and the H⋯O line points nearer the trigonal than the tetrahedral direction at oxygen. A rather similar intermolecular arrangement and interaction pattern was found in the crystal structure of the 1:2 18-crown-6 complex with nitromethane[13] (Fig. 3), where each methyl group of the guest species is in close contact (3.25–3.33 Å) with three alternate oxygens of the host. Interestingly, studies of thermodynamic constants in solution indicated that nitromethane forms both 1:1 and 2:1 complexes with 18-crown-6. The

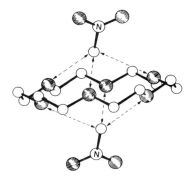

Fig. 3.

former appears to be considerably more important in the solution; nevertheless, only 2:1 complexation occurs in the solid state. Similar observations have been made with respect to complexes with malononitrile.

Coordination of organic guests to crown ether hosts via the methylene ($\gtrless CH_2$) entity is represented by the structure of the 2:1 adduct between malononitrile and 18-crown-6 which has also been determined from X-ray data.[14] Molecules of the ligand are located in the crystal on centres of inversion; the two guest molecules interacting with opposite faces of the crown are related to each other by this symmetry. The three molecular constituents of this complex are held together primarily by hydrogen bonds between the acidic C–H groups of malononitrile and the ether oxygens (Fig. 4). The geometric features of this association are in fact very similar

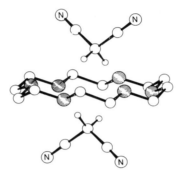

Fig. 4.

to those found in the dimethyl acetylenedicarboxylate complex. In both structures the two coordinating CH moieties are bound to a pair of centrosymmetrically related O atoms. Each one of these oxygens is thus involved in two weak hydrogen bonds and has an approximate tetrahedral arrangement of neighbouring atoms. The convergence of the remaining oxygen sites with respect to the crown centre is, apparently, stabilized by dipole–dipole interactions.

Molecular structures of the hexaether host are characterized by very similar conformational features in the four preceding examples. The 18-crown-6 molecule is centrosymmetric and has approximate D_{3d} symmetry. All torsion angles about the C–C bonds are nearly synclinal (absolute values are between 67 and 74°); those about the adjacent C–O bonds are nearly antiplanar (values between 169 and 180°). In fact, the D_{3d} symmetry of 18-crown-6 has been observed in a large variety of its crystalline complexes, irrespective of the nature of the interacting guests and the adduct stoichiometry (see below).

2.3. Guests with uncharged –NH₂ coordinating entities

Urea, phenyl carbamate, benzenesulfonamide, *p*-nitroaniline, 2,4-dinitro-aniline and 2,4-dinitrophenylhydrazine are substrates that contain uncharged $-NH_2$ group as a potential coordinating entity. Their coordination to 18-crown-6 in the solid has recently been reported.

The overall geometry of the 1:2 complex formed between 18-crown-6 and phenyl carbamate[15] (Fig. 5) is very similar to the structure shown in Fig. 4. The crown ligand occupies a centre of inversion in the crystal and exhibits D_{3d} symmetry. The two centrosymmetrically related guest molecules approach from both sides of the ether moiety, each amino group forming nearly linear N–H···O hydrogen bonds with two nucleophilic sites of the ring. A similar type of association is present in the *p*-nitroaniline complex[16] of 18-crown-6. The crown and the two nitroaniline molecules are held together by four N–H···O bonds (Fig. 6). The geometry of these bonds is slightly distorted, however, due to a partial disorder of one of the ether O atoms in the structure.

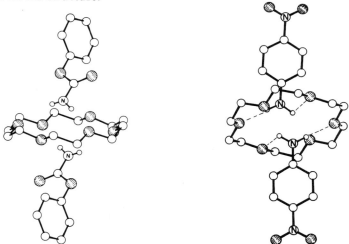

Fig. 5.   **Fig. 6.**

2,4-Dinitrophenylhydrazine also forms a 2:1 adduct with 18-crown-6 (Fig. 7). The complex formed is extremely stable which makes it useful as a potential reagent of phase transfer catalysis in reactions with ketones and aldehydes.[17] The stability of the crystals has been attributed to the acidic nature of the hydrazine protons in this substrate, particularly since the closely related 18-crown-6 complexes with phenylhydrazine and *p*-nitro-phenylhydrazine were found to be less stable. Within the crown the torsion

Fig. 7.

angles about C–C and C–O bonds average 62° and 171°, respectively; these values deviate only slightly from those characteristic of 18-crown-6 with an ideal D_{3d} symmetry. The approach of the N–N hydrazine bonds to the molecular plane of the ligand is at an angle of about 36°. This allows both nitrogen atoms to act as hydrogen donors, while all oxygen atoms of the crown molecule act as hydrogen acceptors. Two of the hydrogen bonds also appear to be bifurcated. Relevant N···O and H···O distances range from 3.0 to 3.2 Å and 2.0 to 2.3 Å, respectively, values which usually represent weak hydrogen-bonding interactions.

The first crystal structure analysis of a crown-ether complex with urea was reported by Harkema *et al.*[18] The crystal structure consists of alternating layers of urea-18-crown-6 adducts and uncomplexed urea molecules. While the overall stoichiometry is one molecule of the polyether and 5 molecules of urea, the stoichiometry of the host-to-guest interaction remains at 1:2 as in the previous examples. Each crown ether ring is linked via four N–H···O hydrogen bonds to two urea molecules which interact from opposite sides of the macrocycle, as shown in Fig. 8. The observed conformation of the ligand differs from the approximate D_{3d} geometry found in other structures (see above); the sequence of torsion angles within the O–CH$_2$–CH$_2$–O units does not correspond exclusively to an *anti-gauche-anti* conformation. In fact, two ethyleneoxy fragments positioned transannularly within the 18-crown-6 rings adopt a *gauche-gauche-anti* arrangement, which imparts an elliptical shape to the ring.

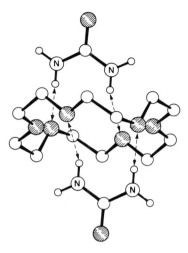

Fig. 8.

A similar biangular conformation of the crown rings was found in the
1 : 2 adduct of 18-crown-6 with benzenesulphonamide where the two guest
molecules related by inversion at the centre of the ligand are bound to it
by two relatively strong and two weak N–H···O hydrogen bonds[19] (Fig. 9).
The corresponding H···O distances are 2.0 and 2.3 Å, respectively. All
torsion angles about the C–C bonds are close to 67°, those about the C–O
bonds are generally near 180° with two exceptions where torsions of 72°
corresponding approximately to a *gauche* arrangement were observed.

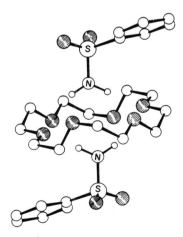

Fig. 9.

The 2:1 complex of 2,4-dinitroaniline with 18-crown-6 provides another example of a structure in which the crown ligands adopt the biangular strained conformation.[20] This crystal structure was found to be partially disordered; nevertheless, the determination of the overall molecular geometry was sufficiently reliable to show that four ethyleneoxy units have *aga* arrangements while two other units in the ring adopt *gga* conformations. The two guest molecules are positioned perpendicularly above and below the plane composed of the ligating oxygens; the intermolecular hydrogen bonding pattern also contains apparently bifurcated N–H···O bridges (Fig. 10).

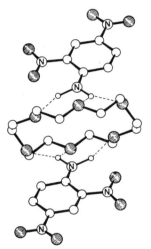

Fig. 10.

2.4. Coordination of metal-ligand assemblies

Several crystalline adducts have been reported between 18-crown-6 and neutral ammine ligands coordinated to transition metals or cations.[21] In the 1:2 complex of 18-crown-6 with *trans*-PtCl$_2$[P(CH$_3$)$_3$]NH$_3$ one molecule of the platinum ammine is bound to each face of the crown (Fig. 11). All six ether oxygen atoms appear to be involved in hydrogen bonds with the two NH$_3$ ligands, the interaction N···O distances ranging from 3.04 and 3.31 Å. Detailed geometries of the anticipated hydrogen bonds have not been reported since the hydrogen positions were poorly resolved in the difference electron density maps. A noteworthy feature in this as well as in the following two examples is the second-sphere type coordination of the crown ether by transition metals.

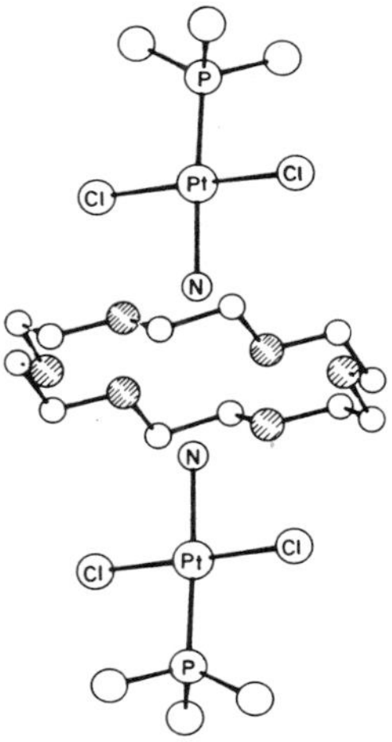

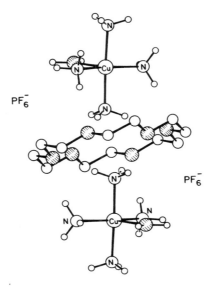

Fig. 11.

The bifunctional role of the 18-crown-6 is also reflected in its $1:1$ adduct with the $[Cu(NH_3)_4(H_2O)](PF_6)_2$ salt. The crystal structure is characterized by high symmetry with the various constituents occupying special positions. The crown molecule is located on a centre of inversion and interacts simultaneously via hydrogen bonding with two alternate $[Cu(NH_3)_4H_2O]^{2+}$ ions related by this symmetry (Fig. 12). The polyfunctional copper–complex

Fig. 12.

cation is in turn located between two adjacent crown entities, thus forming a nearly linear hydrogen-bonded polymeric chain. Of the four ammine ligands, two approach the centre of and interact directly with the adjacent crown effecting two tripods of hydrogen bonds. The other ammine functions donate one proton to each one of the rings; the aquo-ligand does not participate in hydrogen bonding interactions. In this structure the six ether oxygens participate therefore, in ten hydrogen bonds with N⋯O distances between 3.06 and 3.28 Å.

Another chain-polymeric structure formed when 18-crown-6 was interacted with the bis-(ethylenediamine)platinum(2+) hexafluorophosphate salt. The dication and the crown moieties alternate within the chains, each ammine group donating one proton to hydrogen-bond an ether oxygen atom and the other proton to hydrogen bond the PF_6^- counter ion (Fig. 13).

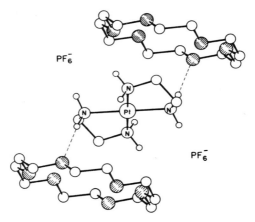

Fig. 13.

Thus, only four of the ring oxygens are involved in direct interactions with the NH_2 groups. Nevertheless, as in preceding structures, the 18-crown-6 preserves its approximate D_{3d} symmetry.

There are several publications in the literature describing coordination of 18-crown-6 to complexes of other transition metals such as hexaaquacobalt(2+)[22] (Fig. 14), hexaaquamanganese(2+)[23] (Fig. 15), pentaaquanitromanganese(2+),[24] and uranyl nitrate tetrahydrate.[25] In all of these structures, the metal species are not directly associated with or located within the crown ring. Instead, the corresponding crystals could be best described as consisting of separate 18-crown-6 and (octahedral) transition-metal-complex entities connected by hydrogen bonding through the ligands. The latter were usually neutral (although probably polarized) water molecules.

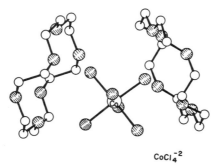

$CoCl_4^{-2}$

Fig. 14. Interaction of octahedral hexaaquacobalt assembly with 18-crown-6.

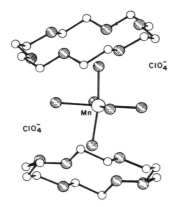

Fig. 15. Interaction of octahedral hexaaquamanganese assembly with 18-crown-6.

Thus, considering only the close surrounding of the polyether macrocycle, the above examples provide structural evidence for a direct association between the 18-crown-6 "host" and water "guest". The presence of the heavy atoms prevented, however, presentation of highly precise geometric data relating to these interactions.

Another example of second-sphere coordination of a transition-metal complex by a crown ether, in a 2:1 adduct of [*trans*-Ir(CO)(CH₃CN)-(PPh₃)₂](PF₆)₂ with 18-crown-6, has most recently been described by Colquhoun, Stoddart and Williams.[26] In this structure the interaction of the two iridium-bound acetonitrile moieties with the crown ring is through the acidic methyl group C–H bonds. The methyl entities approach the crown ring from both sides, being interrelated by inversion at the centre of the ring. Corresponding C⋯O distances of approach vary from 3.24 Å to 3.38 Å;

the distance between the two perching methyl carbons across the cavity is 3.93 Å. When the elongated acetonitrile ligand is replaced by NH_3, a similar association could not be detected (either in solution or in the solid); it appears that the bulky PPh_3 ligands on the metal prevent a close approach of the crown to the centre of the iridium complex.

In a very few cases a transition metal was found to interact directly with 18-crown-6, resembling the well known series of the alkali metal-crown complexes described by Dunitz and coworkers.[8] A suitable example involves inclusion of the UCl_3^+ cation within the centre of 18-crown-6, which is accompanied by a considerable distortion of the crown conformation from D_{3d} symmetry.[27] A similar inclusion of lanthanide cations has been found in the 1:1 complexes of 18-crown-6 with $Nd(NO_3)_3$[28] and with $La(NO_3)_3$.[29] The two complexes are isostructural. The metal cation lies in the centre of the crown ring, being 12-coordinated to the six ether O-atoms and the three bidentate nitrato groups approaching from both sides of 18-crown-6. In these structures the macrocycle adopts an unusual boat conformation with one nitrate ion approaching the more sterically hindered side, and the two remaining nitrates approaching the opposite side. The shape of the ligand cavity is adjusted (by small changes in the conformation) to the slightly different ionic radii of Nd^{3+} (0.99 Å) and La^{3+} (1.06 Å). Insertion of proton-donating aquo ligands to the coordination sphere of another lanthanide ion (Gd^{3+}) leads, however, again to a second-sphere-type coordination of the crown (see above). Thus in the 1:1 adduct of 18-crown-6 with $Gd(NO_3)_3(H_2O)_3$ the metal cation does not penetrate into the ring. Rather, 18-crown-6 forms hydrogen bonds to the coordinated water molecules.

2.5. Guests with partially charged $-NH_2$ coordinating entities

The 1:2 stoichiometric relationship of interacting host and guest species in the crystal has also been found in several compounds in which the coordinating entity of the substrate is partially charged. A complex of 18-crown-6 with 2 mol of guanidinium nitrate was prepared from a mixture of ethanol solutions of both constituents.[30] The guanidinium ion has three equivalent canonical forms and the positive charge is thus equally distributed between the three $-NH_2$ groups. Interestingly, the 1:2 association of the crown with a guanidinium ion observed in the solid phase (Fig. 16) could not be detected in solution. Instead, calorimetric measurements in methanol solution indicated that the host-to-guest interaction ratio is 1:1, even in experiments in which the concentration of the guest exceeded that of the host species by a factor of four.

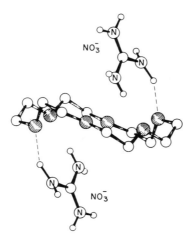

Fig. 16.

Extraction of an aqueous solution of *S-t*-butylthiouronium perchlorate with a solution of 18-crown-6 in chloroform led also to the formation of a 2:1 crystalline complex[31] (Fig. 17). As in many of the previous examples, the molecular complex is centrosymmetric in the crystal, with the 18-crown-6 molecules preserving approximate D_{3d} symmetry. Only one NH_2 group of each guest ion is hydrogen bonded to two oxygens of the crown ring (relevant $N \cdots O$ distances are 2.90 and 2.97 Å); the other $-NH_2$ is bound to the perchlorate anions ($N \cdots O$ distances vary between 3.03 and 3.23 Å). Undoubtedly, the second interaction plays an important role in stabilizing the

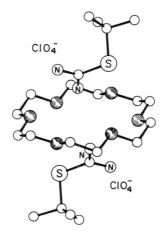

Fig. 17.

2:1 structure by reducing electrostatic repulsions between the two cations attached to opposite faces of a single crown ring. Incidentally, the authors of this work reported in the same paper that by changing concentrations of the host and guest components in the solution mixture it was possible also to form a 1:1 complex of *t*-butylthiouronium perchlorate with 18-crown-6. Unfortunately, detailed structural data have not been published since this structure turned out to be disordered.

The most recent presentation of the Dutch group[32] deals, however, with structures of *crystalline* 1:1 complexes between 18-crown-6 and uronium nitrate or uronium picrate. Preliminary results show that the hydrogen bonding pattern in the structures involves both $-NH_2$ groups which donate three protons to the ether oxygens. Another N–H···O and short O–H···O (2.48 Å) hydrogen bonds are present between cation and anion. The conformation adopted by the crown to optimize the bonding interactions consists of two sets of $ag^+a\,ag^-a\,ag^+g^+$ units and is somewhat unusual. Consequently, the two faces of the macroring seem to have very different complexing properties which makes the coordination of another guest to the same host unfavourable.

2.6. Interaction with ammonium and alkylammonium substrates

Lipophilization of ammonium and alkylammonium salts by crown ethers is principally due to host–guest complexation through a tripod arrangement of NH···O hydrogen bonds. An idealized scheme of this association with a hexaether crown ring, as anticipated from space-filling molecular models, suggests that three alternate oxygens are involved in binding the $-NH_3$ group which centres into the macroring (2). Since the remaining O atoms

(2)

also converge on the guest, it is assumed that their interaction with the ammonium cation is also important. In fact, theoretical calculations on simple model systems indicate that the energy of direct N^+···O interaction is smaller than that of the ^+N-H···O hydrogen bonding only by a factor of three.[33]

In all the observed structures involving ammonium guests and crown penta- and hexaether hosts the stoichiometric ratio of association is 1:1.

In complexes containing 18-crown-6 as the host species, the ammonium nitrogen is displaced between 0.11 and 1.00 Å from the mean plane of the ether oxygen atoms, the macroring conformation having approximate D_{3d} symmetry in each one of these structures. Thus, since the positive charge of the guest is localized mostly on the NH_3 group anchored to the crown, a simultaneous approach of two such cations from opposite sides of the 18-crown-6 would involve strong electrostatic repulsions unfavourable to the structure.

A typical example is provided by the crystal structure of the 1:1 ammonium bromide complex of 18-crown-6[34] (Fig. 18). The ammonium

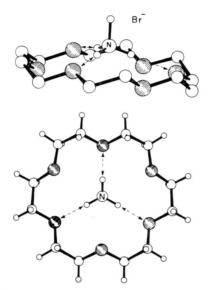

Fig. 18. Two views of the host–guest association between 18-crown-6 and ammonium bromide. Hydrogen bonds are represented by arrows and ion–dipole interactions by dotted lines.

cation is in a perching position with respect to the crown ring, being displaced by 1.00 Å from the mean oxygen plane. The complex is located on a crystallographic mirror plane, and thus shows an ideal tripod of the nearly linear N–H$\cdots$O hydrogen bonds (reported N$\cdots$O distances are within 2.84–2.86 Å). The fourth N–H bond, perpendicular to the plane of the crown, interacts with the nearest bromide ion.

The structure of the 1:1 complex of 18-crown-6 with benzylammonium thiocyanate has been described by Bovill *et al.*[35] In spite of the different shape of the guest species and absence of any crystallographic-symmetry constraints on the molecular structure, the geometry of interaction between the host and the guest (Fig. 19) is remarkably similar to that observed in the previous example (Fig. 18). Thus, the benzylammonium cation forms three linear hydrogen bonds to the macrocycle oxygens (N···O 2.83–2.92 Å), the C–N bond being perpendicular to the ring. The other three

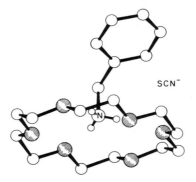

Fig. 19.

oxygen atoms are within the range of 2.93–3.04 Å from the ammonium nitrogen. In the crystal, the thiocyanate counter ion is not associated directly with the complex. Consequently, the binding between the host and the guest appears to be stronger than in the ammonium bromide complex; the nitrogen atom is displaced only 0.86 Å from the median plane of the crown ring, as compared to 1.00 Å in the former structure.

Substitution of the benzyl moiety by a methyl group has a negligible effect on the structure; structural parameters relevant to host–guest interaction in the benzylammonium structure are nearly identical to those found in a 1:1 adduct of methylammonium perchlorate with 18-crown-6.[36] The guest cation in the latter is attached to one face of the ring via three hydrogen bonds, the $-NH_3^+$ group lying 0.84 Å above the mean plane of the ether oxygens (Fig. 20). Furthermore, no specific interactions between the methylammonium moiety and other groups in the crystal have been observed. The

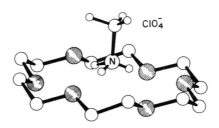

Fig. 20.

three ammonium complexes referred to above, exhibit, therefore, consistent structural features indicating that the guest is bound to the ligand by two different types of interaction: Hydrogen bonding to a triangle of O atoms located on one face of the crown and dipole–dipole attractions involving oxygens positioned on the opposite side of the ring.

Trueblood and coworkers have also analysed crystal structures of the 1 : 1 complexes of 18-crown-6 with hydrazinium perchlorate and hydroxyl-ammonium perchlorate. The latter is characterized by a slightly distorted perching type interaction. This is reflected mainly in the observation that the ammonium nitrogen lies only 0.68 Å from the mean plane of the ligand oxygens (Fig. 21). The more significant penetration of the NH_3^+ group into

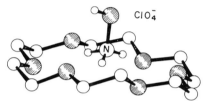

Fig. 21.

the crown ring has also an effect on the geometric details of the host–guest association. While in the previous structures a well defined tripod arrangement of the hydrogen bonds has been found, in this example the ammonium group is involved in bifurcated hydrogen bonds, also exhibiting partial disorder within the macrocyclic cavity. Water molecules present in the crystal lattice of this compound connect the perchlorate anions to the hydroxyl group of the cation by hydrogen bonding simultaneously to both parts of the ion pair. The structural data did not provide enough clues to explain the increased depth of penetration of the NH_3^+ group in this complex.

The hydrazinium perchlorate complex is a unique example, as a consequence of the small size of the guest entity and its potential ability to donate five protons. In this structure, the ammonium nitrogen atom penetrates more deeply into the centre of the crown, lying only 0.11 Å from the median plane of the six ether oxygens (Fig. 22). The ammonium moiety

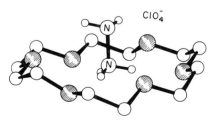

Fig. 22.

binds effectively to the *lower* triangle of the O atoms ($N\cdots O$ distances are within 2.82–2.87 Å) while the $-NH_2$ group forms hydrogen bonds with two oxygens in the upper part of the crown ($N\cdots O$ within 3.05–3.06 Å). Previous structural evidence suggested that the ammonium cation might be too large to fit into the cavity of an 18-membered crown ring and that the interaction geometry at the binding sites is predetermined by the perching arrangement of $-NH_3^+$ (see summarized data in Section 3.4). The above contrasting observation led, however, to the conclusin[36] that

"the primary factor governing the geometry of the interaction at the ether oxygen atoms is the depth of penetration of the $-NH_3^+$ group irrespective of whether hydrogen bonding is involved".

One should keep in mind, nevertheless, that in all other structures analysed so far, a perching rather than a nesting type of association between an ammonium (or alkylammonium) ion and 18-crown-6 derivatives has been observed.

2.7. Interaction with diazonium moieties

Finally, 18-crown-6 also forms another type of crystalline complex with aryldiazonium salts. Haymore[37] has reported the crystal structures of 1:1 complexes with toluenediazonium tetrafluoroborate and phenyldiazonium hexafluorophosphate, showing that the diazonium entity is fully inserted into the hole of the host. The cylindrical diameter of $-N\equiv N^+$ estimated to about 2.4 Å is in fact comparable to the diameter of the crown cavity, allowing such interaction. It is, therefore, somewhat surprising that in solution the phenyldiazonium complex with 18-crown-6 is less stable than with 21-crown-7 where the potential cavity of the crown is obviously larger.[38] The crystallographic data show, indeed, a remarkable similarity between the structures of 18-crown-6 with phenyldiazonium.PF_6^- (Fig. 23) and of

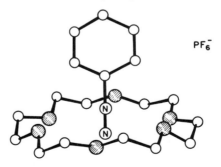

PF_6^-

Fig. 23.

21-crown-7 with *p*-methoxy benzenediazonium.BF$_4$[39] (Fig. 24) in the solid.

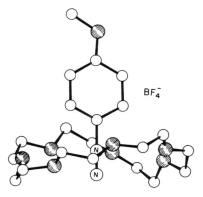

BF$_4^-$

Fig. 24.

Complexation of arenediazonium salts by 18-crown-6 is very sensitive to electronic factors as well as to steric effects due to aromatic substitution.[40] It has been shown, for example, that substitution of benzenediazonium ion by one methyl group in the *ortho* position prevents any effective complexation with the crown ligand. Furthermore, the stability of the potential host–guest complex is severely reduced when the positive charge is delocalized into the aromatic ring away from the diazonium group.

2.8. Conformational features, in relation to the free ligand

The crystal structure of the uncomplexed 18-crown-6 ligand has been analysed at room temperature and more recently[41] also at 100 K. In the solid, the molecules have an elliptical shape and a centrosymmetric (C$_i$) conformation with two transannular methylene groups turning inward. Their structure appears to be stabilized mainly by internal C–H···O attractions (Fig. 25). Clearly, the free ligand does not have a crown shape nor a "cavity". Rather, it has all its dipoles arranged away from each other. Upon complexation with various guests the hexaether host conformationally reorganizes. It forms an open structure with all the oxygens turning inward and converging on the interacting guest. The methylene groups point outward creating a lipophilic skin of C–H bonds, which shields the central guest species from the surrounding solvent.

Conformational analyses of the 18-crown-6 structures by means of ^{13}C NMR and molecular-mechanics calculations have recently been published by several authors. The centrosymmetric conformation of the free

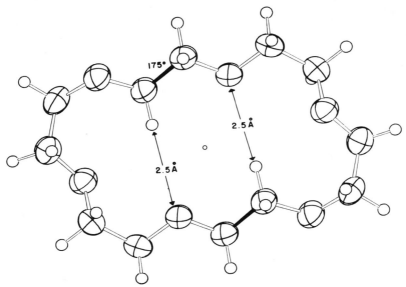

Fig. 25. Molecular structure of the uncomplexed 18-crown-6 host. The *anti* conformation about two C–C bonds is emphasized.

18-crown-6 ligand is of lowest energy due to more favourable intramolecular
electrostatic interactions. Each half of the molecule has the following
sequence of conformational units: $ag^{\pm}a$, aaa and $ag^{\pm}g^{\mp}$. This conformation
was found to be stable not only in the solid phase, but also in nonpolar
solvents.[42]

The D_{3d} conformation occurring in many of the complexed structures of
18-crown-6 is considerably less stable. The ideal crown geometry is characterized by a relatively high steric energy, at about 32.6 kJ mol^{-1} above that
of the free ligand.[35] This is mainly due to larger electrostatic repulsions
between the electronegative oxygen atoms, which essentially directly contact
each other (the nonbonding O···O distances are close to 2.8 Å). However
in complexes of 18-crown-6 with guest cations or polar X–H bonds, these
unfavourable interactions appear to be more than compensated for by
attractions of the guest entities to the oxygen sites. For example, force-field
calculations show that the C_i and D_{3d} structures become similar in energy
in polar solvents. Moreover, in extremely polar media the D_{3d} structure is
even more stable than the C_i conformation, by about 14.2 kJ mol^{-1}.[43]

A biangular centrosymmetric conformation has been observed in
complexes with guests containing the –NH$_2$ binding entity such as urea,
benzenesulphonamide, 2,4-dinitroaniline and guanidine. It is slightly

different from that of the ideal "crown" by having the conformation of two –OCH$_2$CH$_2$O– units changed from *aga* to *gga*. The total ground state energy of the molecule appears to be affected by such a change only to a minor extent. While Bovill and coworkers[35] predict in their calculations that the biangular conformation is slightly more stable than D$_{3d}$, the studies of Wipff *et al.*[43] lead to the opposite conclusion. The latter conclusion finds more support in experimental evidence as the D$_{3d}$ conformation is retained in several adducts of 18-crown-6 with coordination complexes of transition metals, where the apparent interactions between the crown and the other constituents are rather weak. The occurrence of the biangular geometry seems to be confined to those structures which either involve guest entities of a low degree of symmetry or are affected by steric constraints (see below).

Additional conformational modes of the 18-crown-6 ligand were found by force-field calculations to be relatively stable. However, they do not occur frequently and are less important. When one or more of the oxygen sites in 18-crown-6 are replaced by sulphur atoms the conformational features of the ligand, both in the complexed and the free state, become considerably irregular.[9,44]

3. Inclusion of alkylammonium guests within crown ether hosts

Many hydrogen-bonded complexes between alkylammonium ions and macrocyclic polyethers have been reported in recent years, with the host and guest species being subjected to a wide range of structural modifications. The observed geometrical features of intermolecular interaction are, nevertheless, similar in most of the crystalline structures examined. As illustrated in the preceding section, association between an (alkyl) ammonium guest and a crown host of appropriate size is mainly due to complexation through a tripod perching arrangement of $^+$NH$\cdots$O hydrogen bonds on one face of the macrocyclic cavity. Moreover, the C–N bond is usually perpendicular to the mean plane of the ligating O atoms, and the N–H bonds are staggered with respect to those around C. Several examples are shown below.

3.1. Chiral and achiral complexes with penta and hexaether macrocycles

The observed structure of the 1 : 1 complex between dicyclohexyl-18-crown-6 and ammonium perchlorate, which was analysed at −160° C, very much resembles that of 18-crown-6 with ammonium bromide.[37] The ammonium nitrogen is in a perching position about 0.86 Å above the mean plane of the ether oxygen atoms. It forms hydrogen bonds to the upper triangle of oxygen atoms converging on the central guest (Fig. 26). Consequently, the distances between the oxygens in the upper triangle (4.65 Å) become significantly shorter than those between oxygens located on the opposite side

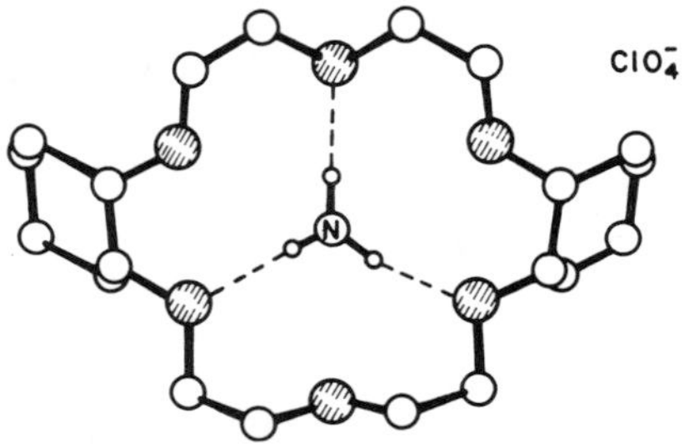

Fig. 26.

of the crown (5.01 Å). The fourth hydrogen atom of NH_4^+, farthest from the ring, is bound to the perchlorate anion. The peripheral cyclohexyl substituents are roughly perpendicular to the hexaether macrocycle, surrounding the complexed ion-pair.

Tertiary-butylammonium salts have often been used as standard guests in complexation reactions with different crown hosts. A recent structural study relates to an inclusion complex of $t\text{-}BuNH_3^+$ ClO_4^- with a cyclic polyether containing a semirigid 1,1'-dinaphthyl unit bound to oxygen in the 2,2'-positions.[45] Formally, this ligand (**3**) contains a 20-membered ring,

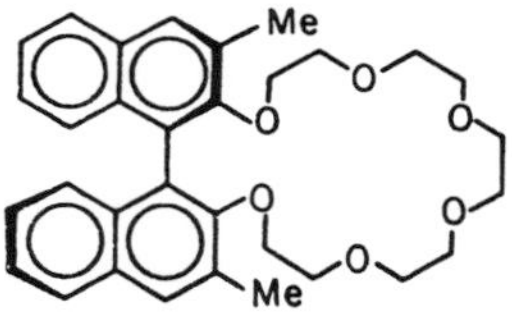

(3)

but due to geometric constraints the size of the cavity which develops upon complexation is roughly similar to that in unsubstituted 18-crown-6. Inspection of the molecular structures of the free and the complexed ligand revealed that its preferred overall conformation is asymmetric, the mean plane of the macroring forming an angle of about 40° with the 1,1'-dinaphthyl bond (Fig. 27). Consequently, one of the methyl groups (substituted on the 3-position of the dinaphthyl unit) covers and directly interacts with one

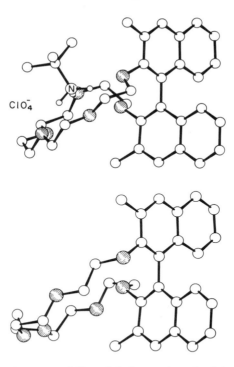

Fig. 27. Molecular structure of ligand **3** (bottom) and of its *t*-butylammonium complex (top).

side of the ring. The ammonium guest approaches the ring from the opposite side and hydrogen bonds to three alternate ether oxygens, the spatial relationships between the host and the guest being almost free from any significant steric hindrance. Intramolecular potential energy calculations confirm that the observed structure represents the most probable overall conformation of this host. It appears therefore that in solution the two faces of the macrocycle are equivalent with respect to a complexation of potential guest species only by virtue of rapidly established equilibria between conformers.

Comparison of the details of molecular conformation for both the free and the complexed ligand illustrates a typical reorganization that usually occurs in macrocyclic polyethers upon complexation. As previously described for the 18-crown-6 system, in the uncomplexed host the conformation about one of the ring C–C bonds is antiplanar rather than *gauche*. As a result, a methylene group turns inward and directs its two hydrogens into the centre of the macroring system to optimize intramolecular van der Waals attractions. On the other hand, the complexed ligand is characterized by an open conformation with synclinal torsion angles about all the C–C bonds, having all oxygens turned inward towards the NH_3^+ group. Also, the free ligand is partially disordered, while the complexed one has a perfectly ordered structure.

It is interesting to note that the above geometric features correlate well with an observation that a similar ligand system with the CH_3 arms replaced by bulkier phenyl substituents exhibits a remarkably selective complexation toward chiral alkylammonium guest moieties.[46] However, attempts at solving a representative structure of one of the diastereomeric complexes formed between phenylglycine methyl ester and a 3,3'-diphenyl-substituted ligand have so far failed.

The similar 9,9'-spirobifluorene pentaether system **4** was also expected to show some discrimination between ions with different radii and enantiomeric selectivity, because of the functional resemblance between the

(4)

spirobifluorene and the dinaphthyl chiral barriers. Nevertheless, only the ammonium thiocyanate complex of this host could be crystallized and subjected to a detailed crystallographic analysis[47] (Fig. 28). It turned out

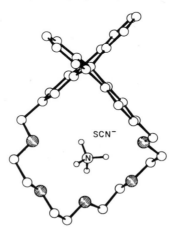

Fig. 28.

that, at room temperature, the complexation of the NH_4^+ ion is not even well defined as the ammonium hydrogen atoms exhibit a statistical disorder. Furthermore, the distances of the central N atom to ether oxygens were found larger than expected (between 2.92 and 3.16 Å), indicating a weak attraction. Most probably, this effect is due to a competing ion-pairing interaction between the ammonium and thiocyanate ions ($N \cdots N = 2.78$ Å), the latter situated on top of the plane of the ring. In spite of the above features the pentaether macroring is characterized by the *gauche* arrangement of all the ethyleneoxy units, resembling remarkably the conformation observed in complexes of unsubstituted 18-crown-6.

An actual example of a chiral host–guest structure is provided by the complex of a phenylglycine methyl ester salt with a chiral host containing two 1,1'-dinaphthyl units separated by a central hexaether macrocyclic binding site[48] (5). As mentioned above the bulky and configurationally stable

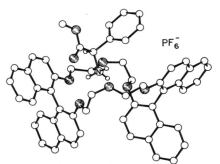

(5)

dinaphthyls play a role of steric and chiral barriers in the process of complex formation, since stereoselectivity depends on the degree of complementary structural relationships between the interacting species. From the two diastereomeric complexes resolved in solution the structure which crystallized corresponds to the less stable (by about $\Delta\Delta G = 8$ kJ mol^{-1}) isomer (Fig. 29). It reflects, in fact, a severe steric hindrance between the alkyl groups of the constituents, as well as the conformational adjustments and reorganization of binding sites that were required to fit the α-amino ester

Fig. 29.

within the host. For example, the naphthalene rings on the interacting face of the cavity are pushed away from each other, and the N–H⋯O hydrogen bonds are far from linear. Furthermore, only five of the ether oxygens are in close contact with N^+, and two of those which accept the hydrogen bonds are adjacent in the ring; in other hexaether ligands alternate oxygen atoms are usually involved in the H-bonding. Apparently, the weak association of the ammonium moiety in this complex is reconciled with additional C–H⋯O and π-acid π-base attractions between the constituents. The PF_6^- counter ions seem to play no role in structuring the host–guest adduct. The inclusion compound crystallizes with 1 mol of chloroform solvent, and the charge separation in this structure is stabilized by delocalization of the negative charge in the relatively large anions, as well as by their hydrogen bonding to chloroform. In the more stable diastereomer, the substituents on the asymmetric centre of the guest are expected to be arranged more favourably with respect to the dinaphthyl walls of the host.

Sousa *et al.*[49] have used the bisdinaphthyl ligand system successfully to achieve a complete optical resolution of racemates of primary amine salts. The crystal structure referred to above turned out to be informative, leading to some interesting interpretations with regard to higher chiral recognition towards phenylglycine methyl ester exhibited by a similar ligand with two CH_3 groups substituted in the 3-positions of one dinaphthyl unit. It has been assumed that increased steric compression between the extended host and the guest as well as between the naphthalene rings on the lower closed side of the macroring contribute to further destabilization of the less stable diastereomer of the modified system. Opposite reasoning could be applied to account in part for the decrease of stereoselectivity in complexation of smaller amino esters by the bisdinaphthyl polyether hosts.

The 1,1'-dinaphthyl residue has also been used as a semirigid support in syntheses of ligands that contain more than a single macroring assembly of binding sites. Such compounds are suitable, for example, as potential hosts to complex efficiently various diammonium guest ions. A model system of this type consists of two 18-crown-6 rings connected by a 2,3 and 2',3'-substituted 1,1'-dinaphthyl group (**6**). Figure 30 illustrates the observed

(**6**)

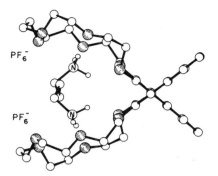

Fig. 30.

structure of this ligand in a complex with a tetramethylenediammonium ion.[50] The two rings are held in a convergent relationship by the aromatic support. Each one of them is hydrogen bonded to an ammonium group of the guest moiety, which is strung between and in contact with the two rings. The geometry of the host–guest interaction is, however, considerably affected by the relatively short dimensions of the $(CH_2)_4$ bridge. The C–N bonds deviate strongly from perpendicularity with respect to the planes of the corresponding rings; in each site the ammonium ions participate in two linear hydrogen bonds to the central O-atoms and one bifurcated H-bond to the aryl O-atoms. Two oxygens in the peripheral region of the 18-crown-6 unit are also oriented towards N^+, but not involved in hydrogen bonding. Correspondingly, the ring framework deviates from the ideal D_{3d} conformation, as the peripheral OCH_2CH_2O fragment adopts an *agg*-type rather than *aga*-type arrangement. In the observed structure, the dihedral angle between the planes of the naphthalene rings attached to one another is about 78°. Crystallographic evidence from related studies indicates however that this angle can vary between 60 and 110°. Such flexibility makes this crown a potential host for a variety of bifunctional guest species.

An introduction of carboxylic residues into the macrocyclic crown rings leads to another class of host compounds with powerful binding abilities toward alkylammonium moieties. One of the first adducts studied is that of 2,6-dimethylylbenzoic acid-18-crown-5 (**7**) with *tertiary*-butylamine

(7)

(Fig. 31a).[51] The carboxylic substituent of the ligand acts, after proton transfer, as an internal counter ion for the complexed guest. The configuration consequently adopted by the crown molecule allows the formation of a hexagonal cavity consisting of the five ether O-atoms and one of the carboxylate O-atoms. Since the carboxylate and ammonium species are on the same side of the ring, a typical coordination of the perching NH_3^+ group can be formed. There is a tripod arrangement of three nearly linear N–H$\cdots$O

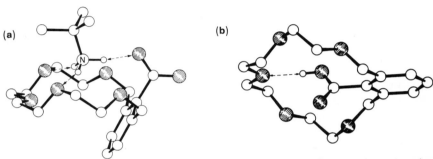

Fig. 31. (a) *t*-Butylammonium complex of host **7**, (b) intramolecular complexation in the free ligand.

hydrogen bonds, the C–N bond being perpendicular to the mean plane of the six ligating oxygens. Evidently, one of the hydrogen bonds is directed at the carboxylate anion and is considerably stronger than the other two bonds. This shows well in the relevant distances: $^+$NH$\cdots$O$^-$ 1.70 Å and $^+$NH$\cdots$O 2.21 Å. Interestingly, the observed geometry of the host–guest complex is characterized by a very high organization and it has a higher degree of symmetry than the constituents in their stable form. The accurate low temperature study led to the location of all the hydrogen atoms with reasonable precision and to define the inherent conformation in the *t*-butylammonium ion which has subsequently been observed in other structures as well. This conformation is characterized by an almost ideally staggered arrangement of the ammonium H-atoms relative to the three methyl carbons. The conformations about the CH$_3$–C bonds are also all staggered.

An interesting correlation can be made between the molecular structures of the complexed and the uncomplexed ligand. As will be discussed later in Section 4.4, the free host represents an intramolecular complex. Almost all of the potential cavity within the molecule is filled with the carboxylic residue occupying the centre of the ring and hydrogen bonding to the transannularly located oxygen (Fig. 31b).[52] Complexation of the alkylammonium guest is thus associated with a major conformational reorganization in the host, the most striking change involving rotation of the benzoic acid fragment by more than 90° about the adjacent C–C bonds.

When four carboxylic groups are attached to a single 18-crown-6 ring, the affinity of the ligand for cationic guests is increased by several orders of magnitude. Indeed, stable crystals of a complex between the tetracarboxy-

R = CO₂H

(8)

macrocyclic receptor **8** and ethylenediamine were obtained from a concentrated aqueous solution containing equimolar quantities of the components.[53] The formation of this 1 : 1 complex is associated with a transfer of two protons from the acid residues to the diamine guest. In the structure, the dication extends roughly perpendicular to the crown ring and parallel to the *pseudo*axial substituents, with one ammonium group fitting tightly into the centre of one face of the crown (Fig. 32). Because of strong

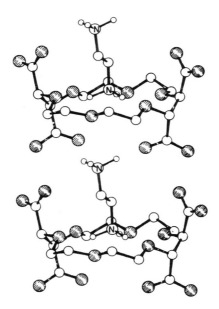

Fig. 32.

electrostatic interactions between the NH_3^+ and CO_2^- groups, the guest penetrates into the macrocyclic cavity more than in most of the previous examples referred to above. In the present complex the ammonium nitrogen lies only 0.56 Å above the mean plane of, and at relatively short hydrogen bonding distances (2.76–2.86 Å) to, all ring oxygens. On the interacting side of the crown the axial carboxy groups are slightly spread apart by steric repulsions with the bound guest; those turned in the opposite direction approach each other, practically covering the lower side of the cavity. The complexed entities are stacked in the crystal one on top of the other, and the ammonium which sticks out from one complex is in contact with the two carboxy residues at the bottom of the adjacent one.

It is remarkable that the substituents and the strong intra- and inter-complex interactions seem to have only a minor effect on the $ag \mp a$ conformational pattern within the ring, which is very similar to that found in other complexes of 18-crown-6. Assuming that such ligands might display an almost constant conformation when complexed with suitable guests, it has been suggested to replace some of the carboxy groups by other functional residues (e.g. $CONHC_6H_5$) in order to achieve a stronger and more toposelective binding of metal cations,[54] or provide secondary interaction sites for a polyfunctional ammonium guest.[55]

3.2. Complexes with larger 24-crown-8 and 30-crown-10 ligands

Large polyether ligands with suitably constrained geometries are also able to form stable complexes with alkylammonium ions. Such an association in the solid as well as in the solution state has recently been reported by Metcalfe *et al.*[56,57] who studied the complexing properties and conformational behaviour of a bisdianhydro-D-mannitolo-30-crown-10 derivative **9**.

(9)

Structures of both the free[56] and complexed[57] molecules were analysed, providing another illustration of conformational changes that occur in polyether ligands during reactions of complexation (Fig. 33). As could have been anticipated from previous studies, the uncomplexed host does not

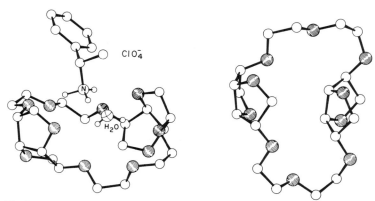

Fig. 33. Structures of the free and the complexed ligand **9**.

contain an open hydrophilic cavity. The molecule exhibits a compressed conformation with several lipophilic C–H bonds turning towards the centre of the ring. Moreover, the optimal relative arrangement of the two conformationally rigid fragments of *cis*-fused five-membered rings, with the convex part of one unit facing the concave side of the other unit, contributes significantly to the stability of the molecular structure. In order to create the essential hydrophilic cavity for complexation of an alkylammonium guest, such as $C_6H_5CHCH_3NH_3^+$, the ligand conformationally reorganizes. The most substantial variation occurs in the conformation of two C–O–C–C fragments connected on both sides to one of the dianhydromannitolo residues. The corresponding torsion angles change from synclinal to antiperiplanar. As a result the ligand unfolds, adopts a face-to-face arrangement for the rigid units of fused rings, and binds simultaneously the ammonium cation and water molecule from the solvent. Seven of the ten oxygen atoms of the ligand are involved in hydrogen bonding. Two ammonium and one water hydrogen form bifurcated hydrogen bonds to three pairs of the oxygens, and the second H-atom of water binds to another oxygen. The third hydrogen of the ammonium group provides a hydrogen-bonding bridge to the oxygen atom of the H_2O molecule. All relevant $N^+\cdots O$ and $H_2O\cdots O$ distances are within the normal range between 2.83 and 3.03 Å.

A recent publication reports an interesting second sphere coordination of the dibenzo-30-crown-10 host to a cationic platinum complex having a square planar environment of one 2,2'-bipyridyl and two monodentate NH_3 ligands.[58] Evidently, in this structure both *cis* NH_3 ligands on the transition metal are involved in $N-H\cdots O$ hydrogen bonds to three of the ten oxygens in the crown ether. The observed $N\cdots O$ interaction distances vary between 2.90 and 3.02 Å. In addition, the macrocyclic ether envelopes the transition metal complex to allow an effective charge transfer interaction between the two benzene substituents (π-donors) of the crown host and one of the pyridine rings (π-acceptor) in the guest ligand. The corresponding conformational features of the large crown reflect the departure of two ethyleneoxy units from the most commonly observed $ag \mp a$ to an agg arrangement.

The above examples indicate that the 30-membered polyether ring is too large to form a monomolecular inclusion complex with a single ammonium entity. Instead, such ligands tend to interact with two (or more) guest species. It has also been observed that as the size of the ring, and consequently the ligand flexibility, increases the conformational patterns of the crown rings become less regular.

In the characteristic configuration of the benzylammonium cation the C–N bond is roughly perpendicular to the plane of the phenyl ring. Correspondingly, inclusion complexes of this substrate with 18-crown-6-type ligands are not expected to be very stable mainly due to potential steric hindrance between the macroring and the aromatic substituent. Nevertheless, a rather unique complexation was found to occur between benzylammonium perchlorate and a larger 24-crown-8 ligand containing a furan-2,5-diketo unit as part of the polyether ring.[59] The overall geometry of this compound is depicted in Fig. 34. Although all the torsion angles about C–C

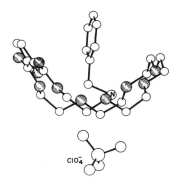

Fig. 34.

bonds are close to *gauche* and those about C–O bonds are nearly *anti* as in complexed 18-crown-6 moieties, the large ligand is not pseudoplanar but rather resembles a cradle. The guest ion penetrates into the ring from its concave side in such a manner that the phenyl moiety is perpendicular and the C–NH$_3^+$ bond is roughly parallel to the mean plane of the ligating oxygens, in obvious contrast with previous observations. As a result, both the methylene and the ammonium groups interact directly with the polyether macrocycle. The former occupies one part of the cavity with two weak

C–H···O interactions at 2.48 and 2.64 Å. The ammonium ion strongly hydrogen bonds to two ether oxygens (with H···O distances of 1.9 Å) in another part of the ring. A most interesting structural feature is associated however, with the third ammonium hydrogen which penetrates deeply into the cavity and hydrogen bonds to the perchlorate counter ion located on the opposite face of the macrocycle. Two oxygens of the perchlorate are involved, at H···Cl = 2.27 and 2.36 Å. This is the only example of an alkylammonium-crown ether complex known so far, where the charged species ion-pair *through* the crown cavity.

3.3. Compounds with nitrogen-containing hosts

There are several examples of alkylammonium complexes with crown ethers which incorporate heteroatoms other than oxygen in the macrorings. Substitution of a pyridine ring for one CH_2OCH_2 unit in 18-crown-6 creates a host with an electron-donating nitrogen site (**10**), and a powerful binding

(10)

ability towards cations. A stable crystalline complex forms between the monopyrido-18-crown-6 and the standard *t*-butylammonium perchlorate guest. The structural details of this compound have been revealed by a low-temperature analysis[60] (Fig. 35). As in other compounds of 18-crown-6,

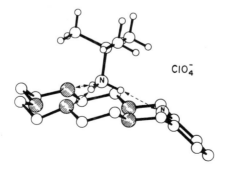

ClO_4^-

Fig. 35.

the complexed macrocycle has a roughly planar relaxed conformation. The undistorted host–guest association is stabilized by a tripod of linear hydrogen bonds, with all bonding distances between the ammonium nitrogen and the heteroatoms contained within the range 2.94–3.03 Å. The six ligating atoms have their electron pairs directed towards the centred guest, which is displaced about 1.14 Å from the mean plane of the crown ring. Since the *t*-butylammonium cation was found to form hydrogen bonds with two ether O-atoms and the pyridine N rather than with three ether oxygens, it has tentatively been concluded that the pyridino nitrogen provides a better hydrogen-bonding site than an ether oxygen. This is in agreement with earlier indications obtained from a comparative analysis of the free energies of complexation.[33] The pyridine-containing ligand has also been found to possess unusual characteristics of binding metal cations.[61]

The last crystallographic determination presented in this section relates to a rather unique *2:1* complex between an alkylammonium guest and a monocyclic crown host. The ligand is a diaza-24-crown-6 containing two transannularly located N–CH₃ groups (**11**). Although in methylene chloride

(11)

this ligand forms a 1:1 complex with a methylammonium thiocyanate salt (spectroscopic evidence), in the presence of benzylammonium thiocyanate in the solution a 1:1 complex crystallizes.[62] In order to avoid electrostatic ion–ion repulsion, the ammonium ions do not penetrate into the centre of the cavity from opposite sides. Rather, the guest species lie roughly parallel to the average plane of the diaza-24-crown-6 ligand (Fig. 36). Each benzylammonium cation interacts only weakly with the host macrocycle via a single hydrogen bond, being attached at the same time to two neighbouring thiocyanate ions. Again, the nitrogen rather than oxygen heteroatoms are preferred in hydrogen-bonding interactions with the guest cations.

Formation of complexes between aza derivatives of crown ethers and primary alkylammonium salts is well documented in the literature. Thus, both monoaza and diaza analogues of 18-crown-6 were found to form complexes with alkylammonium ions in methylene chloride solutions.[63,64] NMR investigations of these structures have indicated clearly that the

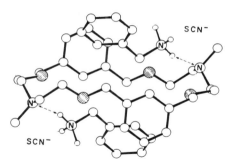

Fig. 36.

hydrogen-bonding association involves the nitrogen atom of the crown. The additional bonding valency of nitrogen (in comparison to oxygen) allows many interesting structural variations. For example, it is possible to connect two monocyclic ligands into one bifunctional paracyclophane-like host by a suitable bridge. The number of bridging strands can vary according to the number of the nitrogen atoms in each macrocyclic ring. Correspondingly, bis(monoaza) crown ethers with one connecting bridge, consisting of either rigid aromatic rings or a flexible $(CH_2)_n$ chain, have been synthesized by Johnson *et al.*[65] Doubly bridged diaza-18-crown-6 ligands have been reported by Kintzinger *et al.*[66] and triply bridged triaza-18-crown-6 rings have been described by Kotzyba-Hibert *et al.*[67] Complexation studies with these dicrown ligands have indicated that they are able to bind either two primary alkylammonium substrates or a single diammonium cation. In the latter case the structure of the adduct that formed appears to be stabilized by interaction of each $-NH_3^+$ group with one of the two 18-crown-6 units of the polycyclic ligand. It has also been shown that some of these hosts exhibit high selectivity in complexation towards $^+NH_3-(CH_2)_n-NH_3^+$ dications characterized by different chain lengths. The different possible modes of interaction between the bifunctional hosts and the alkylammonium guests have so far been investigated only in solution, by NMR spectroscopy; full structural details from crystallographic determinations are not yet available. An extensive review of the ammonium macropolycyclic complexes is given by B. Dietrich in the following chapter.

Naturally, all structural examples of the alkylammonium-crown ether adducts described above refer to macrocycles containing "hard" heteroatoms only. "Soft" heteroatoms such as S or P are ineffective for this purpose; they are used frequently, however, in complexation reactions with heavier cations. Several examples involving S were given in the preceding section. An additional illustration is provided by the cobalt complexes with

4, 7, 13, 16-tetraphenyl-1, 10-dioxa-4, 7, 13, 16-tetraphosphacyclooctadecane ligands[68] (18-membered crown-like macrocycles containing four phosphino and two etheral groups).

3.4. Geometry of binding interactions

Evidently, all complexes between ammine or ammonium species and macrocyclic polyethers that have been presented above are stabilized by N–H$\cdots$X (X = O or N) hydrogen bonds. Characteristic parameters of the interaction of an ammonium ion with uncharged ether oxygen are N$\cdots$O 2.95 ± 0.10 Å, H$\cdots$O 2.1 ± 0.1 Å and N–H$\cdots$O $160 \pm 15°$. Largest deviations from the mean values occur in the following cases. When carboxylate oxygens are involved as acceptors the respective hydrogen bonding distances become considerably shorter (Figs. 31 and 32). On the other hand, bifurcated hydrogen bonds are usually much weaker and far from linear (Fig. 30). Bonds of distorted geometries have been found in structures showing imperfectly complementary relationship of host and guest due to steric effects (Figs. 29 and 30). Structural data on alkylammonium adducts also suggest that the O-atoms that are not hydrogen bonded seem to be involved in stabilization of the host–guest complex through direct pole–dipole interactions with the positive charge of the guest ion. These oxygens are usually held in a convergent form, turned toward the centre of the crown ring, at about 3 Å from the electrophilic N (see below).

Optimal geometric relationships exist between the –NH$_3^+$ group and 18-crown-6 type ligands in which the heteroatoms lie alternately above and below their mean plane and form a cavity about 2.8 Å in diameter. Thus any comparison of fine structural details associated with the binding interactions should be confined to such systems. Survey of the crystallographic data available so far (excluding the transition metal complexes) with respect to interaction geometries at the binding sites reveals some interesting details. Since tabulations of relevant data for individual structures have already been published elsewhere,[5,45] only the systematic trends are described below. In most complexes the –NH$_3^+$ group is in a perching position with respect to the macroring, lying between 0.9 and 1.2 Å above the mean plane of the ligating heteroatoms. Substrates containing additional sites for hydrogen bonding to the ligand, such as hydroxylammonium or hydrazinium ions, tend to penetrate more deeply into the macrocyclic cavity. To allow efficient binding with all three hydrogen atoms, the ammonium substrate is centered above the crown cavity with the R–NH$_3^+$ (R = H or C) bond nearly perpendicular to the best plane of the ligating heteroatoms. Significant deviations from the tripod arrangement of hydrogen bonds to alternating oxygen sites around the cavity were found to occur only in severely strained complexes (Figs. 29 and 30).

In *perching* adducts the hydrogen bonding involves heteroatoms which are situated on the complexed face of the macrocycle; those located on the opposite side participate in dipolar attractive interactions. The latter are characterized by substantially longer N···O distances (average 3.10 ± 0.20 Å), as compared to the NH···O contacts (2.95 ± 0.10 Å) observed for the hydrogen bonding attractions. In each one of the individual structures the organization of binding sites is differently affected by steric constraints and other features of molecular geometry. However, the overall pattern of host-guest interaction is very similar in all compounds. For example, we observe a consistent tendency for the adoption of a trigonal rather than tetrahedral approach of the hydrogen bonds to the accepting heteroatoms. Correspondingly, the angle of approach of N···O vectors which represent the dipolar interactions to the respective COC planes are preferably tetrahedral (ca. 55°).

The above described geometric pattern of NH···O and N···O interactions is shown schematically in Fig. 37. It illustrates also (for the model alkyl-ammonium complex of 18-crown-6) the relative orientation of oxygen lone-pair orbitals with respect to the ammonium nitrogen, assuming that the

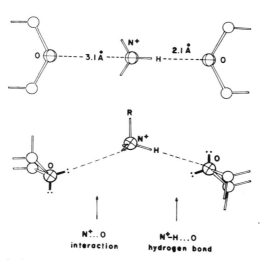

Fig. 37. Geometric features associated with the two types of interaction between a perching ammonium ion and 18-crown-6. For clarity, only two transannular oxygens are shown; approximate orientations of their lone-pair orbitals are marked by dark lines.

symmetry of valent orbitals on O is ideally tetrahedral. *In general*, however, the geometry of interaction at the ligating heteroatoms is primarily dependent on the depth of penetration of the $-NH_3^+$ group into the crown ring.[36]

4. Other types of complexes with monocyclic hosts

4.1. Interaction of hydronium ions with crown ethers

Crown ether macrocycles were found to form stable crystalline complexes not only with ammonium ions (see above) or with neutral water molecules (see below) but also with hydronium ions as isolated monomeric species. The first detailed report has recently been published,[69] relating to the inclusion compound formed by a tetracarboxylic-18-crown-6 ligand **8** with hydronium chloride. Crystallographic analysis (suitable crystals were obtained by slow evaporation of a 1 M HCl solution of **8**) shows that the hydronium cation is indeed included as a discrete entity within the 18-crown-6 cavity (Fig. 38). Because of its apparently smaller size the H_3O^+

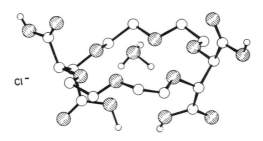

Fig. 38.

ion penetrates deeper into the centre of the crown ring than the ammonium ion in related compounds, hydrogen bonding to the lower triangle of the ether O-atoms. The mean distance of the guest oxygen from the plane of the ligating oxygens is 0.61 Å; that from the upper triangle of the ring oxygen atoms is 0.34 Å. The inclusion complex is thus stabilized by three hydrogen bonds (to "lower" oxygens at 2.67–2.74 Å) and three ion-dipole interactions (to "upper" oxygens at 2.76–2.82 Å) from the central guest to the ligand. The four neutral carboxylic groups are all planar extending above and below the pseudoplanar macroring; they do not interact directly with the central guest moiety. Paralleling to what has been found in the ethylenediamino complex of this ligand (Fig. 32), the face complexing the cation is slightly more open. The opposite face is essentially covered by the two carboxy groups which approach each other, shielding the lower side of the ring from the Cl⁻ anions. The latter are located in between the complexed entities, forming relatively short contacts with carboxylic residues of neighbouring molecules. The hydronium cation is probably too small to stabilize a fully unfolded conformation of the host; the geometry of a small fragment of the ring is therefore irregular and disordered.

A very similar type of host–guest association has been observed between the dicyclohexyl-18-crown-6 ligand and a hydronium perchlorate salt.[37] The hydronium cation is centred above the polyether cavity more in a perching than in a nesting position, forming a tripod arrangement of hydrogen bonds to three of the ether oxygens (Fig. 39). Thus, in both structures the guest

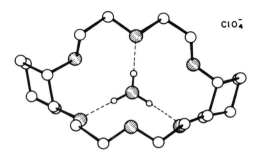

ClO_4^-

Fig. 39.

ion is characterized by a pyramidal geometry. The cyclohexyl rings are on the same side of the ring to which H_3O^+ is bound, thus extending the lipophilic envelope around the complexed cation. There is no direct contact between the positive and the negative ions in the crystal.

The above two examples illustrate that the overall geometric features of host-to-guest binding in crown ether complexes with oxonium cations are very similar to those observed in complexes with the isoelectronic ammonium guest entities.

4.2. Exclusive inclusion of water molecules

The complex of the 30-crown-10 derivative referred to in the preceding section is composed of more than two components, the large cavity containing side-by-side an alkylammonium cation and an H_2O molecule. Similarly, several other complexes have been reported in which metal or ammonium

ions and water molecules are bound simultaneously to the same ligand.[4]
The co-inclusion of water in these compounds appeared to be essential
mainly because the crown host was too large for the formation of a stable
1 : 1 complex. Recently, several stoichiometric water complexes of polyether
ligands have also been reported, where the H_2O molecules occupy exclusively the centre of the macrocyclic cavity.

The 1 : 1 water adduct of the 3,3'-(1,1'-bi-2-naphthol)-21-crown-5 (**12**) provides the first example.[70] In this complex the water molecule is however

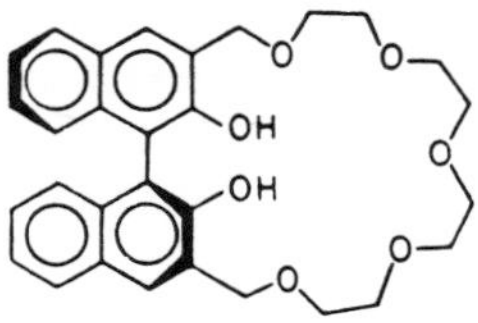

(12)

too small to fit into the ligand cavity and thus occupies only a part of it
(Fig. 40). Nevertheless, the guest moiety is not entirely nested within the
ring, but rather slightly displaced from the mean plane of the oxygen-ligating
atoms. The complexation involves two relatively weak hydrogen bonds with

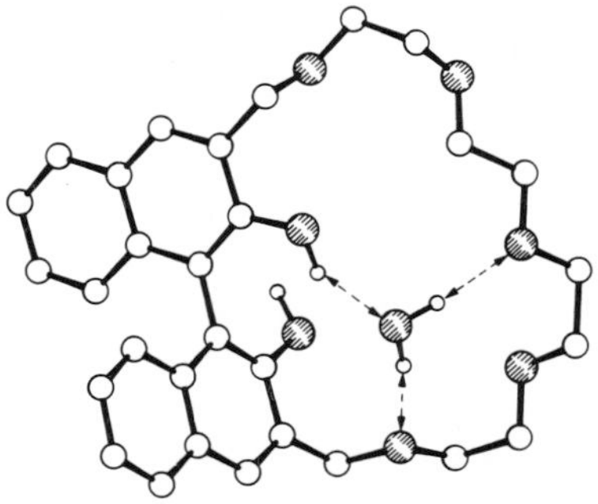

Fig. 40.

the water hydrogens donated to two ether oxygens (relevant distances are:
$O_w \cdots O$ 2.93–2.97 Å and $H_w \cdots O$ 2.0 Å), and a stronger H-bond from one of
the acidic phenolic groups to the water oxygen ($O \cdots O_w = 2.69$ Å, $H \cdots O_w =$
1.9 Å). While the former are nearly linear ($O–H_w \cdots O$ 171–173°), the latter
is considerably bent ($O–H \cdots O_w$ 136°). The second phenolic group participates in interactions between two centrosymmetrically related entities of
the host–guest complex.

Incidentally, because of the imperfect fit between the component moieties, several different features of molecular conformation are assembled in this structure. The complexed part of the macroring is well ordered and consists of two *aga* conformational units. On the other hand, the uncomplexed fragment of the ligand, which is not involved in specific interactions with the guest molecule, has an irregular and even partially disordered conformation. Moreover, one of the OCH_2CH_2O torsion angles becomes antiplanar in order to fill the empty part of the intramolecular cavity by an inward-turning methylene group.

A polyethereal bis-lactone 18-membered macrocycle **13**, analogous to the 2,6-dimethylylpyridine-18-crown-6 ligand (see above) but with carbonyl

(13)

groups replacing the methylyl units, readily forms a complex with two moles of water.[71] In fact, this ligand crystallizes as a dihydrate from organic solvents (such as hexane) in the presence of only trace amounts of water. The structural analysis revealed that only one water molecule is bound directly to the macrocycle (Fig. 41), forming hydrogen bonds to two of the ether oxygens at O···O distances of 2.92 and 3.05 Å. This molecule is in a

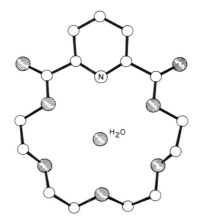

Fig. 41.

perching arrangement, being displaced by about 1.49 Å from the mean plane of the ligand; the second water molecule is bound to the O-atom of the first well above the macroring. All the six potential donor atoms form a typical crown cavity, the aliphatic part of the ring being characterized by a regular relaxed conformation. Although the pyridine N-atom was shown to be a very good hydrogen-bond acceptor, it is not involved in hydrogen-bonding interactions; in the present structure the electron density at nitrogen is reduced by the carbonyl electron-withdrawing groups.

An exclusive inclusion of water by a larger tetraethylene glycol bis-(2-pyridyl) ketone macrocycle **14** has also been reported.[72] The stoichiometry

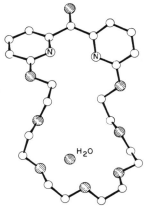

(14)

of the corresponding crystalline complex is 1 : 1. As in the previous examples the water molecule is too small to fill the whole macrocyclic cavity (Fig. 42). It is therefore coordinated preferentially to one oxygen atom in the ring at $O \cdots O = 3.00$ Å, and only weakly to other oxygen sites. Due to such a weak association, the water hydrogens are most probably disordered within the cavity. Again, the rigid fragment of the ligand including the pyridine rings does not interact directly with the water species. Rather, the guest molecule is bound to the extreme portion of the polyether ring, where conformational adjustments can easily be made to encircle as efficiently as possible the small substrate moiety.

Fig. 42.

4.3. Encapsulated guanidinium and uronium cations

The crystal structures of complexes of 18-crown-6 with urea, uronium nitrate and guanidinium nitrate have already been referred to (Section 2). In these compounds the cavity of the host is too small to encapsulate the guest species, and therefore perching-type structures occur. Consequently, the guests associate *via* hydrogen bonds not only with the 18-crown-6 host but also with other surrounding moieties. It has been anticipated however from molecular models that the urea-like species may fit well in the cavity of a 27-crown-9 macrocyclic framework; with suitable hosts it became possible to achieve a complete transfer of uronium and guanidinium salts into organic media.

Only two crystal structure determinations of such complexes have been published so far by the Dutch group.[73] In the complex between benzo-27-crown-9 and guanidinium perchlorate (Fig. 43) the guanidinium ion is

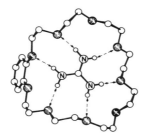

Fig. 43.

indeed completely encapsulated within the macrocyclic cavity. The structure is stabilized by six roughly linear hydrogen bonds, extending from each $-NH_2$ group to the nearest ether oxygens. The corresponding $N \cdots O$ distances are within the expected range from 2.84 Å to 3.08 Å. The three non-hydrogen bonded oxygens are also turned inward converging on the guest. Torsion angles along the molecular framework have fairly irregular values, the ligand adopting a bent structure.

A very similar complex is formed between this ligand and uronium nitrate. The uronium ion is well accommodated within the macroring, all its hydrogen atoms being involved in hydrogen bonding to the non-aryl ether O atoms. The $O-H \cdots O$ bond (2.53 Å) is much shorter than the $N-H \cdots O$ bonds, indicating that the positive charge is mostly localized on the urea oxygen. Since the aryl oxygens of the host are not forced to participate in the hydrogen bonds, the uronium complex has a considerably flatter structure than the guanidinium complex. In the former the mean plane of the guest ion nearly coincides with that of the crown. A less efficient hydrogen bonding pattern has reportedly been observed in a similar complex of uronium with a larger dibenzo-30–crown-10 host.

4.4. Intramolecular complexes

Crown ether hosts containing 2-carboxy-1,3-xylyl units tend to form "intramolecular complexes" in which the carboxy groups occupy potential cavities of the macrocycle. It is well known that carboxylic acids almost invariably form intermolecular carboxy-dimers in the solid. Therefore, a stable intramolecular complex will occur only if a complementary arrangement of potentially binding sites within the cavity can be achieved.

We have already referred to the structure of the 2-carboxy-1,3-xylyl-18-crown-5 ligand as an intramolecular complex (Fig. 31). The carboxylic group is centred within the macrocycle, and takes part in a short and almost linear hydrogen bond with the transannularly located ether oxygen. The geometric details of this association are $O \cdots O$ 2.71 Å, $H \cdots O$ 1.8 Å, $O-H \cdots O$ 174°. Additional attractive dipole–dipole interactions occur between the two benzyl O-atoms and the carboxylic carbon atom, as indicated by the observed $O(benzyl) \cdots C(carboxyl)$ distances of 2.75 and 2.77 Å. These distances are considerably shorter than the normal van der Waals distance (~3.1 Å). Moreover, the relative arrangement of the respective atoms suggests that one of the lone pairs of each benzyl oxygen points towards the carboxyl C, resembling the characteristic geometry of $O \cdots C{=}O$ type attractive interactions observed in many other compounds.[74]

A remarkably similar configuration has been found[75] in the crystal structure of 2,6-dimethylylbenzoic acid-2,6-dimethylylpyridine-18-crown-5 (**15**).

(15)

Again, the molecular structure of this ligand is uniquely stabilized by an internal hydrogen bonding which involves the carboxylic acid group located in the centre of this ligand (Fig. 44). The planes of the nearly parallel phenyl and pyridine rings are displaced by about 1.2 Å to optimize the $O-H \cdots N$ interaction from the carboxyl to the pyridine nitrogen; $O \cdots N$ 2.66 Å, $H \cdots N$

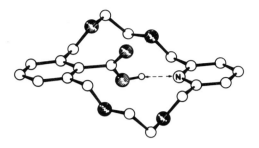

Fig. 44.

1.8 Å, O–H$\cdots$N 169°. The intramolecular O–H$\cdots$N hydrogen bond in this ligand appears to be slightly stronger than the O–H$\cdots$O bond in the previous example, in agreement with the determinations of corresponding pK_a values; in water, the pK_a value of the pyridine-containing compound is by one unit weaker than that for the benzoic-18-crown-5 acid.[76] Furthermore, the stronger bonding to the transannular nitrogen leads to somewhat weaker O(benzyl)$\cdots$C=O attractions. This is reflected in the corresponding O$\cdots$C distances of 2.84 and 2.87 Å, which are systematically longer by about 0.1 Å than those observed in the former ligand. The molecular framework in both carboxylic structures shows approximate C_2 symmetry and *gauche* conformations for the peripheral ethyleneoxy units.

The structure of a larger bis(2'-carboxy-1',3'-xylyl)-24-crown-6 ligand **16** also represents an intramolecular complex with the two carboxy groups

$$\text{(16)}$$

converging on each other to provide an ideal hydrogen-bonding arrangement (Fig. 45).[77] The molecule has D_2 symmetry. The two coplanar carboxylic functions form an intramolecular cyclic pair of equivalent linear O–H$\cdots$O hydrogen bonds; relevant structural details, including the dimensions of the hydrogen bonds O$\cdots$O 2.69 Å, H$\cdots$O 1.9 Å and O–H$\cdots$O 169°, are very similar to previously observed geometries of *inter*molecular carboxy dimers in crystals of carboxylic acids. The potential cavity in the present ligand is thus efficiently filled with the dicarboxylic pair. As an apparent result of the dominant function of the intramolecular H-bonding interaction, the potential dipole–dipole attractions between benzyl O-atoms and carboxylic

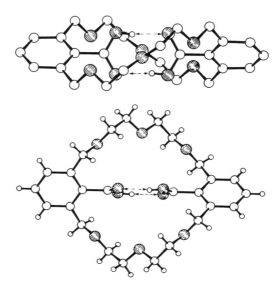

Fig. 45. Two perpendicular views of the intramolecular complex **16**.

carbons is much less pronounced than in the other compounds. The corresponding O···C=O distance is 2.97 Å as compared to previously observed mean values of 2.86 and 2.76 Å. Actually, this structure provides a rare example of a non-rigid monocyclic crown whose geometry is characterized by a very high organization in the crystal due to the well defined pattern of intramolecular interactions as well as the partly constrained configuration inherent to the 2-carboxy-1,3-xylyl residues.

In the three carboxylic acid-crown ligands all polar groups are hidden within the macrocycle and covered by a lipophilic skin of C–H bonds. The corresponding crystal structures consist of discrete molecules of the acid, being stabilized mainly by ordinary van der Waals forces. When these ligands are treated with a suitable alkylamine or alkylammonium substrate stable intermolecular host–guest-type complexes can be formed; the multiple NH_3^+···O interactions between ligand and substrate will then compensate for the loss of the intramolecular hydrogen bonds. One such complex with *t*-butylamine has been described above. However, no good quality crystals suitable for crystal structure determination could be prepared with the other ligands.

The 20-crown-6 ligand containing two pairs of methyl and benzamido geminal substituents **17** reveals a rather poor binding capability towards ammonium and metal cations. On the other hand, structural data suggest that this compound may also be regarded as an intramolecular complex

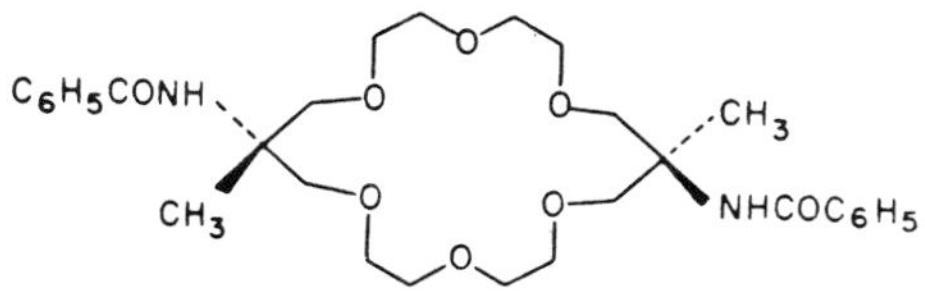

C_6H_5CONH, ─ O   O ─ ,CH_3

CH_3 ─ O   O ─ NHCOC_6H_5

(17)

(Fig. 46).[78] In the observed configuration of the free ligand, the two bulky benzamido groups are *trans* with respect to the macroring and approach its centre from opposite sides. The crown molecule adopts a chair-like conformation which is stabilized mainly by linear hydrogen bonds from the nitrogen atoms to two of the ether O-atoms. Consistently, the macroring is contracted along a direction roughly parallel to that of the N–H···O bonds.

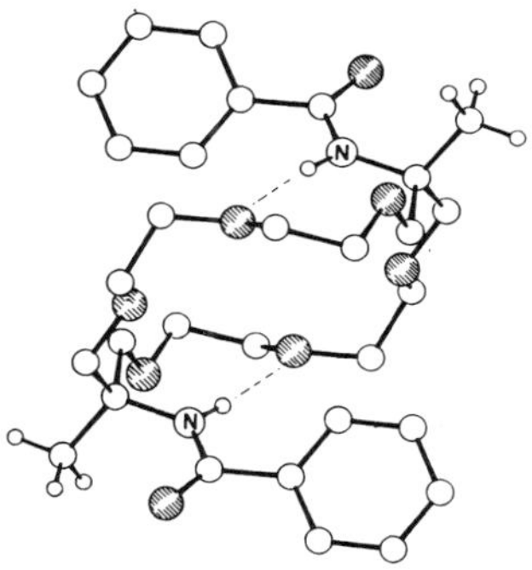

Fig. 46.

The centrosymmetrically related phenyl groups are nearly perpendicular to the mean plane of the macrocycle, providing steric inhibition for a potential association of this ligand with other species. The molecular structure is slightly disordered in those parts of the ring which are not involved in hydrogen bonds or van der Waals interactions with the benzamido substituents. This correlates well with previous observations of irregular conformational features in ligands lacking a sufficiently extended pattern of stabilizing interactions.

4.5. Complexes of hemispherands and spherands, ligands with sterically enforced cavities

Incorporation of rigid aromatic groups into the crown ether macrocycles has already been illustrated in previous sections. For example, biphenyl and binaphthyl substituents are commonly used as steric barriers in order to effect selective binding of chiral guests. Smaller groups such as 2-carboxy-1, 3-xylyl units or even phenyl rings also impose some rigidity on the crown ring, of which they are part. Nevertheless, in many of the ligands described, the polyether macrocycle remained sufficiently flexible to undergo conformational reorganization (when required) upon complexation with a potential substrate.

Hemispherands and spherands are unique examples of synthetic polyethers which possess either partially or fully enforced cavities. Hemispherand **18** contains one rigid *m*-teranisyl unit as part of the macroring.

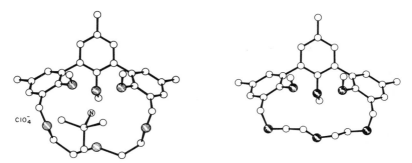

(18)

The oxygen-ligating sites of anisole units are held in sterically enforced conformations; yet, this ligand remains conformationally flexible to some extent mainly in the diethylene glycol bridge. In the crystal structure, the cavity within the empty ligand is filled (partially) with two inward-turning methylene groups of the aliphatic fragment while the unshared electron pairs of the non-aryl oxygens are oriented outward. The O–CH$_3$ groups are directed along axes roughly normal to the mean plane of the macroring and away from its centre, those attached to adjacent aryls pointing in opposite directions[79] (Fig. 47).

Fig. 47. Molecular structures of hemispherand **18** and its complex with t-BuNH$_3^+$.ClO$_4^-$.

Upon complexation with *t*-butylammonium perchlorate, the ligand cavity becomes wider to accommodate the interacting guest. Similar to the conformational reorganizations found in other hosts, this is accomplished mainly by a change of the $O-CH_2-CH_2-O$ torsions within the bridge from an *anti* to a *gauche* conformation. In the observed structure of the complex (Fig. 47) the unshared electrons of the six oxygens are thus turned inward, and they interact with the guest ammonium ion that centres into the macrocycle.[5] The alkylammonium cation is bonded to the ligand via three hydrogen bonds. Those involving oxygens of the diethylene glycol bridge are linear; the third is a bifurcated hydrogen bond directed to the two (*pseudo-meta*) aryl oxygens which lie on the complexed side of the ring. The Inner $O-CH_3$ substituent is centred on the opposite face of the ligand. Comparison of average molecular dimensions of the free and complexed hemispherand also shows minor conformational changes within the *pseudo*-rigid teranisyl group. This includes some excessive folding of the outer aryl groups in the complex, and a slight increase (*c.* 0.4 Å) of the distance observed between the *pseudo-meta* oxygen sites.

Trueblood *et al.*[80] and Cram *et al.*[81] have recently reported the syntheses and structural characteristics of cyclohexametaarylene spherands. These hosts appear to be *fully organized* during synthesis (which is templated by Li^+), and their overall conformation remains unchanged upon complexation with guest species. In spherand **19** the intramolecular cavity is defined by

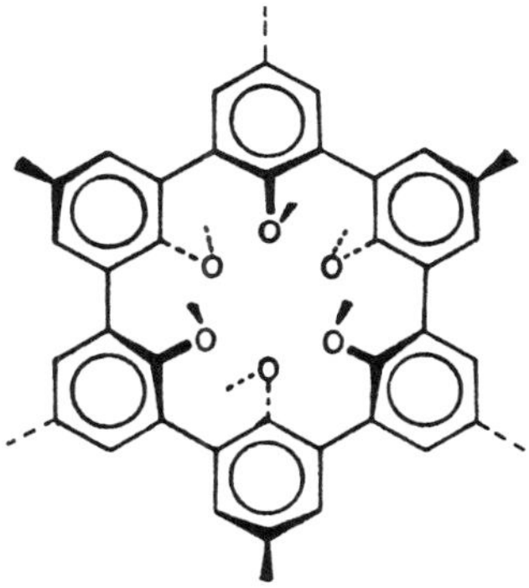

(19)

the six aryl oxygens which are octahedrally arranged around the molecular centre (Fig. 48). The entire molecule has approximate D_{3d} symmetry and is covered by a lipophilic skin of C–H bonds which shields the unshared electrons on oxygen from interaction with the solvent. Crystalline complexes

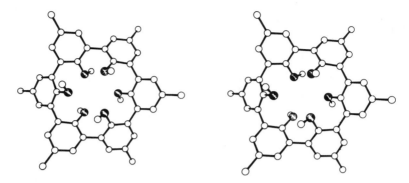

Fig. 48. Stereoview of empty spherand **19**.

of spherand **19** with LiCl and NaCH$_3$SO$_4$ salts have been reported.[80] Figures 49 and 50 show their respective structures in which the cation guests are fully encapsulated by the host (in the crystal, the corresponding anions are spread between the complexed lithiospherium and sodiospherium entities).

The crystallographic analyses showed that similar conformational organization characterizes the ligand in all three structures, the molecular conformation being severely strained due to electron–electron repulsions between the inward-turning oxygen lone-pairs. Upon insertion of the Li$^+$ cations between the oxygens this strain is somewhat reduced. As a result of the attractive electrostatic interactions present in the lithiospherium complex the cavity diameter (1.48 Å) is smaller and the oxygen–oxygen distances are shorter in this structure than in the uncomplexed host. On the

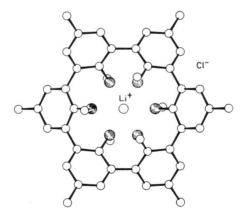

Fig. 49.

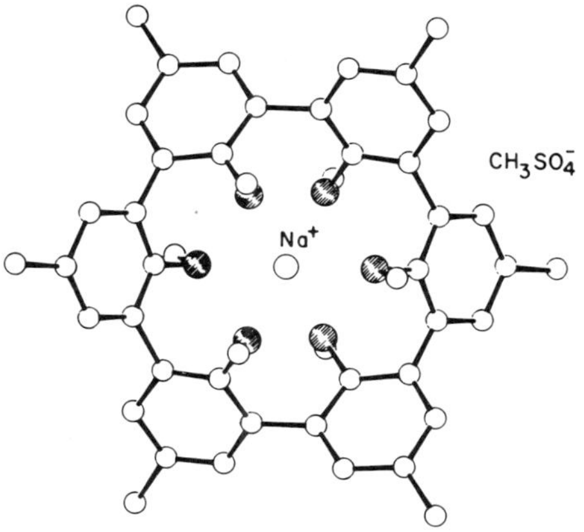

$CH_3SO_4^-$

Na^+

Fig. 50.

other hand, insertion of the larger Na^+ cation into the cavity is associated with an increased conformational strain of the system. The ligand cavity expands from 1.62 Å (in diameter) in the free spherand to about 1.76 Å in the sodiospherium complex, and the twist angle between adjacent aryl groups increases by about 8° (from 52 to 60°).

Additional examples of rigid and highly strained spherands are provided by compounds **20** and **21** in which two pairs of the aryl oxygens are bridged. Both ligands formed crystalline complexes with lithium salts and their structures were analysed in detail, leading to interesting observations.[81] The relative orientation of the aryl groups in hosts **20** and **21** is different from that observed in spherand **19**. The bridging groups and four of the aryl oxygens are located on the same side of the macrocycle, while the remaining

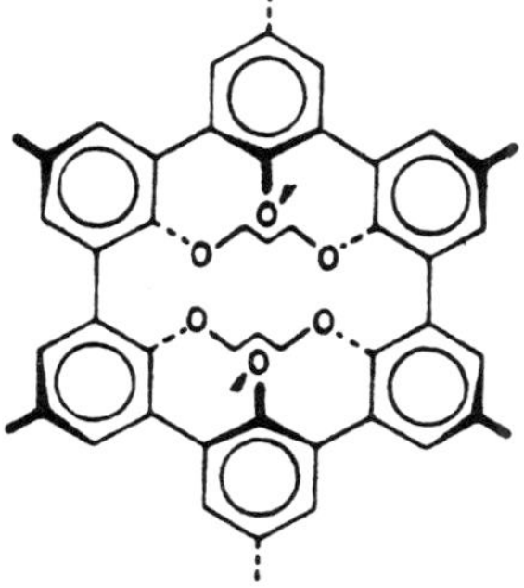

(20) 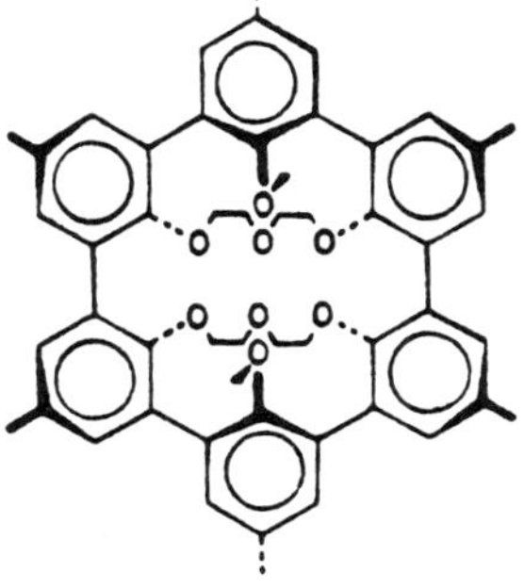   **(21)**

unbridged *pseudo-para* methoxyls point in the opposite direction. Consequently, the distribution of the oxygens around the cavity is uneven. In the lithiospherium complex of **20** (Fig. 51) the Li^+ cation is best described as being five-coordinated, with $Li\cdots O$ distances ranging from 2.00 to 2.08 Å.

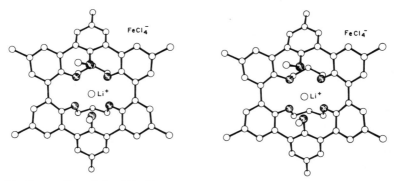

Fig. 51. Stereoview of the $LiFeCl_4$ complex of spherand **20**.

Steric strain in this structure is reflected mainly in a significant distortion of oxygen atoms from the mean plane of their attached aromatic rings, irregular folding of the aryls, and rather short nonbonding $O\cdots O$ distances (2.50–2.72 Å).

Host **21** contains eight oxygen-ligating sites which define an elliptically shaped cavity lined with 32 electrons. The conformation of this ligand also appears to be extremely strained. In the complex with lithium chloride (Fig. 52) only seven oxygens are effectively liganded to Li^+ with distances

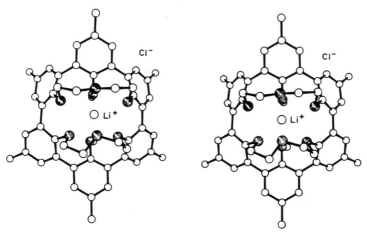

Fig. 52. Stereoview of the LiCl complex of spherand **21**.

varying from 2.03 to 2.42 Å. The structural data presented in this section thus indicate that Li^+ is five-coordinated by spherand **20**, six-coordinated by **19** and seven-coordinated by **21**, the respective cation diameters calculated from the observed Li-to-O distances being 1.27, 1.48 and 1.71 Å. Crystalline complexes of the bridged spherands with sodium salts have not yet been reported.

Measurements of the free energies of complexation have indicated that spherands **19** and **20** have more binding power toward Li^+ than Na^+, while host **21** binds Na^+ better than Li^+. No significant complexation of larger cations by these spherands has been detected. Cram and coworkers have however discussed the results of an extensive study of complexation reactions between a large variety of hosts and lithium and sodium picrates in a $CDCl_3$ solution. They showed that the spherands have generally more binding power toward Li^+ and Na^+ salts than the corresponding hemispherands, which in turn are better binders than monocyclic crown ethers containing an intramolecular cavity of a comparable type. This observation led to the conclusion that

"the larger the number of host ligating sites organized for maximum binding during synthesis rather than during complexation, the higher the free energy of complexation".

A more detailed discussion of this subject is given by Cram and Trueblood.[5]

The spherands exhibit enhanced selectivities toward metal cations.[82] On the other hand, a fascinating design of even more complex ligands, "cavitands", has recently been reported.[83] Molecular models show that these ligands have rigid cavities large enough to accommodate simple neutral guest molecules of suitable size.

5. Inclusion compounds of open-chain polyether ligands

5.1. Complexes of metal cations

Noncyclic crown-type polyethers, consisting of several oxygen atoms connected by ethylene groups, are also capable of binding guest molecules and ions. Metal ion complexes of the open ligands, as those of the cyclic crowns, have been of particular interest to many chemists because they seem to provide model compounds for studies of cation transport across membranes. (See Chapter 16, Volume 3.)

One of the first systematic structural investigations of such systems relates to a series of complexes formed between ethylene oxide oligomers and cadmium or mercuric chloride.[84,85] Detailed structural data are available for the molecular complexes of tetraethylene glycol dimethyl ether (TEGM) with $CdCl_2$ and $HgCl_2$, as well as of tetra- and hexa-ethylene glycol diethyl ethers (TEGE and HEGE) with $HgCl_2$. Less precise data have been published for the mercuric chloride adduct with a polymer of oxyethylene. Experimental evidence in the solid and in solution indicates that a helical conformation represents the lowest energy form for a polyethylene oxide chain.[86] Similarly, in the 1:2 complex between HEGE and $HgCl_2$ the molecular structure of the ligand resembles a helix with two turns. Each Hg atom is coordinated to four oxygen sites in either half of the polyether, the central oxygen being thus coordinated simultaneously to the two metal atoms. All Hg···O distances are within 2.7–2.9 Å. The shorter TEGM and TEGE ligands form only 1:1 complexes with mercuric chloride. Their conformation, composed exclusively of $ag^{\pm}a$ units, is nearly the same in the two structures except for the terminal groups. The metal atom was found to be almost fully encircled by the open chain molecule, all the ligating oxygens adopting a coplanar arrangement. Interatomic distances between the Hg and O atoms are relatively short (2.8–3.0 Å). Iwamoto had concluded from these studies[84,85] that the $ag^{\pm}a$ conformation for CH_2–O–CH_2–CH_2–O–CH_2, which appeared also in either half of the larger HEGE ligand, seems

"very important and favorable for coordination between the O and Hg atoms".

The observed interaction modes are shown schematically in Fig. 53.

Cadmium chloride has an ionic nature and is differently associated with TEGE. The stoichiometric ratio is 2:1. The ligand adopts an extended rather than a convergent type of conformation with *agg* as well as $ag^{\pm}a$ conformational units. The different coordination radius of Cd is reflected in shorter

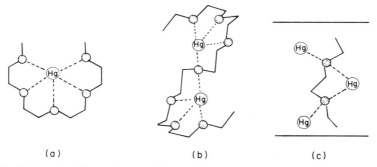

(a)   (b)   (c)

Fig. 53. Schematic illustrations of the mercuric complexes with (a) TEGM, (b) HEGE, and (c) polyethylene oxide.

Cd···O distances (as compared to the Hg···O bonds) of 2.4–2.5 Å. The differences in molecular conformation of the various ligands correlate well with the variations observed in infrared spectra.[85]

More recently a rather extensive series of open-chain polyethers capable of forming stable complexes with alkali metal ions have been reported by Vögtle (synthesis), Saenger (structure) and coworkers. Characteristically, these molecules contain, in addition to the open chain (of varying size) of ethylene oxide units, rigid aromatic donor end groups.[87]

The first example involves the hexadentate 1,8-bis(2-methoxy-phenoxy)-3,6-dioxaoctane polyether **22** which resembles the 18-crown-6 system in its

(22)

size and number of ether oxygens. It forms a 1:1 complex with sodium isothiocyanate salt in methanol/ethyl acetate solutions[88] (Fig. 54). As in many other structures involving metal salts the cation, apart from being coordinated to the oxygen nucleophiles, remains connected to its counter ion (via the N atom). The ligand has a helical structure to allow efficient interaction of all six oxygens with the Na$^+$ ion at 2.33 Å $<$ Na$^+$···O $\leqslant$ 2.54 Å, without any collision between the terminal oxyanisole residues. The coordination number of the sodium ion in this structure is therefore 7.

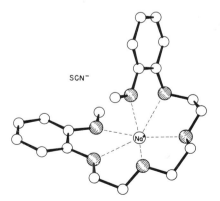

Fig. 54.

Coordination of potassium cations in their complexes with open-chain polyethers, described below, involves either seven or eight ligands. In the 1:1 complex of KNCS with 1,11-bis(2-nitrophenoxy)-3,6,9-trioxaundecane (**23**) the cation ligating sites consist of the five ether oxygens, one O-atom of

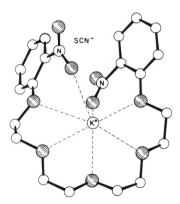

(23)

each nitro group and N of the anion. Again, the ligand adopts a helical structure to optimize all interactions, avoiding at the same time steric repulsions between the terminal nitrophenoxy groups[89] (Figure 55). In the observed conformation, which is composed almost entirely of $ag^{\pm}a$ units, the aromatic fragments are stacked intramolecularly in an antiparallel mode. All $K^{+}\cdots O$ interaction distances range from 2.79 to 2.91 Å with one exception of 3.22 Å.

Fig. 55.

Replacement of the terminal nitrophenoxy groups by 2-acetylamino-phenoxy substituents (**24**) changes the structure of the 1:1 complex significantly (Fig. 56). The cation is sevenfold coordinated to four of the five ether oxygens ($K^{+}\cdots O$ at 2.74–3.09 Å), two acetyl groups of adjacent ligands ($K^{+}\cdots O$ at 2.61–2.87 Å) and either the thiocyanate anion or water (there are two entities of the complex in the asymmetric unit). Consequently a

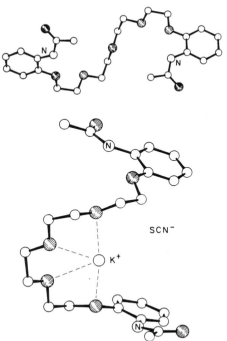

(24)

polymeric structure is formed within the crystal lattice, where every K$^+$ ion is bound to three adjacent ligands.[90] In a parallel manner, every ligand molecule wraps circularly around one cation but binds its two acetyl substituents to two other cations. As in the previous example the terminal aromatic groups of neighbouring ligand moieties are stacked in an antiparallel mode. NMR investigations suggested that the polymeric structure of this compound observed in the solid also occurs in the liquid phase.

Crystallographic analysis of the free ligand has shown an S-shaped molecular structure with a regular scheme of $ag^{\pm}a$ conformational units

Fig. 56. Molecular structures of ligand **24** (top) and its complex with KSCN (bottom).

322

I. Goldberg

along the oligoether chain (Fig. 56). Interestingly, upon K^+ complexation the conformational pattern within the open ligand remains exclusively $ag^{\pm}a$, but with reversed signs for some of the *gauche* torsions. Therefore, it has been suggested by Saenger and coworkers that the main flexibility of these particular noncyclic polyether molecules is around the C–C bonds.

A similar type of association between K^+ and an open-chain ligand is formed upon complexation of 2,2'-di-*o*-carboxymethoxyphenoxydiethyl ether (**25**) with potassium picrate.[91] The ligand adopts a nearly circular

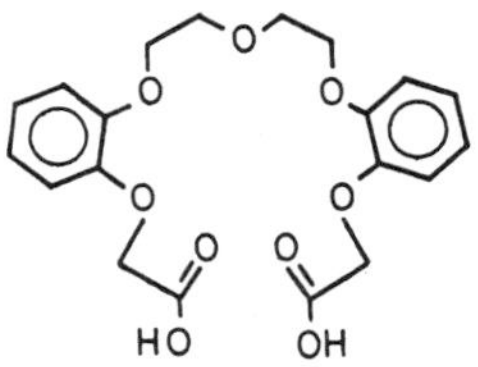

(25)

conformation around the cation, which is provided with an irregular eight-fold coordination. To avoid intramolecular repulsions the terminal carboxylic groups are displaced in opposite directions from the average plane of the molecule. Seven $K^+\cdots O$ interactions involving all ether and carbonyl oxygen sites are confined to one unit of the complex cation. In addition, each K^+ is bound to the carbonyl group of the adjacent ligand, thus joining two units of the complex into one dimer. All $K^+\cdots O$ binding distances are within a typical range of 2.75–2.94 Å.

With larger chain-ligands the stoichiometric ratio of the complex formation changes from 1:1 (mononuclear) to 1:2 (binuclear). A suitable example is that of a binuclear complex between 1,5-bis{2-[5-(2-nitrophenoxy)-3-oxapentyloxy]phenoxy}-3-oxapentane (**26**) and potassium isothiocyanate

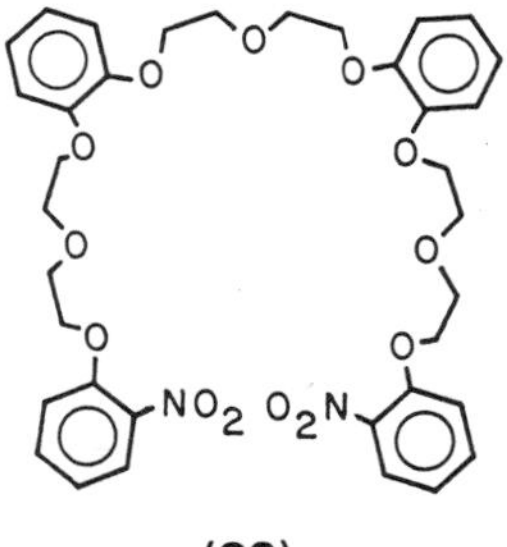

(26)

(Fig. 57).[92] The ligand contains nine ether oxygens and two terminal NO_2 groups which can be bound directly to the metal cation. It has an S-shape configuration with C_2 symmetry. One cation is situated in the centre of each S-loop, being coordinated to seven O-atoms of the ligand as well as ion-paired with the counter ion. In the observed structure each nitro group and

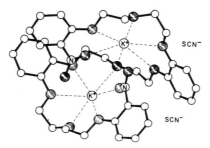

Fig. 57.

the central ether oxygen are bound simultaneously to both cations, stabiliz-ing the spiral configuration of the ligand. There are relatively large variations in the observed $K^+\cdots O$ distances (from 2.70 to 3.30 Å), probably because of the complex spatial relationships, different types of oxygen ligating sites, and multiple coordination of some of them in this structure.

Rubidium salts also form crystalline complexes with open-chain poly-ethers of varying size. The complex between bis(8-quinolyloxyethyl) ether **27** with rubidium iodide represents the smallest structure[93] (Fig. 58). There

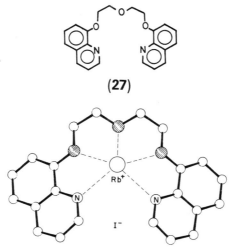

Fig. 58.

are only five heteroatoms ligating the Rb^+ ion with $Rb^+\cdots O$ and $Rb^+\cdots N$ distances of 3.07–3.17 Å and 2.97 Å, respectively. Two symmetry related I^- ions at 3.69 and 3.90 Å complement the coordination sphere around the rubidium. The heteroatoms are nearly coplanar, the cation being displaced by about 1.5 Å from this plane.

When the polyether chain is extended by two ethyleneoxy units to give **28**, the ligand contains enough nucleophilic sites to satisfy the coordination requirements of Rb^+. Indeed in the crystal structure of the bis[(8-quinolyloxy)-ethoxyethyl] ether rubidium iodide 1:1 complex the Rb^+ is coordin-

(28)

ated only to the seven heteroatoms of the ligand at characteristic distances between 2.9 and 3.1 Å. There is no direct interaction between the cation and I^-. The conformation of the host deviates slightly from the usual $ag^{\pm}a$ pattern, allowing the formation of a helical complex and leading to nearly perpendicular orientations of the terminal rings with respect to each other (Fig. 59).[94]

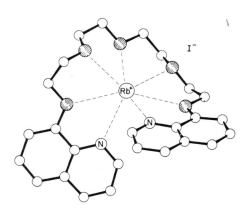

Fig. 59.

Addition of methyl substituents to the quinoline rings at 2-positions leads to changes in the ligand configuration and its coordination around the complexed cation. This is illustrated by the structure of bis(2-methyl-8-quinolyloxyethoxyethyl) ether with RbI. In the crystal the host species is wrapped around the central cation in a configuration resembling one turn of a helix, with the terminal heterocycles stacked on top of each other. The cation interacts with all seven donor heteroatoms but, in contrast to the previous structure, it is also coordinated to the I⁻ anion (Fig. 60).[95] The

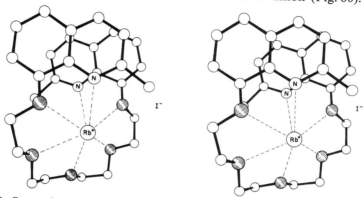

Fig. 60. Stereoview of the complex between rubidium iodide and the dimethyl-derivative of host **28**.

additional attraction to I⁻ leads to slightly weaker interactions between Rb^+ and the heteroatoms. Furthermore, the geometry of the coordination sphere around Rb^+ is not entirely symmetric, the $Rb^+ \cdots N$ or O distances varying from 2.84 to 3.12 Å. The molecular structures of the previous quinoline ligand and the present 2-methylquinoline host differ mainly in two torsion angles about C–O bonds and relative disposition of the terminal rings, which in the latter case impose some stereochemical constraints on the ligand.

Replacement of the 2-methylquinoline residues by tropolone rings giving **29** does not affect the total number of potentially ligating sites on the ligand

(29)

(apart from changing the nitrogen donors to oxygen ones). Therefore, the overall pattern of association between the various constituents remains nearly unchanged as in the complex of 1,11-bis(tropolone)-3,6,9-trioxaun-decane with RbI (Fig. 61).[96] The rubidium cation is wrapped by the ligand in a circular manner, interacting directly with all seven O-atoms at 2.81–3.18 Å. The cation is further coordinated by I$^-$ at nearly 3.67 Å. The two

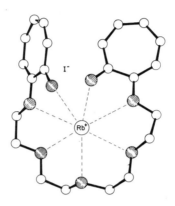

Fig. 61.

approaching tropolone rings are approximately at right angles to each other, allowing formation of an intramolecular $O \cdots C{=}O$ dipolar interaction between the two rings. The latter seems to stabilize the circular configuration of the ligand without any significant distortion of the C–C and C–O torsion angles from their characteristic values.

    Further extension of the polyether chain to contain eight oxygen sites in addition to the terminal aromatic groups leads to formation of a helical structure with one and a half turns. A suitable example is provided by the 1:1 complex of 1,20-bis(8-quinolyloxy)-3,6,9,12,15,18-hexaoxaicosane (**30**)

(30)

with rubidium iodide. The cation is in the centre of the complex surrounded by all the heteroatoms which are arranged on a sphere (Fig. 62).[97] It has a unique tenfold coordination, being completely shielded by the ligand from any direct interaction with the I^- ion. The average $Rb^+ \cdots N,O$ distance is 3.07 Å. In spite of the large number of conformational degrees of freedom in this ligand, only one C–O–C–C torsion angle (127°) was found to deviate largely from the expected value of 180°.

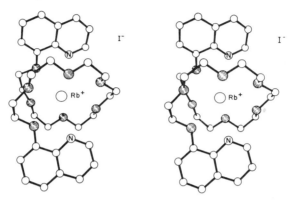

Fig. 62. Stereoview of the RbI complex of **30**.

Many of the structures of linear polyethers described above are characterized by common features. The conformation of the polyether chain is usually fitted as efficiently as possible to the coordination sphere of the complexed cation. In some structures the metal cations are further coordinated directly to their counter ions. As in complexes of cyclic crown ethers, the observed conformations about the C–C bonds are almost exclusively *gauche*, conformational flexibility of the open ligands being most frequently associated with variations of the C–O–C–C torsion angles within the entire *trans → gauche* range. The observed interaction distances between the ligand and the enclosed metal cation very often reflect the relative electronegativity of the ligating heteroatoms. As expected, the less electronegative atoms (e.g., oxygens adjacent to aromatic rings) have usually larger coordination distances.[98]

5.2. Complexes with non-metal guests

The number of publications which relate to crystalline complexes of linear oligoethers with guest species other than metal ions is very small. Thiourea was found to complex easily with one of the ligands referred to above: the

bis[(8-quinolyloxy)ethoxyethyl] ether **28**. The stoichiometry of the complexation is 1:1. The ligand in this structure adopts an S-like configuration,
each S-loop interacting via hydrogen bonds with one NH_2 group of thiourea
(Fig. 63).[99] Particularly strong hydrogen bonds involve the quinolyl nitrogen

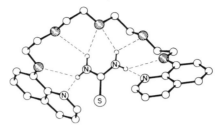

Fig. 63.

atoms. The remaining $NH_2\cdots O$ bridges are relatively long and indicate weak
interactions. An analogous complex containing one thiourea molecule
wrapped by a polyether chain is that between bis[2-(*o*-methoxyphenoxy)-
ethoxyethyl] ether and thiourea. In this structure the guest molecule is
coordinated to all seven donor sites of the ligand with some of the hydrogen
bonds being bifurcated (Figure 64).[100]

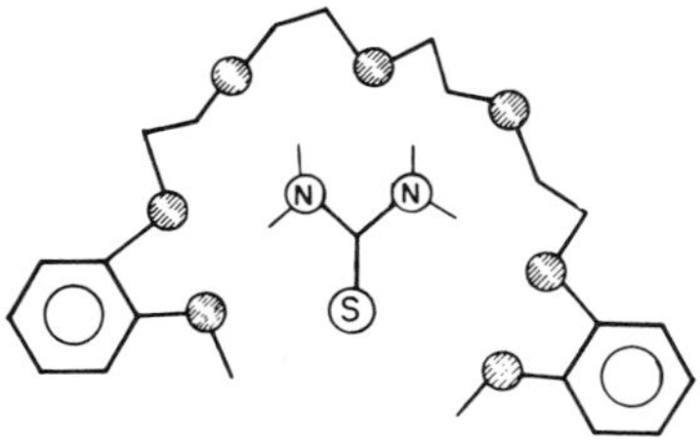

Fig. 64.

It should be noted in this context that stoichiometric complexation of
urea by various open chain polyether ligands has also been reported (for
a review see Vögtle *et al.*[4]). Nevertheless, no detailed structural data are yet
available. Other potential host molecules for urea and thiourea are the
pyridine N-oxides. The only crystal structure analysis reported so far relates
however to the complex between oxybis(2-methylene-6-methylpyridine *N*-
oxide) (**31**) and 2 moles of water. This host appears to be too short and rigid

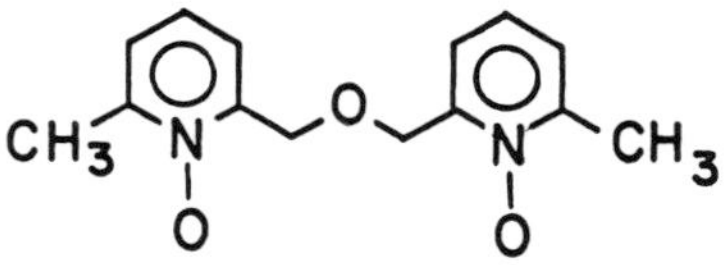

(31)

to wrap around the water molecule. Instead, a dimeric structure is formed by two molecules of the ligand with a pair of water molecules included in it (Fig. 65).[101] Two other H_2O molecules hydrogen-bond to the *N*-oxide oxygen atoms from both sides of the cluster. Interestingly, the ether oxygens point outward and are not involved in coordination. The hydrogen bonding

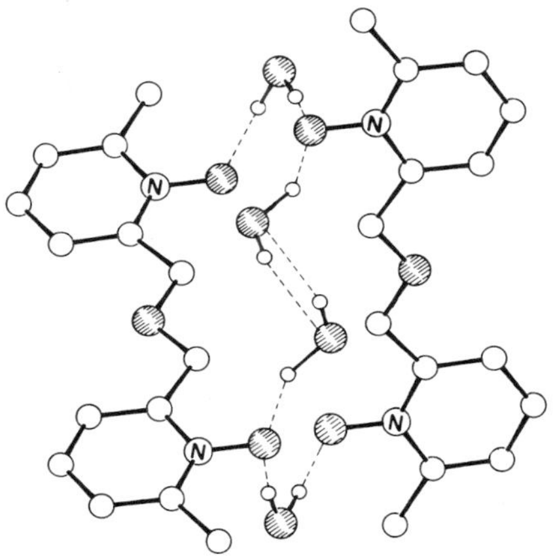

Fig. 65.

interactions, which are not particularly strong (all relevant O···O distances are between 2.78 and 3.06 Å), extend over the entire crystal. This leads to the formation of a layered-type crystal structure with alternating layers of the organic ligand and water. The interesting tripod ligand, tris(2-methyl-8-quinolyloxy)ethylamine, was found to form a monomolecular host–guest complex with water. In this structure the water molecule is attached to one quinolyl chain of the ligand.[102]

A large number of complexes between cyclic crown ether receptors and alkylammonium substrates have been surveyed in a previous section of this chapter. The affinity of open chain polyether ligands for alkylammonium guests is usually much weaker, and the complexation process thus much less effective. In this respect the final example is a unique one as it refers to a stable crystalline host–guest complex of an acyclic ligand and an amino acid ester salt (Fig. 66).[103] The synthetic ligand contains a 7-membered linear

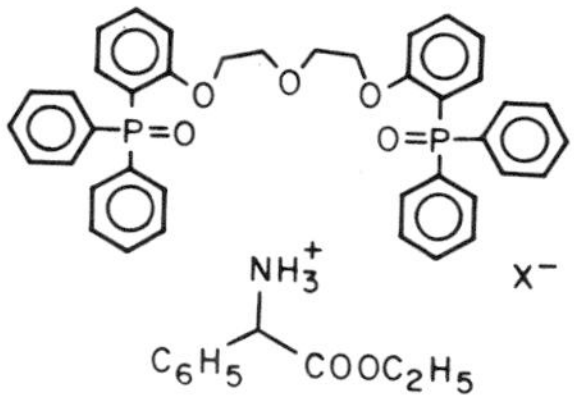

Fig. 66.

chain including three ether oxygens, and two terminal triphenylphosphine oxide groups. The latter are known, in fact, for their very high formation constants in hydrogen bonding. In the crystal, the ligand with its five oxygen ligating sites is folded around the ammonium group. Three hydrogen bonds of the ammonium nitrogen, extended in a tripod arrangement to the two phosphine oxide groups and the central ether oxygen atom (N···O 2.65– 2.82 Å), stabilize the structure. Further stabilization is provided by π–π interactions involving the planar phenyl and ester functions of the guest and the aromatic rings of the host. The present ligand exhibits a remarkable degree of selectivity towards ammonium compounds in extraction experiments. It appears that a significant complexation is achieved only with those substrate moieties that can bind simultaneously to the diethyleneglycol and the two triphenylphosphine oxide residues.

The structural properties of other noncyclic oligoether hosts have recently been reviewed by Vögtle and coworkers.[4] In fact, the open-chain ligands usually tend to exhibit lower selectivities in binding metal cations than their cyclic and less flexible analogues. Considerable effort is continuously being made however in the design of better hosts in order to achieve a more effective and selective complexation.

6. Concluding remarks

Throughout this chapter we have attempted to illustrate the various types of structures that characterize host–guest complexes of macrocyclic and acyclic crown ethers. It has been emphasized that many geometric features occur systematically in these systems; the accumulated data could thus be very useful in predicting new structures.

However, the list of references included in this review is by no means a complete one. Because of the limited space available, discussion of complexes with ligands smaller than 18-crown-6 (e.g., 12-crown-4[104]) or with hosts containing fused crown ether rings[105,106] has been omitted. Little attention has been paid to complexes containing more than two components. These include adducts in which ionic guests and water molecules are accommodated, side by side, within a single crown ether cavity as well as crystal structures of ternary solids. The latter consist of layers of ordinary metal ion or alkylammonium crown ether complexes separated by layers of other organic molecules; two representative examples involve complexes of dibenzo-18-crown-6 with KI and thiourea[107] and with *t*-butylammonium perchlorate and DDQ.[108] Several macrocyclic binuclear complexes with metal cations have also been reported.[109,110] However, the amount of data available is too small to define any general trends that might be characteristic of the structure of such compounds. Another subject related to the binding of *anions* by macrocyclic polyammonium crown-like ligands is only at the initial stages of research, and its evaluation should perhaps be better postponed to the future. Although several hosts suitable for complexing mono- as well as polyfunctional anionic substrates have already been synthesized (e.g., Hosseini and Lehn[111]), a very limited amount of structural work has appeared in the literature as yet.

The possibilities of structural variations in potential receptor molecules are perhaps as wide as is the creative imagination of a synthetic chemist. Even within the limited field of monocyclic and noncyclic hosts new developments are continuously being made. Recent examples are the flexible octopus-type molecules with several oligoether chains attached to a single support[4] and the rigid cyclic cavitands.[83] Undoubtedly, this field is still not exhausted and further progress, mainly in improving the complexation selectivities of hosts toward specific guests, is expected in the future.

Acknowledgement

The author wishes to thank F. R. Fronczek, S. Harkema, B. L. Haymore, R. Hilgenfeld, G. R. Newcome, W. Saenger, I. O. Sutherland, K. N. Trueblood, J. W. H. M. Uiterwijk, F. Vögtle and the Cambridge Crystallographic Data Centre for kindly providing relevant reprints, structural data or manuscripts before publication. He is also grateful to Mrs R. Magen for her help with the diagrams.

References

1. M. R. Truter, *Struct. Bonding* (*Berlin*), 1973, **16**, 71.
2. N. K. Dalley, in *Synthetic Multidentate Macrocyclic Compounds*, (eds. R. M. Izatt and J. J. Christensen), New York Academic Press, 1978, pp. 207–243.
3. R. Hilgenfeld and W. Saenger, *Top. Curr. Chem.*, 1982, **101**, 1.
4. F. Vögtle, H. Sieger and W. M. Müller, *Top. Curr. Chem.*, 1981, **98**, 107.
5. D. J. Cram and K. N. Trueblood, *Top. Curr. Chem.*, 1981, **98**, 43.
6. I. Goldberg, in *The Chemistry of Ethers, Crown Ethers, Hydroxyl Groups and their Sulfur Analogues.* Supplement El (ed. S. Patai), London, J. Wiley and Sons, 1980, pp. 175–214.
7. C. J. Pedersen, *J. Am. Chem. Soc.*, 1967, **89**, 7017.
8. J. D. Dunitz, M. Dobler, P. Seiler and R. P. Phizackerley, *Acta Crystallogr.*, 1974, **B30**, 2733.

9. M. L. Campbell, N. K. Dalley, R. M. Izatt and J. D. Lamb, *Acta Crystallogr.*, 1981, **B37**, 1664.
10. J. D. Lamb, R. M. Izatt, P. A. Robertson and J. J. Christensen, *J. Am. Chem. Soc.*, 1980, **102**, 2452.
11. I. Goldberg, *Acta Crystallogr.*, 1975, **B31**, 754.
12. J. A. Bandy, M. R. Truter and F. Vögtle, *Acta Crystallogr.* 1981, **B37**, 1568.
13. J. A. A. de Boer, D. N. Reinhoudt, S. Harkema, G. J. van Hummel and F. de Jong, *J. Am. Chem. Soc.*, 1982, **104**, 4073.
14. R. Kaufmann, A. Knöchel, J. Kopf, J. Oehler and G. Rudolph, *Chem. Ber.*, 1977, **110**, 2249.
15. J. Galloy, W. H. Watson, F. Vögtle and W. M. Mueller, *Acta Crystallogr.*, 1982, **B38**, 1245.
16. G. Weber, *Z. Naturforsch. Teil B*, 1981, **36**, 896.
17. W. Saenger and R. Hilgenfeld, *Z. Naturforsch. Teil B*, 1981, **36**, 242.
18. S. Harkema, G. J. van Hummel, K. Daasvatn and D. N. Reinhoudt, *J. Chem. Soc., Chem. Commun.*, 1981, 368.
19. A. Knöchel, J. Kopf, J. Oehler and G. Rudolph, *J. Chem. Soc., Chem. Commun.*, 1978, 595.
20. G. Weber and G. M. Sheldrick, *Acta Crystallogr.*, 1981, B37, 2108.
21. H. M. Colquhoun, J. F. Stoddart and D. J. Williams, *J. Chem. Soc., Chem. Commun.*, 1981, 847, 849, 851.
22. T. B. Vance, Jr., E. M. Holt, C. G. Pierpont and S. L. Holt, *Acta Crystallogr.*, 1980, **B36**, 150.
23. T. B. Vance, Jr., E. M. Holt, D. L. Varie and S. L. Holt, *Acta Crystallogr.*, 1980, **B36**, 153.
24. A. Knöchel, J. Kopf, J. Oehler and G. Rudolph, *Inorg. Nucl. Chem. Lett.*, 1978, **14**, 61.
25. P. G. Eller and R. A. Penneman, *Inorg. Chem.*, 1976, **15**, 2439.
26. H. M. Colquhoun, J. F. Stoddart and D. J. Williams, *J. Am. Chem. Soc.*, 1982, **104**, 1426.
27. G. Bombieri, G. de Paoli and A. Immirzi, *J. Inorg. Nucl. Chem.*, 1978, **40**, 1889.
28. J. C. G. Bünzli, B. Klein and D. Wessner, *Inorg. Chim. Acta*, 1980, **44**, L147.
29. J. D. J. Backer-Dirks, J. E. Cooke, A. M. R. Galas, J. S. Ghotra, C. J. Gray, F. A. Hart and M. B. Hursthouse, *J. Chem. Soc., Dalton Trans.*, 1980, 2191.
30. J. A. Bandy, M. R. Truter, J. N. Wingfield and J. D. Lamb, *J. Chem. Soc., Perkin Trans. 2*, 1981, 1025.
31. J. W. H. M. Uiterwijk, S. Harkema, G. J. van Hummel, J. Geevers and D. N. Reinhoudt, *Acta Crystallogr.*, 1982, **B38**, 1862.
32. J. W. H. M. Uiterwijk, S. Harkema, D. N. Reinhoudt, K. Daasvatn, H. J. den Hertog, Jr. and J. Geevers, *Angew. Chem. Suppl.*, 1982, 1100.
33. J. M. Timko, S. S. Moore, D. M. Walba, P. C. Hiberty and D. J. Cram, *J. Am. Chem. Soc.*, 1977, **99**, 4207.
34. O. Nagano, A. Kobayashi and Y. Sasaki, *Bull. Chem. Soc. Jpn.*, 1978, **51**, 790.
35. M. J. Bovill, D. J. Chadwick, I. O. Sutherland and D. Watkin, *J. Chem. Soc., Perkin Trans. 2*, 1980, 1529.
36. K. N. Trueblood, C. B. Knobler, D. S. Lawrence and R. V. Stevens, *J. Am. Chem. Soc.*, 1982, **104**, 1355.
37. B. L. Haymore, 1982, Private communication; results also presented at the Symposium on Macrocyclic Compounds, Provo (Utah), 1980.
38. J. Krane and T. Skjetne, *Tetrahedron Lett.*, 1980, 1775.

39. P. Groth, *Acta Chem. Scand., Ser. A*, 1981, **35**, 541.
40. R. M. Izatt, J. D. Lamb, B. E. Rossiter, N. E. Izatt, J. J. Christensen and B. L. Haymore, *J. Chem. Soc., Chem. Commun.*, 1978, 386.
41. E. Maverick, P. Seiler, W. B. Schweizer and J. D. Dunitz, *Acta Crystallogr.*, 1980, **B36**, 615.
42. J. Dale, *Isr. J. Chem.*, 1980, **20**, 3; and references cited therein.
43. G. Wipff, P. Weiner and P. Kollman, *J. Am. Chem. Soc.*, 1982, **104**, 3249.
44. N. K. Dalley, S. B. Larson, J. S. Smith, K. L. Matheson, R. M. Izatt and J. J. Christensen, *J. Heterocycl. Chem.*, 1981, **18**, 463.
45. I. Goldberg, *J. Am. Chem. Soc.*, 1980, **102**, 4106.
46. D. J. Cram, 1979. Private communication.
47. M. Dobler and K. Neupert-Laves, *Acta Crystallogr.*, 1978, **A34**, S101.
48. I. Goldberg, *J. Am. Chem. Soc.*, 1977, **99**, 6049.
49. L. R. Sousa, G. D. Y. Sogah, D. H. Hoffman and D. J. Cram, *J. Am. Chem. Soc.*, 1978, **100**, 4569.
50. I. Goldberg, *Acta Crystallogr.*, 1977, **B33**, 472.
51. I. Goldberg, *Acta Crystallogr.*, 1975, **B31**, 2592.
52. I. Goldberg, *Acta Crystallogr.*, 1976, **B32**, 41.
53. J. J. Daly, P. Schönholzer, J. P. Behr and J. M. Lehn, *Helv. Chim. Acta*, 1981, **64**, 1444.
54. J. P. Behr, J. M. Lehn, D. Moras and J. C. Thierry, *J. Am. Chem. Soc.*, 1981, **103**, 701.
55. J. P. Behr and J. M. Lehn, *Helv. Chim. Acta*, 1980, **63**, 2112.
56. J. C. Metcalfe, J. F. Stoddart, G. Jones, T. H. Crawshaw, A. Quick and D. J. Williams, *J. Chem. Soc., Chem. Commun.*, 1981, 430.
57. J. C. Metcalfe, J. F. Stoddart, G. Jones, T. H. Crawshaw, E. Gavuzzo and D. J. Williams, *J. Chem. Soc., Chem. Commun.*, 1981, 432.
58. H. M. Colquhoun, J. F. Stoddart, D. J. Williams, J. B. Wolstenholme and R. Zarzycki, *Angew. Chem. Int. Ed. Engl.*, 1981, **20**, 1051.
59. N. K. Dalley, J. S. Bradshaw, S. B. Larson and S. H. Simonsen, *Acta Crystallogr.*, 1982, **B38**, 1859.
60. E. Maverick, L. Grossenbacher and K. N. Trueblood, *Acta Crystallogr.*, 1979, **B35**, 2233.
61. R. M. Izatt, J. D. Lamb, R. E. Asay, G. E. Maas, J. S. Bradshaw, J. J. Christensen and S. S. Moore, *J. Am. Chem. Soc.*, 1977, **99**, 6134.
62. M. J. Bovill, D. J. Chadwick, M. R. Johnson, N. F. Jones, I. O. Sutherland, and R. F. Newton, *J. Chem. Soc., Chem. Commun.*, 1979, 1065.
63. M. R. Johnson, I. O. Sutherland and R. F. Newton, *J. Chem. Soc., Perkin Trans. I*, 1979, 357.
64. L. C. Hodgkinson, M. R. Johnson, S. J. Leigh, N. Spencer, I. O. Sutherland and R. F. Newton, *J. Chem. Soc., Perkin Trans. I*, 1979, 2193.
65. M. R. Johnson, I. O. Sutherland and R. F. Newton, *J. Chem. Soc., Perkin Trans. I*, 1980, 586.
66. J. P. Kintzinger, F. Kotzyba-Hibert, J. M. Lehn, A. Pagelot and K. Saigo, *J. Chem. Soc., Chem. Commun.*, 1981, 833.
67. F. Kotzyba-Hibert, J. M. Lehn and K. Saigo, *J. Am. Chem. Soc.*, 1981, **103**, 4266.
68. M. Ciampolini, P. Dapporto, N. Nardi and F. Zanobini, *J. Chem. Soc., Chem. Commun.*, 1980, 177.
69. J. P. Behr, P. Dumas and D. Moras, *J. Am. Chem. Soc.*, 1982, **104**, 4540.
70. I. Goldberg, *Acta Crystallogr.*, 1978, **B34**, 3387.

71. G. R. Newkome, F. R. Fronczek and D. K. Kohli, *Acta Crystallogr.*, 1981, **B37**, 2114.
72. G. R. Newkome, H. C. R. Taylor, F. R. Fronczek, T. J. Delord, D. K. Kohli and F. Vögtle, *J. Am. Chem. Soc.*, 1981, **103**, 7376.
73. J. W. H. M. Uiterwijk, S. Harkema, J. Geevers and D. N. Reinhoudt, *J. Chem. Soc., Chem. Commun.*, 1982, 200.
74. H. B. Bürgi, J. D. Dunitz and E. Shefter, *Acta Crystallogr.*, 1974, **B30**, 1517.
75. I. Goldberg and H. Rezmovitz, *Acta Crystallogr.*, 1978, **B34**, 2894.
76. T. W. Bell, P. G. Cheng, M. Newcomb and D. J. Cram, *J. Am. Chem. Soc.*, 1982, **104**, 5185.
77. I. Goldberg, *Acta Crystallogr.*, 1981, **B37**, 102.
78. I. Goldberg, *Acta Crystallogr.*, 1978, **B34**, 2224.
79. I. Goldberg, *Cryst. Struct. Commun.*, 1980, **9**, 1201.
80. K. N. Trueblood, C. B. Knobler, E. Maverick, R. C. Helgeson, S. B. Brown and D. J. Cram, *J. Am. Chem. Soc.*, 1981, **103**, 5594.
81. D. J. Cram, G. M. Lein, T. Kaneda, R. C. Helgeson, C. B. Knobler, E. Maverick and K. N. Trueblood, *J. Am. Chem. Soc.*, 1981, **103**, 6228.
82. G. M. Lein and D. J. Cram, *J. Chem. Soc., Chem. Commun.*, 1982, 301.
83. J. R. Moran, S. Karbach and D. J. Cram, *J. Am. Chem. Soc.*, 1982, **104**, 5826.
84. R. Iwamoto, *Bull. Chem. Soc. Jpn.*, 1973, **46**, 1114, 1118, 1123.
85. R. Iwamoto and H. Wakano, *J. Am. Chem. Soc.*, 1976, **98**, 3764.
86. M. Yokoyama, H. Ishihara, R. Iwamoto and H. Tadokoro, *Macromolecules*, 1969, **2**, 184.
87. F. Vögtle and H. Sieger, *Angew. Chem. Int. Ed. Engl.*, 1977, **16**, 396.
88. I. H. Suh, G. Weber and W. Saenger, *Acta Crystallogr.*, 1978, **B34**, 2752.
89. I. H. Suh, G. Weber, M. Kaftory, W. Saenger, H. Sieger and F. Vögtle, *Z. Naturforsch, Teil B*, 1980, **35**, 352.
90. I. H. Suh, G. Weber and W. Saenger, *Acta Crystallogr.*, 1980, **B36**, 946.
91. D. L. Hughes, C. L. Mortimer, D. G. Parsons, M. R. Truter and J. N. Wingfield, *Inorg. Chim. Acta*, 1977, **21**, L23.
92. G. Weber and W. Saenger, *Acta Crystallogr.*, 1980, **B36**, 61.
93. W. Saenger and B. S. Reddy, *Acta Crystallogr.*, 1979, **B35**, 56.
94. W. Saenger and H. Brand, *Acta Crystallogr.*, 1979, **B35**, 838.
95. G. Weber and W. Saenger, *Acta Crystallogr.*, 1979, **B35**, 1346.
96. K. K. Chacko and W. Saenger, *Z. Naturforsch. Teil B*, 1980, **35**, 1533.
97. G. Weber and W. Saenger, *Acta Crystallogr.*, 1979, **B35**, 3093.
98. W. Saenger, I. H. Suh and G. Weber, *Isr. J. Chem.*, 1979, **18**, 253.
99. G. Weber and W. Saenger, *Acta Crystallogr.*, 1980, **B36**, 424.
100. I. H. Suh and W. Saenger, *Angew. Chem. Int. Ed. Engl.*, 1978, **17**, 534.
101. G. Weber and W. Saenger, *Acta Crystallogr.*, 1980, **B36**, 207.
102. G. Weber and G. M. Sheldrick, *Acta Crystallogr.*, 1980, **B36**, 1978.
103. A. H. Alberts, K. Timmer, J. G. Noltes and A. L. Spek, *J. Am. Chem. Soc.*, 1979, **101**, 3375.
104. E. Mason and H. A. Eick, *Acta Crystallogr.*, 1982, **B38**, 1821.
105. M. Czugler and E. Weber, *J. Chem. Soc., Chem. Commun.*, 1981, 472.
106. D. M. Walba, R. M. Richards, S. P. Sherwood and R. C. Haltiwanger, *J. Am. Chem. Soc.*, 1981, **103**, 6213.
107. R. Hilgenfeld and W. Saenger, *Angew. Chem., Int. Ed. Engl.*, 1981, **20**, 1045.
108. J. A. A. de Boer, D. N. Reinhoudt, J. W. H. M. Uiterwijk and S. Harkema, *J. Chem. Soc., Chem. Commun.*, 1982, 194.

109. M. G. B. Drew, M. McCann and S. M. Nelson, *J. Chem. Soc., Chem. Commun.,* 1979, 481.
110. N. A. Bailey, M. M. Eddy, D. E. Fenton, G. Jones, S. Moss and A. Mukhopad-hyay, *J. Chem. Soc., Chem. Commun.,* 1981, 628.
111. M. W. Hosseini and J. M. Lehn, *J. Am. Chem. Soc.,* 1982, **104**, 3525; and references cited therein.

10 · CRYPTATE COMPLEXES

B. DIETRICH

Université Louis Pasteur, Strasbourg, France

1. Introduction

From time to time we all take a "random-walk" around the library picking up books from the shelves just to have a look. If our hand is well guided, our reaction may be: "crown ethers again". Checking more carefully, one may discover from a volume of *Justus Liebigs Annalen*, for the year 1937, the compound dibenzo-20-crown-6 (the only difference from the classical one being that the phenolic ethers are *meta* and not *ortho*);[1] while another careful choice, *J. Chem. Soc.*, 1959, might reveal a methylated 12-crown-4, together with unambiguous evidence for its ability to solubilize alkali metals*.[2]

It is a fascinating question for all of us, why a new area starts to develop only at one moment, even if some compounds or facts exist long before. In our case, respectively thirty, and eight years before the 1967 report of the synthesis of the crown ethers by Pedersen and the discovery of their ability to complex alkali and alkaline earth cations.[3] It may appear that a germination period is required.

* Other examples can be found in the book by Gokel and Korzeniowski (see further reading at the end of this chapter).

INCLUSION COMPOUNDS
ISBN 0-12-067101-8

However, another path, originating in the chemistry of natural substances, was at the same time and independently, leading towards the same target.

In biochemistry Moore and Pressman[4] demonstrated, as early as 1964, the ionophore properties of valinomycin. This gave an Ariadne's thread to the puzzling problem of ion transport across biological membranes. Mainly three groups—located in the Soviet Union (Shemyakin, Ovchinnikov *et al.*), Switzerland (Prelog, Dunitz, Simon, Dobler *et al.*) and in the USA (Pressman *et al.*)—worked over years on the natural ionophores. In the late sixties, synthetic analogues of valinomycin were made in order to elucidate the structural and functional characteristics of the carrier. A vast number of other cation ionophores were discovered and studied: enniatin, nonactin, monensin, nigericin, etc.[5] From the chemist's point of view, the elucidation of the structure of the potassium complex of nonactin by X-ray crystallography was very important because of the spatial surrounding of the cation.[6] (For an X-ray discussion see ref. 7.)

The binding properties of both the natural and the synthetic macrocycles attracted many people to the challenge of synthesizing new complexing compounds. The "design of complexing agents"[8] led to hundreds of representatives of this new class. Along the way, and to subdivide a fast growing family, new names flourished: crown ether, cryptand, coronand, host–guest, podand, spherand, etc.

Applications spread all over chemistry: salt separation, purification of actinides, isotope separation, removal of harmful metals, ion selective electrodes, alkali anions, anionic polymerization, reaction mechanisms, phase transfer catalysis, bio-organic and inorganic modelling, analytical chemistry, chelatotherapy, and others.

In this chapter we will focus our attention only on *cryptates* in the etymological sense: $\kappa\rho\upsilon\pi\tau\sigma\sigma$ = hidden. In other words only macrobicyclic and macropolycyclic systems will be mentioned with a few exceptions which will be explained. This rule will not hold for the short discussion of *anion complexation*, still a poorly explored area.

Due to space limitations the references will not be exhaustive. This implies that in most cases the reader should take note of the "references cited therein". We will also try to give all the major reviews and books which have appeared on this and related subjects.

2. Macrobicyclic ligands—cryptates

2.1. Diaza-polyoxa macrobicycles

The natural substances and the crown ethers hold the bound cation in their macrocyclic cavity. Complete inclusion of the bound cation is achieved by macrobicyclic ligands.

2.1.1. Synthesis

The synthesis of the first macrobicycle, compound **1**, was achieved in 1968 by Lehn's group[9] (Fig. 1).

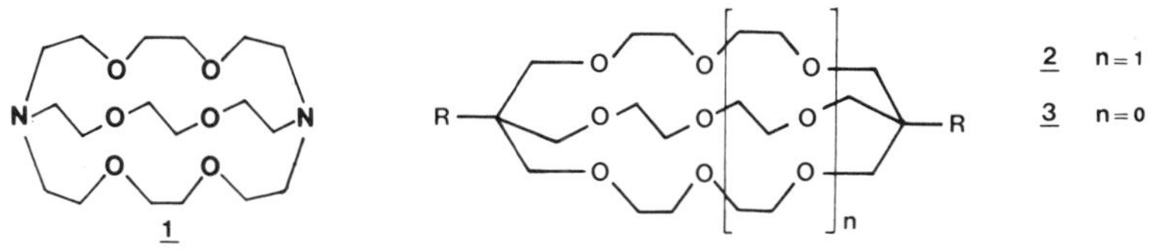

Fig. 1. Macrobicyclic polyethers.

Prior to the synthesis, several other potential complexing agents for alkali and alkaline earth cations had been considered, such as, for example, systems of types **2** and **3** (see Fig. 1)*. However, since bridgehead carbons would lead to difficult isomerism problems (this aspect has been demonstrated recently[10]) nitrogen atoms were chosen instead. They have two major advantages, (i) there are no longer isomerism problems (because of nitrogen inversion) and (ii) they can participate in the coordination of the cation. The successive steps for the preparation of compound **1** are described in Fig. 2.

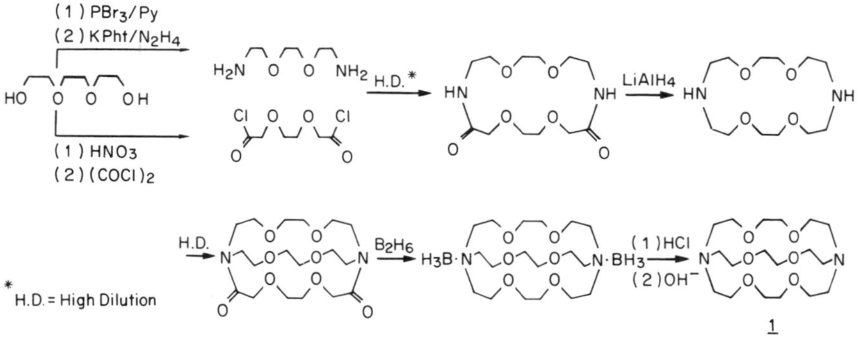

*H.D. = High Dilution

Fig. 2. Synthetic scheme for the first diaza–polyoxa-macrobicycle: [2.2.2.]

This synthesis was not straightforward, and several steps were time consuming. For example, the apparently trivial nitric oxidation of triethylene glycol had to be performed under unusually strict conditions of temperature. The reduction of the macrobicyclic diamide was also a critical step. Using LiAlH$_4$ as reducing agent we cleaved the diamide strand leading back to

---

* Using another synthetic scheme the larger analogue **2** ($n = 1$) was later obtained by Coxon and Stoddart.[11]

the starting monocycle. Twice we thought that we had the final compound in hand. In one case we had the very stable diamino-borane derivative; only very drastic conditions of hydrolysis, 6M hydrogen chloride at 100°C for several hours, gave us the dihydrochloride of **1**. The second case was encountered when we tried to obtain the free ligand **1** by treating the dihydrochloride with KOH; the compound produced was not the free ligand but the potassium complex. Using basic ion-exchange resin, however, we finally obtained **1**.

We immediately began a simple ¹H NMR experiment which gave clear evidence that this bicyclic polyether forms complexes with alkali (AC) and alkaline-earth (AEC) cations: addition of AC or AEC salts to a solution of **1** in CDCl₃ gave a set of new peaks growing progressively while at the same time the peaks of the starting material disappeared. Quite important was the fact that the final spectrum had the same symmetry as that of the starting compound **1**. This led to the presumption that the cation was included in the cavity. Further studies using selective electrodes, pH-metric titration and X-ray analysis[12] showed: (i) that the complexes formed with many cations are very stable and (ii) that the cation is located within the central cavity. This last characteristic suggested to us the name of *cryptate*.[13]

This first macrobicyclic ligand had, as expected, a net preference for the potassium cation. Changing the size of the cavity would then lead to ligands showing selectivity of complexation for all members of a group (AC or AEC). Having this goal in mind, we achieved the synthesis of the compounds represented in Fig. 3.[14] The synthetic pathways are similar to the original

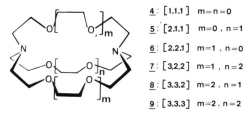

4: [1.1.1] m=n=0
5: [2.1.1] m=0 , n=1
6: [2.2.1] m=1 , n=0
7: [3.2.2] m=1 , n=2
8: [3.3.2] m=2 , n=1
9: [3.3.3] m=2 , n=2

Fig. 3. Generalized structure of diaza–polyoxa-macrobicycles.

method described in Fig. 2. Thanks to the powerful high dilution technique, appreciable overall yields are obtained in all cases except for compound **4** (for this bicycle see details in Section 3.2.1.1 on macrotricyclic ligands).

The IUPAC nomenclature applied to compound **1** is 4,7,13,16,21,24-hexaoxa-1,10-diazabicyclo[8,8,8]hexacosane. To avoid this correct but cumbersome description we proposed trivial names. This coding, by now generally used, is very simple and is formed by three numbers, each designating the numbers of oxygen in each chain. Compound **1** in this system will then be described simply as 222 (see nomenclature of all compounds in Fig. 3). This coding applies only if two successive heteroatoms are separated by two carbon atoms. For the uncomplexed compounds in Fig. 3 the name *cryptand* was proposed. To indicate their bicyclic character we call them [2]-*cryptands*. Tricyclic and tetracyclic ligands will respectively be designated as [3]-*cryptands* and [4]-*cryptands*. Finally for the monocyclic precursor of [2.2.2] the coding is simply [2.2]. To express the complexation of a cation (K^+ for example) with a cryptand the compact description $[K^+ \subset 2.2.2]$ will be used.

Other macrobicyclic polyethers were synthesized by us and other groups and will be mentioned later in this chapter.

2.1.2. Complex formation—stability and selectivity

The complexation of a metal cation M^{n+} by a ligand L is expressed by Equation 1

$$L + M^{n+} \rightleftharpoons [L, M^{n+}] \tag{1}$$

The stability constant is defined in Equation 2

$$K_s = \frac{[L, M^{n+}]}{[L][M^{n+}]} \tag{2}$$

As the activity coefficients of all the species are unknown the thermodynamic stability constants are not given here but rather concentration stability constants.

Since all the ligands of Fig. 3 are diamines the acid-base equilibria have to be considered:

$$L + H^+ \rightleftharpoons LH^+ \tag{3}$$

$$LH^+ + H^+ \rightleftharpoons LH_2^{2+} \tag{4}$$

$$K_1 = \frac{[LH^+]}{[L][H^+]} \tag{5}$$

$$K_2 = \frac{[LH_2^{2+}]}{[LH^+][H^+]} \tag{6}$$

From Equation 1 stability constants can be obtained by using cation selective electrodes[15] (direct measurement of uncomplexed cation concentration). This method is unfortunately limited to the range $1 < \log K_s < 5$. Using the pH-metric titration method[15] equations 1, 3 and 4 have to be taken into account. Addition of a metal cation M^{n+} to a solution of the ligand L affects the titration curves (see Fig. 4). Analysis of these curves leads to the stability

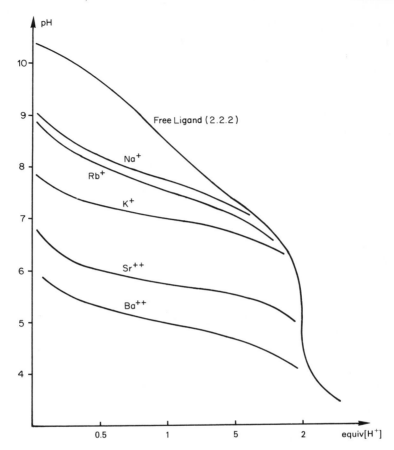

Fig. 4. Titration curves of [2.2.2] and several of its complexes.

constants. This method is well adapted to high stability constants. The results of these measurements are given in Table 1.

High *stability* constants are generally observed. The most attractive property of this class of substances is their *selectivity* of complexation. The selectivity of a ligand in the presence of two cations, M_1^+ and M_2^+ is defined as the ratio $K_{sM_1^+}/K_{sM_2^+}$. Some notable selectivities are expressed in Table 1. The *cavity size* is the determining factor for this behaviour and one can note the correlation between cavity size and ion size in Table 1. A small ligand (such as [2.1.1], [2.2.1] or [2.2.2]) with a small cavity forms the most stable complexes and shows the highest selectivity, i.e. *peak selectivity*. A larger ligand ([3.2.2], [3.3.2] or [3.3.3]) forms less stable complexes and has low selectivity, i.e. *plateau selectivity*.

Table 1. Stability constants in log $K_s^{(a)}$ (solvent, water)—Selectivities[15]

Ligand	Cavity radius in Å	Ionic radius (Å)							
		Li+	Na+	K+	Rb+	Cs+	Ca2+	Sr2+	Ba2+
		0.60	0.95	1.33	1.48	1.69	0.99	1.13	1.35
4 [1.1.1]	~0.5	2.2	—	—	—	—	—	—	—
5 [2.1.1]	0.8	5.5	3.2	<2	<2	<2	2.5	<2	<2
6 [2.2.1]	1.1	2.5	5.40	3.95	2.55	<2	6.95	7.35	6.30
1 [2.2.2]	1.4	<2	3.9	5.4	4.35	<2	4.4	8.0	9.5
7 [3.2.2]	1.8	<2	1.65	2.2	2.05	2.0	2.0	3.4	6.0
8 [3.3.2]	2.1	<2	<2	<2	<0.7	<2	2.0	2.0	3.65
9 [3.3.3]	2.4	<2	<2	<2	<0.5	<2	2.0	2.0	<2
9 [3.3.3](b)	2.4	—	2.7	5.4	5.7	5.9	—	—	—
2(b)	—	—	1.1	2.2	1.9	1.6	—	—	—

Selectivity: [2.1.1] Li+/Na ~ 200; [2.2.1] Na+/Li+ ~ 800; Na+/K+ ~ 30; [2.2.2] K+/Na+ ~ 30; Sr2+/Ca2+ ~ 4000.
(a) Determined by pH-metric titrations.
(b) Solvent: methanol, determined by ion selective electrodes.

If the cation to be complexed does not fit exactly into the internal cavity of the ligand the stability constant of the complex decreases dramatically. This explains the relatively low K_s observed for ligands 7, 8 and 9. This point is even better illustrated by Stoddart's bicyclic compound 2 which shows a stability constant (in methanol) with a potassium cation equal to log $K_s = 2.2$.

The large difference in stabilities of the complexes with K+ observed for ligands 2 and 9 also emphasizes the contribution of the nitrogen atoms in the complexation process. Further investigation of this contribution would be of interest; studies on compound 10 (Fig. 5) which, to our knowledge, has not yet been synthesized, would elucidate this point (see also Section 2.2.4 below). Recently Cox et al.[16] did a comprehensive study of the solvent dependence of cryptate stabilities ([2.1.1], [2.2.1], [2.2.2]) with many

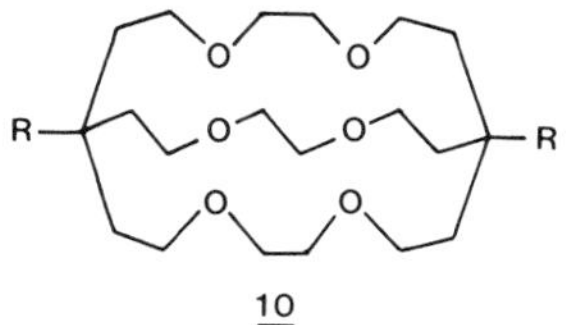

10

Fig. 5. Carbon-bridgehead analogue of [2.2.2].

cations. Many solvents were used and the general tendencies are a large variation with solvent of the stability constants but no major changes are observed in the order of selectivity.

2.1.3. Proton cryptates

Due to its small cavity, compound 4 [1.1.1] forms a weak complex with the lithium cation and shows no complexation at all with larger cations.

On the other hand, fascinating proton cryptates can be obtained. The [1.1.1] bicycle binds one or two protons inside its intramolecular cavity[17,18] (external protonation also occurs) as shown in Fig. 6. The presence of cryptate type structures is indicated both by NMR spectral analysis and by the high resistance of the complex to deprotonation. If 11 is heated at 60° C in 5M KOH for 80 h, it gives only partially the monoprotonated form 12.

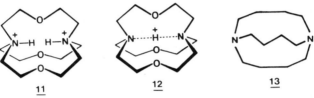

Fig. 6. Proton cryptates.

This last compound could not be deprotonated; all attempts decomposed the molecule. By skilful NMR experiments, internal–external isomerism and the rate of proton transfer were studied.[17,18] Bicycle [1.1.1] shows an extremely slow rate of proton transfer. The smaller bicycle containing bridgehead nitrogens 13 (Fig. 6), synthesized by Alder *et al.*,[19] exhibits similar properties.[20,21] Bicycle [1.1.1] and compound 13 have in common a troubling and singular characteristic; they both behave thermodynamically as strong bases, and yet kinetically, both bases exhibit slow protonation.

2.1.4. Structures

The first structural study of [2.2.2]-RbSCN complex in 1970 by Weiss *et al.*[12] confirmed that the cation was located in the central cavity of the macrobicyclic system. For several years this group made intensive structural studies on many cryptates: free ligand [2.2.2] and complexes formed with Na^+, K^+, Cs^+, Tl^+, Ag^+, Ca^{2+}, Ba^{2+}, Pb^{2+}; [2.2.1] complexes with Na^+, K^+, Co^{2+}; [2.1.1] complexes with Li^+, and [3.2.2] complexes with Ba^{2+} (see compilation and discussion in ref. 7 p. 177). In the cryptate [$K^+ \subset 2.2.2$], all eight heteroatoms are coordinated to the cation. The conformation of the ligand in this complex has minimal strain and adopts D_3 symmetry. Symmetries close to D_3 are found in most of the [2.2.2] cryptates. For other cations (Na^+, Rb^+, Cs^+) the ligand has to change its cavity size. This is clearly illustrated by the modification of the $N \cdots N$ distances which are: 5.50 Å for Na^+, 5.75 Å for K^+, 6.01 Å for Rb^+ and 6.07 Å for Cs^+. In [$Na^+ \subset 2.2.2$] the ligand reaches the limit of compression and in [$Cs^+ \subset 2.2.2$] the upper limit of expansion. For non-symmetric cryptands, like [2.2.1], the cation can either occupy a central position, as for the Na^+ complex, or, as in the case of the K^+ complex, move closer to the macrocyclic moiety having the longest strands ([2.2] moiety).

In many cases anions or solvent molecules have no direct interaction with the complexed cations. For doubly-charged cations (Ca^{2+}, Ba^{2+}, Pb^{2+}) this is often not the rule and water molecules or anions participate in the cation coordination.

2.1.5. Cryptate effect

Comparing polyaza[14]ane N_4 macrocycle (cyclam) to its open-chain tetramine analogue, Cabbiness and Margerum[22,23] observed that the macrocyclic Cu(II) complex is much more stable (by a factor of 10^4) than the open-chain Cu(II)-complex. They termed this enhancement the *macrocyclic effect*. The same effect is observed with polyether complexing agents[24] (see Fig. 7a).

In the cryptate series it has been established that a similar effect takes place.[15] One can compare (Fig. 7b) the [$K^+ \subset 2.2.2$] complex and its "one-strand-opened" analogue.* The stability increases by a factor of 10^5. This effect was called the *macrobicyclic* or *cryptate effect*.

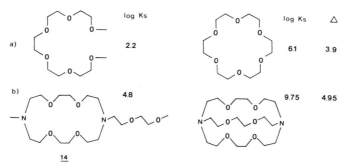

Fig. 7. Stability enhancement by macrocyclic and macrobicyclic effects. Solvents: (a) methanol, (b) methanol–water, 95:5.

* Incidentally note that **14** is an earlier example of the new class of complexing agents called by Gokel "Lariat Ether".[25]

Thermodynamics can explain this effect (see ref. 29) but an intuitive interpretation can also be stated. In the macrobicyclic compound many of the unfavourable interactions (electron–electron repulsion of neighbouring oxygens, steric hindrance) exist, at least in part, before the complexation of the cation, these interactions are built-in during the synthesis. In compound **14**, energy must be consumed in order to overcome these unfavourable interactions in the complexation process. These above considerations, not always taken sufficiently into account in the design of ligands, are well illustrated in the "spherand host".

In the elegant chapter by Cram and Trueblood[26] the importance of ligand shape design is pointed out. In the following excerpt all the fundamental facts are expressed:

"the larger the number of host ligating sites organized for binding during synthesis rather than during complexation, the greater the standard free energy change that accompanies complex formation".[27]

In some cases this strict synthetic building is pushed so far that complexation brings about a decrease in the unfavourable interactions. Reference 26 should be consulted for details. (See also Chapter 9, Volume 2.)

2.1.6. Thermodynamics

The thermodynamics of cryptate formation have been studied in order to assign enthalpy, ΔH, and entropy, ΔS, contributions to the free energy of complexation, ΔG[28,29]. Cryptands, being uncharged ligands, belong to the enthalpic type of complexing agents; but as the complexed cations are "hard", they are of the entropic type. This ambivalent situation leads to several possibilities: enthalpic stabilized complexes, entropic stabilized complexes, and both entropic and enthalpic stabilization.

In Table 2 the three types are encountered. The cryptates of Na^+, K^+, Rb^+ (except $[Na^+ \subset 2.2.1]$ are purely enthalpic, the entropic terms being unfavourable ($\Delta H > \Delta G$). The Ca^{2+} complexes with [2.1.1] and [2.2.2] are entropic type cryptates. Finally [2.2.1] complexes with Na^+, Ca^{2+}, Sr^{2+} and Ba^{2+} are entropically and enthalpically stabilized.

It is of interest to note that the purely enthalpic origin of cryptate formation found in many cases, is markedly different from that of chelate formation which usually shows a strong positive entropy of complexation. A detailed discussion of this matter is given in ref. 29. A more general article on thermodynamics of macrocyclic complexes appeared recently.[30]

2.1.7. Other physicochemical studies

Editorial requests for a *brief* review on Cryptates, and the scope of this book (overviews on all inclusion compounds), do not allow further develop-

Table 2. Free energy ΔG, *enthalpy,* ΔH, *and entropy,* ΔS, *of complexation*[29]

Ligand	Thermodynamic parameter	Na$^+$	K$^+$	Rb$^+$	Ca^{2+}	Sr^{2+}	Ba^{2+}
5 [2.1.1]	$-\Delta G_c^{(a)}$	18.8	—	—	14.2	—	—
	$-\Delta H_c^{(a)}$	22.6	—	—	0.4	—	—
	$T\Delta S_c^{(a)}$	−3.8	—	—	13.8	—	—
	$\Delta S_c^{(b)}$	−12.5	—	—	46.4	—	—
6 [2.2.1]	$-\Delta G_c^{(a)}$	30.1	22.6	14.42	39.7	41.8	35.9
	$-\Delta H_c^{(a)}$	22.36	28.4	22.6	12.1	25.5	26.3
	$T\Delta S_c^{(a)}$	7.74	−5.8	−8.18	27.6	16.3	9.6
	$\Delta S_c^{(b)}$	25.9	−19.6	−27.2	92	54.8	32.2
1 [2.2.2]	$-\Delta G_c^{(a)}$	22.1	30.1	24.7	25.1	45.5	53.9
	$-\Delta H_c^{(a)}$	30.9	47.6	49.3	0.8	43.0	58.9
	$T\Delta S_c^{(a)}$	−8.8	−17.5	−24.6	24.3	2.5	−5.0
	$\Delta S_c^{(b)}$	−29.3	−58.9	−82.8	81.5	8.4	−16.7

(a) In units of kJ mol^{-1}.
(b) In units of kJ mol^{-1} K^{-1}.

ment on this vast subject and only a few references on different topics can be included. I should apologize then, to all the workers in various fields for citing only briefly some of the studies.

All the references indicated below have to be taken in the sense: "and references cited therein". A review by Popov and Lehn including many physicochemical studies appeared recently and should be consulted for detailed information.[31] References on the following aspects of cryptate complexes are: *kinetics*;[30,32–36] *NMR* (including some kinetics measurements) ¹H and ¹³C NMR,[37,43] ⁷Li,[38,39] ²³Na,[40–43] ¹³³Cs,[44,45] ¹⁷O,[46] ¹⁵N;[47] *conductance*;[48] *electrochemistry*;[49] *electronic and vibrational spectroscopy*;[50,51] *Raman spectroscopy*;[52] *electron spectroscopy.*[53]

2.2. Structural modifications

In the diaza-polyethers described in Section 2.1 many changes can be made in order to allow wider use of cryptates. These changes can affect either the *lipophilicity* of the ligand or the *nature* of the *heteroatoms*. Cryptands of these different types will be considered in this part.

2.2.1. Lipophilic cryptands
There were two major reasons for deciding to make these compounds: (i) the cryptates formed with more lipophilic ligands would be more soluble

in organic solvents (application to anionic reactivity), and (ii), according to Simon,[54] lipophilic complexes would show a different selectivity of complexation of monovalent versus divalent cations (a more antibiotic-like behaviour was then expected). The lipophilic cryptands shown in Fig. 8 were synthesized[55] by the original method described in Fig. 2 (see nomenclature under the compounds).

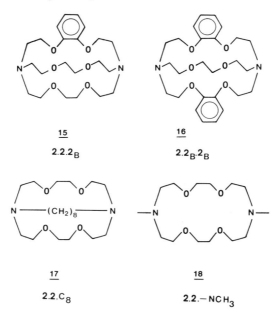

Fig. 8. Lipophilic cryptands.

Complex formation with these compounds was observed by NMR and UV-spectroscopy, or isolated complex analysis. Owing to the small water solubilities of most of these cryptands, we measured the stability constants in a 95:5 methanol:water mixture. Selected results are indicated in Table 3.

The following discussion is based on a K^+/Ba^{2+} comparison: this makes sense since the ionic radii are very close ($K^+ = 1.33$ Å, $Ba^{2+} = 1.35$ Å); only minor size effects can then occur. Several comments can be made: (i) cryptand [2.2.2] **1** has a large preference for Ba^{2+}; (ii) compound **16** [$2.2_B.2_B$] forms complexes of identical stability with Ba^{2+} and K^+; (iii) **17**, [$2.2.C_8$] has a marked selectivity for K^+, even higher than nonactin.* These changes

_____

* Compound [$2_B.2_B.2_B$] has been synthesized[56] but, only the stability in methanol of the potassium complex is indicated: log $K_s \sim 7$.

Table 3. *Selectivity of monovalent cations versus divalent cations*[a] (*solvent; methanol–water 95:5*)

Ligand	K^+	Ba^{2+}	K^+/Ba^{2+}
1 [2.2.2]	9.45	11.5	1/110
15 [2.2.2$_B$]	9.05	11.05	1/100
16 [2.2$_B$.2$_B$]	8.6	8.5	~1
17 [2.2.C$_8$]	4.35	<2	>200
18 [2.2–NCH$_3$]	4.4	6.65	1/200
Nonactin	3.6	1.7	80

[a] Determined by pH-metric titrations.

of Ba^{2+}/K^+ selectivity can be explained in the following way. In **16**, the thickness of the organic layer separating the complexed cation from the solvent, which still has favourable interactions with the cation, is much larger than in **1**. The overall result is a destabilization of the complexes with K^+ and Ba^{2+}, but this effect is more pronounced for the divalent cation than for the monovalent one. This behaviour can be rationalized by the Born equation.[8,54] The change between **1** and **17** has another origin: **1** has eight donating atoms and **17** only six; as K^+ and Ba^{2+} have respectively $n = 6$ and $n = 8$ for hydration numbers, it is again the divalent cation which is much more affected by the decrease in donating atoms. Finally, the crucial role of the solvent is very well illustrated with compounds **17** and **18**. Both have identical numbers of donating atoms but **17** has the $-(CH_2)_8-$ strand which partially shields the cation from solvent approach. This shielding does not exist in compound **18** in which the solvent can participate on both sides in the coordination of the ion. Again K^+, which has not such a large requirement for extra coordinating solvent sites, has the same stability constant in **17** and **18**. But for Ba^{2+} there is a large change between the monocycle **18** and bicycle **17**. Vögtle recently reported a similar shielding effect[57] (see Section 2.2.7). The control over AEC/AC selectivity can then be achieved by changing either the lipophilicity or the number of donors in the ligands. Compounds **16** and chiefly **17** show, with respect to M^+/M^{2+} selectivity, a similarity with a large number of natural macrocyclic ligands.

2.2.2. Nitrogen containing cryptands

Several macrobicycles containing an increasing number of nitrogens have been synthesized by Lehn and Montavon[58] (see Fig. 9).

By pH-metric titration, stability constants have been obtained[59] and these are given in Table 4.

Table 4. Stability constants in log K_S^{59} (solvent, water)[a]

Ligand	Li+	Na+	K+	Rb+	Mg2+	Ca2+	Sr2+	Ba2+	Ag+	Tl+	Co2+	Ni2+	Cu2+	Zn2+	Cd2+	Hg2+	Pb2+
1 [2.2.2]	<2	3.9	5.4	4.3	<2	4.4	8.0	9.5	9.6	6.3	<2.5	<3.5	6.8	<2.5	7.1	18.2	12.7
19 [2.2.2$_{ON}$]	1.5	3.2	4.2	3.0	1.9	4.6	7.4	9.0	10.8	6.3	5.2	5.0	9.7	6.3	9.7	21.7	14.1
20 [2.2.2$_N$]	2.4	2.5	2.7	2.3	2.6	4.3	6.1	6.7	11.5	5.5	4.9	5.1	12.7	6.0	12.0	24.9	15.3
21 [2.2$_N$·2$_N$]	—	—	1.7	—	—	1.5	1.5	3.7	13.0	4.1	5.2	5.7	12.5	6.8	10.7	26.1	15.5

[a] Determined by pH-metric titration.

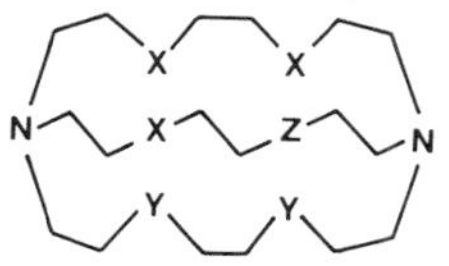

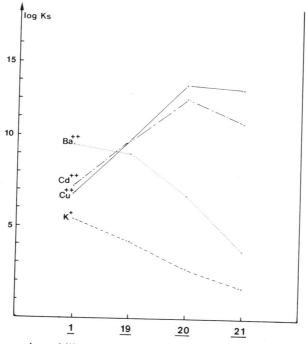

19 X=Y=O   Z=NCH$_3$

20 X=Z=O   Y=NCH$_3$

21 Y=O   X=Z=NCH$_3$

Fig. 9. Poly-nitrogen containing cryptands.

For AC and AEC a stepwise decrease in stability is observed going from
1 ([2.2.2]) to 21 (change of "hard" to "intermediate" base). Not surprisingly
the reverse holds true for transition metal cations; this is clearly illustrated
in Fig. 10.

Many types of selectivity are of importance, and potential applications
to the detoxication of heavy metals can be imagined (on this problem, see
ref. 60). For this purpose a good ligand has to bind the toxic metal (Pb^{2+},
Cd^{2+}, Hg^{2+}) strongly but, equally important, disregard the biologically
essential cations (Na^+, K^+, Mg^{2+}, Ca^{2+}, Zn^{2+}). This is achieved in many
cases. It is noteworthy that the selectivities for toxic metals versus AC and

Fig. 10. Change in stability constant as a function of the number n of nitrogen
donating atoms, where $n_1 = 2$, $n_{19} = 3$, $n_{20} = 4$, $n_{21} = 6$.

Mg^{2+} are always high but versus Ca^{2+} and Zn^{2+}, all ligands do not exhibit the required characteristics. Note in Table 4 some remarkable selectivities: **20**: $Cd^{2+}/Zn^{2+} = 10^6$, $Hg^{2+}/Zn^{2+} = 10^{19}$, $Pb^{2+}/Zn^{2+} = 10^9$, $Cd^{2+}/Ca^{2+} = 10^8$; **21**: $Cd^{2+}/Zn^{2+} = 10^4$, $Cd^{2+}/Ca^{2+} = 10^9$.

2.2.3. Sulphur containing cryptands

Four macrobicycles, having different numbers of atoms of the "soft" donor, sulphur, have been synthesized[61] (see Fig. 11). As the number of sulphur atoms increases, the solubilities in water, and in many other solvents, drop dramatically. This is a severe limitation to the use of this type of compound.

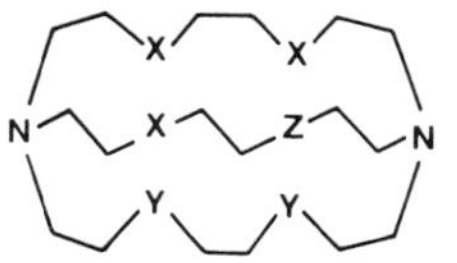

22	X=Z=O	Y=S
23	X=O	Y=Z=S
24	Y=O	X=Z=S
25	X=Y=Z=S	

Fig. 11. Sulphur containing cryptands.

Nevertheless we were able to form many complexes with these ligands and to do some stability measurements in methanol using cation specific electrodes[62] (see Table 5).

Bicycle **22** retains a reasonable stability with AC but in **23** there is a drop by a factor of 10^5 in stability with K^+ (**22** has still six "hard" or "intermediate" binding sites, only five are left in **23**). No extensive investigations in transition metal cations have been carried out so far; however, it has been observed that bicycle **24** forms a very stable complex with Cu(I).[63]

Most of the work described in the preceding sections was done by Lehn's group over a period of 15 years. The stepwise changes introduced in each new type of cryptand allowed the evolution of a deeper understanding of all the parameters which control complexation: cavity size, lipophilicity, and the nature of the heteroatoms. A similar approach was taken, in various

Table 5. Stability constants[a] *in* log K_s[62] *(solvent, methanol)*

	Li$^+$	Na$^+$	K$^+$	Rb$^+$	Cs$^+$	Tl$^+$	Ag$^+$
1 [2.2.2]	2.65	>8	>8	6.4	4.4	—	—
22 [2.2.2$_S$]	2.2	6.0	7.0	4.4	2.3	7.5	9.5
23 [2.2$_{OS}$.2$_S$]	2.3	2.8	2.4	<2	<2	5.2	8.0

[a] Determined by cation-sensitive electrodes.

related areas by other groups, on "host–guest" chemistry,[64] on carbohydrate crown ether derivatives,[65] on podands[66] etc. In the following parts different new types of cryptand are illustrated, including their complexation ability. As we will see, many important contributions have appeared in recent years.

2.2.4. Bridged macrocyclic polyethers

In 1978 Parsons published the synthesis of the two macrobicycles[67] shown in Fig. 12.

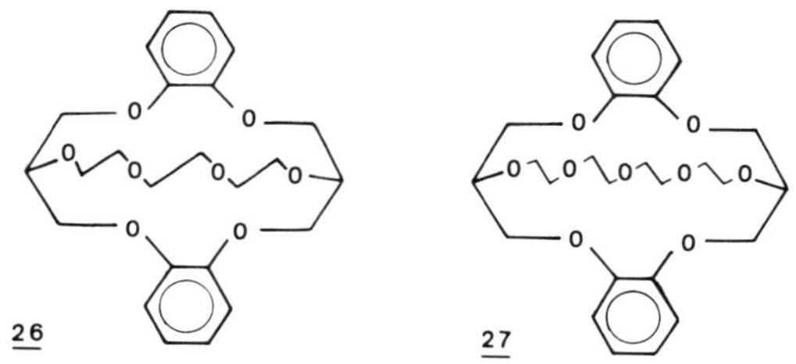

Fig. 12. Carbon-bridgehead macrobicyclic polyethers.

These compounds constitute a recent advance in the chemistry of macrobicyclic cryptands. As shown in Table 6, the stability constants, measured in water using cation sensitive electrodes, are very high and in the same range as those of [2.2.2] cryptates, measured by pH-metric titration.[15] [Na$^+$ ⊂ 26] has the same high stability as [Na$^+$ ⊂ 2.2.1]. Quite noticeable is the uncommonly high selectivity of K$^+$/Rb$^+$ by ligand 26.

The larger macrobicycle 27 forms less stable complexes with Na$^+$ and K$^+$ but, owing to the cavity size, a more stable complex with Rb$^+$. The high stabilities can be explained by two factors: (i) the large number of oxygen atoms included in the bicycles, eight for 26 and nine for 27, which is large

Table 6. Stability constants in log K_s[67] (solvent, water)

	Na$^+$	K$^+$	Rb$^+$
26[a]	5.4	5.7	3.8
27[a]	3.5	4.3	4.4
1 [2.2.2][b]	3.9	5.4	4.35
6 [2.2.1][b]	5.4	3.95	2.55
18-Crown-6[a]	<0.3	2.05	—

[a] Determined by cation selective electrodes.
[b] Determined pH-metric titrations.

enough for AC; (ii) in **26** and **27** two successive oxygens are separated only by two carbon atoms even on the bridgehead. This was not the case in Stoddart's compound **2** discussed earlier (Section 2.1.2). Later on, other analogous macrobicycles were synthesized (see Fig. 13); the tribenzo **29**[68] and three isomeric macrobicyclic polyethers (**30**, **31**, **32**).[69] Compound **29**

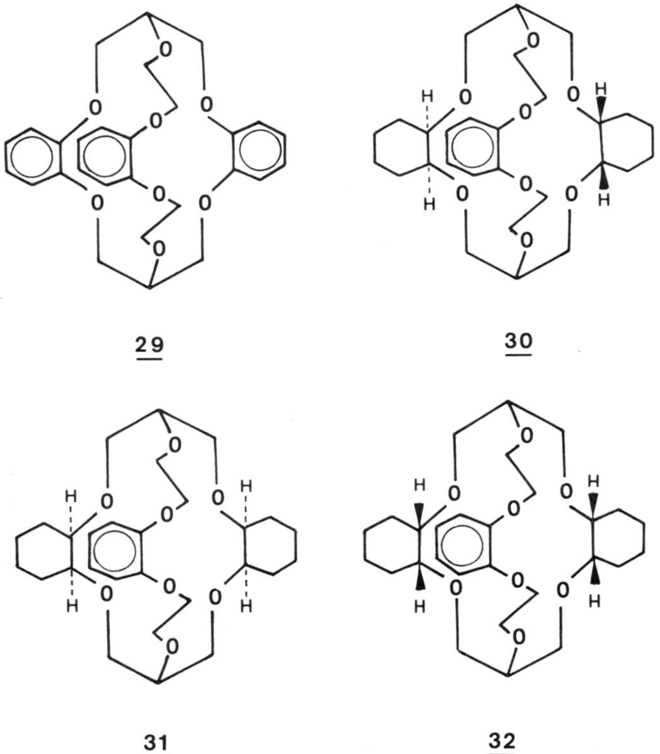

Fig. 13. Carbon-bridgehead macrobicyclic polyethers.

forms very stable complexes in methanol, with K^+ (log $K_s \sim 8.7$) and Cs^+ (log $K_s = 4.2$ for the 1:1 complex; log $\beta_2 = 7.7$ for the 2 ligand:1 metal complex). In compounds **30, 31** and **32**, large differences in stability constants are observed,[69] i.e., all isomers do not have the same ability to complex. A few X-ray structures of both the free ligands and their complexes have been studied for the compounds described in this part.[70–72] Another important fact is the non-sensitivity to pH change of these compounds. This new class of cryptands has certainly a promising future.

2.2.5. *Pyridino–cryptands*

In 1976 Vögtle *et al.* synthesised[73] the first cryptand having a pyridine in one chain, compound **33** in Figure 14. The complexes, formed with many cations, are very stable; and similar in stability to those of [2.2.1]. For example, in water, expressed in log K_s: $[Na^+ \subset \mathbf{33}] = 5.3$ ($[Na^+ \subset 2.2.1] = 5.4$). For Ca^{2+}, Sr^{2+}, Ba^{2+} the stabilities with **33** are, respectively, log $K_s = 7.8$, 8.6, 7.9 compared to 6.9, 7.3, 6.3 with [2.2.1].

More recently Newkome *et al.* reported a cryptand containing three pyridines (**34** in Fig. 14). X-ray analysis showed that the bridgehead nitrogen possesses a planar configuration.[74] This group synthesized, using a fast but

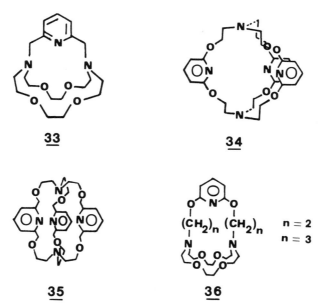

33   **34**

35   **36**

Fig. 14. Pyridine containing cryptands.

low-yielding route, the larger analogue **35**. The X-ray structure revealed that the free ligand possesses a peculiar characteristic: one of the pyridine rings is included in the cavity![75] This compound seems to form (based on elemental analysis) binuclear complexes with $CoCl_2$ and $CuCl_2$. Later this group proposed a more convenient synthesis of pyridino-cryptand **36** by a quaternization–dealkylation procedure.[76]

Bipyridino-cryptands have also been described;[77] complexes with Li^+ and Na^+ cations are mentioned.

2.2.6. Photoresponsive cryptands

The cryptand **37**, containing the photosensitive azobenzene moiety has been described.[78] By irradiation, **37** is transformed into the *cis* azobenzene **37'**. As easily seen in Fig. 15, this *trans-cis* isomerism leads to a change of cavity

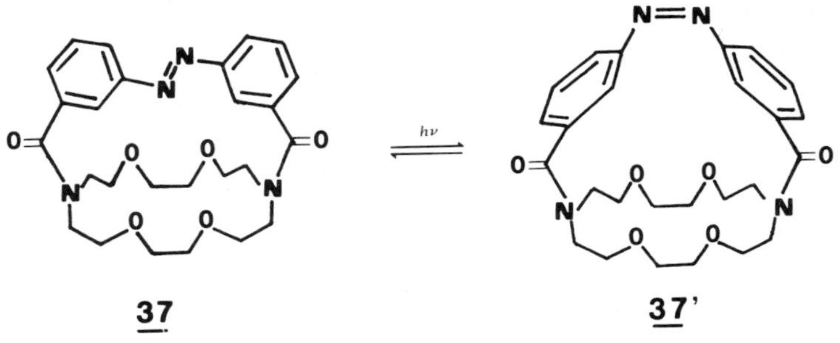

37   **37'**

Fig. 15. Photoresponsive cryptands.

size, *i.e.* a change in cation selectivity. In Table 7 the extraction percentages are given. From these results it can be concluded that K^+ binds preferentially to the "*cis*" cryptand; the smaller Li^+, Na^+ cations are relatively better extracted by the "*trans*" cryptand. Applications to ion transport are suggested by the authors. The same group recently published several papers on this interesting subject.[79-81]

Table 7. Extraction[a] of alkali metal salts[78] (water, benzene)

	Li^+	Na^+	K^+	Cs^+
Trans 37	1.1	14.8	15.8	<0.1
Photo-irradiated 37	<0.1	13.5	27.2	<0.1
Cis 37'	—	12.6	34.8	—

The photo-irradiated **37** is a 60/40 mixture of *cis/trans*. Values for pure *Cis-37'* are then calculated.
[a] Expressed in percentage of metal salt extracted.

2.2.7. Various types of macrobicycles

Many other cryptands, which will be mentioned shortly, have been synthesized in recent years. A comprehensive review up to mid 1979 on crowns and cryptands was published[82] and should be consulted for more general information. The various cryptands are illustrated in Fig. 16. Compound **38** forms complexes with AC and AEC; the number, *n*, of carbons in the chain

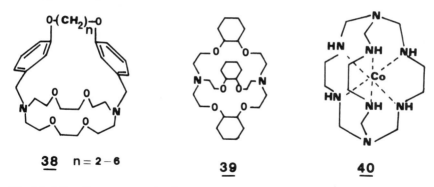

Fig. 16. Miscellaneous cryptands.

does not greatly affect the stability constants.[57] As noted in Section 2.2.1, the divalent cations are weakly complexed due to the shielding effect. The lipophilic cryptand **39** is an efficient phase transfer catalyst.[83] The complex **40** is formed in one step and in very high yield by the action of formaldehyde and ammonia on $Co(en)_3^{3+}$. This synthesis occurs with retention of chirality. The complex was called a sepulchrate by the authors.[84,85] Similar types of complex were described by the same group.[86]

Several other cryptands containing various side chains have also been described: Ferrocene cryptands;[87–89] Urethane-cryptands;[90] Terphenylene cryptands;[91,92] Chiral binaphthyl-cryptands.[93] All these compounds display varying complexation abilities.

Finally, naturally occurring macrocycles have been modified in order to create cavities, or to enhance the cavity character. This was realized for cyclodextrins which are transformed into "capped" cyclodextrins.[94–100] See Chapters 13 and 14, Volume 3. In recent years numerous results have been obtained in the modified porphyrin series. There are "crowned",[101] "pocket",[102] "face-to-face",[101–108] "bridged" or "capped" porphyrins.[109–119] This new area involves interests ranging from theoretical through biological to practical.

3. Macrotricycles, macrotetracycles, mononuclear and dinuclear cryptates

The preceding section was largely concerned with the recognition of a single simple entity, a cation. The various ligands described allow a large variety of specific complexation of specific cations. In this part more sophisticated receptors will be treated. Apart from those described in Section 3.1, they are designed to be receptors for (i) several substrates leading to *dinuclear complexes*, and (ii) multifunctional substrates providing an entry to *molecular complexes*. To achieve these goals the ligands must incorporate in their structures two separate receptor sites. A gradual rise in complexity results in the synthesized systems bearing, more and more, the characteristics of biological receptors.

References 8, 121, 122 and 149 should be consulted on ligand topology.

3.1. Spheroidal macrotricycles

One of the most aesthetically pleasing cryptands, compound **41**, was synthesized by Graf and Lehn.[120] The elaborate synthetic pathway is described in ref. 121. In Fig. 17 this macrotricycle and analogous ones are illustrated.

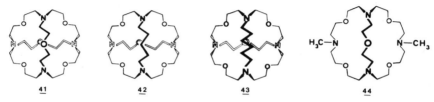

Fig. 17. Spherical macrotricycles.

Cryptand **41** possesses a broad range of binding properties. It complexes, depending on the pH, a cation, a water molecule or an anion. The stability constants of the complexes formed with AC and AEC are listed in Table 8. It is noteworthy that (i) for **41** the stability constants are high, (ii) the rubidium cation complexes with **41** and **42** have the highest stability con-

Table 8. Stability constants[a] in log K_s[123] (solvent, water)

	Na$^+$	K$^+$	Rb$^+$	Cs$^+$	Ca^{2+}	Sr^{2+}	Ba^{2+}
41	1.6	3.4	4.2	3.4	4.1	6.7	8.2
42	1.8	2.5	3.3	2.8	2.4	2.8	5.3
43	<2.0	<2.0	<2.0	<2.0	<2.0	~2.0	3.7
44	—	1.3	1.3	—	—	—	—
1 [2.2.2]	3.9	5.4	4.35	<2	4.4	8.0	9.5

[a] Determined by pH-metric titrations.

stants, in contrast to those of the smaller (K^+) and the larger (Cs^+) cations, and (iii) structural changes in 41, giving 42 and 43, bring about a large drop in K_s. In addition by comparing 41 to 44, a large spherical *cryptate effect* can be detected.[121] The number of donating atoms, (6 oxygens, 4 nitrogens) in 41 explains the observed high stability constants. NMR kinetic measurements showed that the rate of complex formation is not as fast as that in the [2] cryptand series. The same studies demonstrated that the dissociation process is slow.[121] These results can both be explained by the increasing difficulty encountered by the cation in entering, or leaving, the well-hidden cavity.

Based on NMR studies (^1H, ^{13}C, ^{17}O) the diprotonated and triprotonated ligand 41 appears to form complexes with a water molecule.[122] Unambiguous proof, from, for example, an X-ray analysis, has not yet been obtained.

Complexes with the ammonium cation, NH_4^+, have also been observed. This was achieved by various intensive studies:[123] ^1H, ^{13}C, ^{14}N nuclear magnetic resonance and I.R. spectroscopy. The X-ray structure of [$NH_4^+ \subset$ 41] makes clear that the ammonium cation is held inside the cavity, and the bridgehead nitrogens, arranged at the corners of a tetrahedron, form linear $N^+–H\cdots N$ hydrogen bonds with NH_4^+.[124] The binding is completed by bent $N^+–H\cdots O$ hydrogen bonds formed between NH_4^+ and the six oxygen atoms of the ligands. Detailed analysis of X-ray data (interatomic distances) indicated that the cavity is slightly too large for NH_4^+. Nevertheless, as shown in Table 9, exceptionally high stability constants and selectivities

Table 9. Stability constants[a] in log K_s[123] (*solvent, water*)

	NH_4^+	K^+	Rb^+	Selectivity NH_4^+/K^+
41	6.1	3.4	4.2	500
42	4.3	2.5	3.3	63
44	1.7	1.3	1.3	2.5

[a] Determined by pH-metric titrations.

(versus AC) are observed for compound 41; they are the highest known to date. As already mentioned above, a structural modification of 41, e.g. compound 42, or the opening of one chain, e.g. compound 44, has a very large effect on the stabilities of the complexes.[123]

Finally, in their tetraprotonated form these macrotricycles display anionic complexation which will be discussed in Section 5.

3.2. Cylindrical macrotricycles

3.2.1. Polyaza-polyoxa-macrotricycles

3.2.1.1. Macrotricycles based on [12] N_2O_2 monocycle. As mentioned in Section 2.1.1 the synthesis of compound **4** [1.1.1] gives a low yield (by an improved synthesis of this cryptand by another method, was published in a recent paper[125]). This was due to the competitive formation of the dimer **45***.[126] (See Fig. 18.) Ligand **45** forms complexes of low stability with AC (in water log $K_s < 2$); and the stoichiometries are 1:1. With AEC, more stable 1:1 complexes are obtained (log $K_s = 6.5, 7.0, 8.0$, for Ca^{2+}, Sr^{2+}, Ba^{2+}). The divalent cryptates display an intramolecular cation exchange process which has been studied by ^{13}C NMR.[127] There is an intramolecular cation jump (at a slow rate) from one monocycle to the other as shown in Fig. 18. An

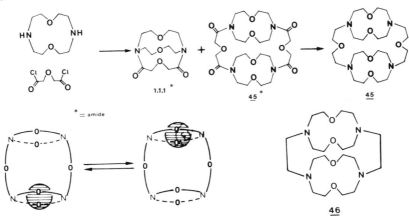

Fig. 18. Cylindrical macrotricycles: synthesis and intramolecular exchange.

intermolecular exchange also occurs but at an even slower rate. Cryptand **45** forms complexes with the silver cation Ag^+. The 1:1 complex has a stability constant in water of log $K_s = 6.0$. A complex of stoichiometry 1 ligand: 2 Ag^+ was also demonstrated by NMR experiments. This was the first binuclear complex formed in the cryptate series. X-ray structural studies of $[2 Ag^+ \subset 45]$ indicated that the two cations are held close to the two macrocycles.[128] A bis-Cu(II) complex $[2 Cu^{2+} \subset 45]$ can also be formed.[129]

A macrotricycle, compound **46**, in which the two monocycles [1.1] are linked by two ethylene bridges has been described by Calverley and Dale[130] (see Fig. 18). The X-ray structure determination of the Na^+ complex shows a cubic arrangement of the eight donor atoms.[131]

3.2.1.2. *Macrotricycles and macrotetracycles based on [18]N₂O₄ macro-*
cycle. Larger macropolycyclic (tri and tetra) molecules have been described
(Fig. 19).[132] The strategies of synthesis are given in ref. 133. With these
cryptands, formation of metal cation complexes have been observed by
both ¹H and ¹³C NMR. Mononuclear and binuclear complexes can be
formed,[134] for example: $[2K^+ \subset \mathbf{49}]$, 2SCN⁻; $[2Ba^{2+} \subset \mathbf{49}]$, 2SCN⁻; $[2Na^+ \subset \mathbf{49}]$, 2I⁻. A crystal structure of this last complex shows the location of the

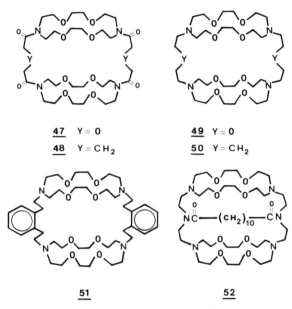

Fig. 19. Macrotricycles **47–51** and macrotetracycle **52**.

sodium cations.[135] The $Na^+ \cdots Na^+$ distance is quite large, 6.4 Å. Consequently, no major interaction between the two cations takes place, i.e. the
two macrocycle units are almost independent. Based on ¹³C data a hetero-
nuclear bimetallic cryptate $[Ag^+, Pb^{2+} \subset \mathbf{49}]$ was postulated. Intramolecular
cation exchange in mononuclear complexes has also been studied.[134] The
stabilities of the complexes (see Table 10) do not differ greatly from those
observed for the individual monocyclic unit. Selectivities are not very high.
For dinuclear cryptates the stabilities of the complexes are even smaller;
selectivities are similar to the 1:1 complex selectivities.

Fluorescence experiments have established other interesting properties
of the macropolycyclic ligands.[132] A strong increase in fluorescence was
observed when the potassium salt of 6-*p*-toluidinonaphthalene-2-sulfonate

Table 10. Stability constants[a] of mononuclear ($\log K_{S1}$) and dinuclear ($\log K_{S2}$) cryptates (solvent, methanol: water 95:5)

		Na$^+$	K$^+$	Rb$^+$	Cs$^+$	Ca^{2+}	Sr^{2+}	Ba^{2+}
49	$\log K_{S1}$	3.6	4.8	3.7	4.4	4.0	5.5	6.7
	$\log K_{S2}$	3.2	3.9	3.3	3.0	—	5.5	6.3
50	$\log K_{S1}$	3.2	4.0	3.5	3.5	—	—	—
	$\log K_{S2}$	1.5	3.2	3.0	2.5			
51	$\log K_{S1}$	3.0	3.6	3.0	—	3.6	4.9	5.9
	$\log K_{S2}$	2.9	2.7	2.8	—	—	5 ± 1	6 ± 1
18[b]	$\log K_S$	3.3	4.4	4.3	4.1	4.4	6.1	6.7

[a] Determined by selective electrodes.
[b] See formula in Fig. 8.

(TNS$^-$) was added to an aqueous solution of compound **49**. It is well known that the fluorescent probe TNS$^-$ is very sensitive (giving an enhancement of fluorescence) to lipophilic proximity.[136] The above observation was explained in the following way: K$^+$ complexed by the macrotricycle **49** leads to the lipophilic cation [K$^+ \subset$ **49**] which associates with the lipophilic anion TNS$^-$ involving enhancement of fluorescence. Molecular complexes have also been obtained; compound **52**, with its large lipophilic pocket, interacts with TNS$^-$ (giving a large fluorescence increase) in the absence of any cation.[132]

Macrotricyclic ligand **53**, formed with two [18]–N$_2$O$_4$ macrocycles linked by squaric acid, forms complexes with K$^+$, Rb$^+$, Cs$^+$.[137]

Finally, two chiral macrotricyclic ligands **54** and **55** have been reported[138] (see Fig. 20).

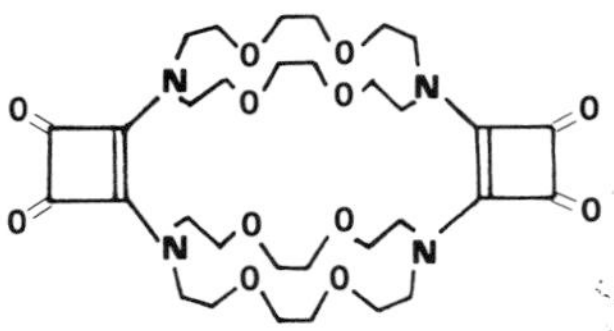

<u>**53**</u> 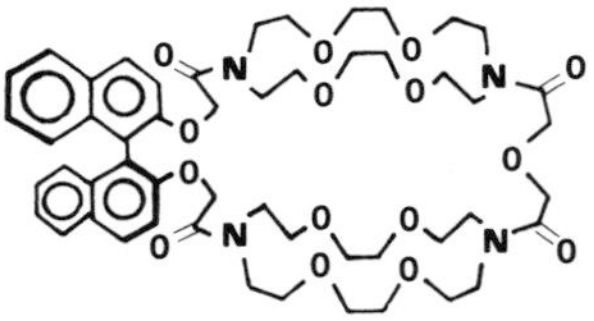   <u>**54**</u>

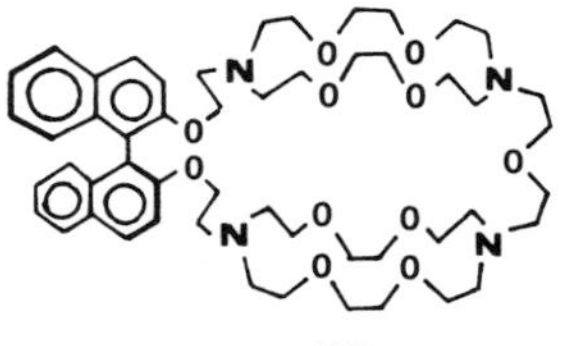

<u>**55**</u>

Fig. 20. Macrotricycle **53**. Chiral macrotricycles **54**, **55**.

Using both extraction methods and transport experiments, partial resolution of racemic substrates has been obtained with these compounds.[138]

3.2.1.3. Macropolycycles linked by lipophilic chains; Molecular complexes. A further step towards "abiotic receptors"[139] has been made in the last few years. Complexing systems for organic guest molecules have been specifically designed, mainly by two research groups, those of Sutherland and of Lehn.

In Fig. 21 some selected representatives of this new class are illustrated.

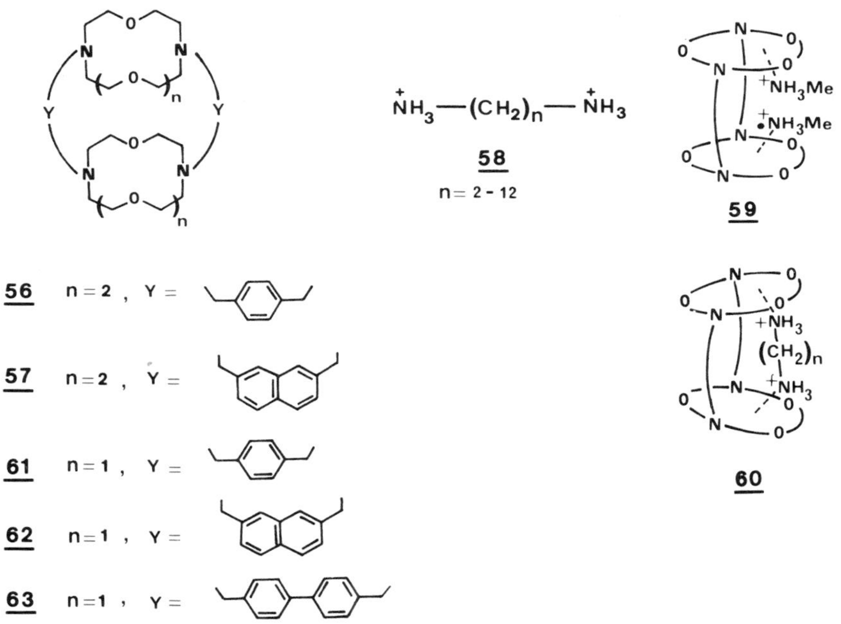

$$\overset{+}{N}H_3 - (CH_2)_n - \overset{+}{N}H_3$$

58

n = 2 - 12

Fig. 21. Macropolycycles designed for complexation of diammonium salts.

Compounds **56** and **57** form complexes with methylammonium thiocyanate. NMR experiments seem to indicate that 1:2 inclusion complexes **59** are formed.[140] Using as substrate the bis primary alkylammonium salt **58** ($n = 6$), there was no doubt that an inclusion complex **60** was formed. Indeed, in the NMR spectrum, high-field chemical shifts for the methylene groups of the diammonium salt were observed, i.e. these groups lie in the centre of the cavity and in the shielding zone of the aromatic rings. Another set of macrotricycles, compounds **61–63**, having various lengths of the two lateral bridges, were reported soon after.[141] Each of these ligands binds preferentially one of the substrates 58 ($n = 2-12$). The selection is based on the fit between cavity size and substrate length. For example, 63 selects the $n = 5$ and $n = 6$ salts 58. By intensive studies the authors were able to state that the optimal guest 58 for a given host should be about 2 Å less in length than the $CH_2-Ar-CH_2$ bridges.[142]

Synthesized by Kotzyba-Hibert and Lehn,[143] macrotricycles 64–66, 104 (Fig. 22) and the previously described ligand 49 are very efficient complexing agents of the salts 58. For example, in the case of the complex $[H_3N^+-(CH_2)_5-NH_3^+ \subset 64]$, the α, β, and γ methylene signals of the substrate have

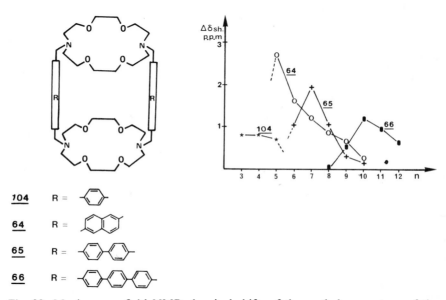

104	R =	(para-phenylene)
64	R =	(naphthalene)
65	R =	(biphenylene)
66	R =	(terphenylene)

Fig. 22. Maximum upfield NMR chemical shifts of the methylene protons of the complexed diammonium salts: $H_3\overset{+}{N}-(CH_2)_n-\overset{+}{N}H_3,2X^-$.

upfield shifts of 0.84 ppm, 1.85 ppm and 2.56 ppm, respectively. A systematic study seems to show that the better the fit between cavity size and substrate length, the larger the upfield shift.[144] This is illustrated in Fig. 22. Molecular dynamics have also been investigated. Experimental data show that the strong complexes (steric fit) display strong dynamic coupling.[144]

Comparative studies have demonstrated that the monocycle $[18]-N_3O_3$ forms much more stable ammonium complexes than monocycle $[18]-N_2O_4$.[145] This enhancement was explained by the fact that in $[18]-N_3O_3$

three $\overset{+}{N}$–H$\cdots$N hydrogen bonds are formed (see Fig. 23a). This favourable situation is not encountered in the $[18]$–N_2O_4 cycle. Incorporating this symmetric macrocycle into polycycles should, therefore, improve the complexation of ammonium salts. This was realized in compounds **67** and **68**.[146] In a similar fashion to that described above, NMR experiments detected the inclusion complexes of diammonium salts **58**. The shielding effects due to the aromatic groups are particularly great in $[H_3N^+-(CH_2)_6-\overset{+}{N}H_3 \subset \mathbf{67}]$ and reach a maximum for the γ-CH_2 protons; an upfield shift of 3.2 ppm was observed. These macrotetracycles having a more defined cavity size, display higher selectivities of complexation. Compound **67** for example, binds the $n = 5$, $n = 6$ **58** substrates but not the smaller ($n = 4$) or the larger ($n = 8$) salts.

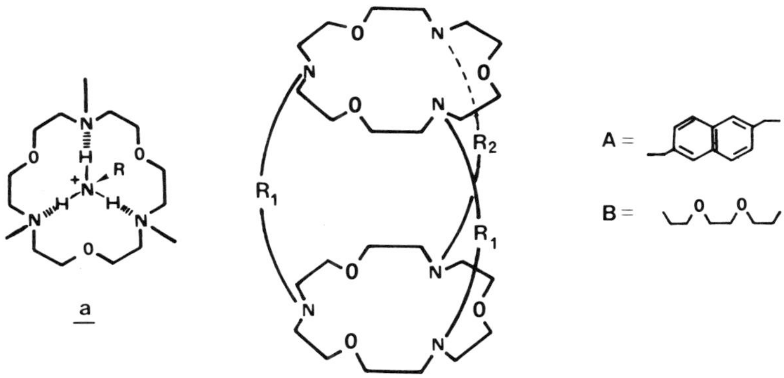

a

67 $R_1 = R_2 = A$

68 $R_1 = A$; $R_2 = B$

g. 23. Macrotetracyclic receptors for diammonium salts.

Definitive proof that these molecular complexes are cryptate-like was recently obtained from the X-ray structure analysis of both free ligand **64** and the complex $[H_3N^+-(CH_2)_5-\overset{+}{N}H_3 \subset \mathbf{64}]$.[147] The cadaverine dication occupies the centre of the macrotricyclic system, in which the aliphatic chain of the substrate is fully extended. To accommodate the diammonium substrate large changes in the ligand shape occur, the salient feature being the moving apart of the parallel naphthalene groups. The separation of these groups varies from 2.91 Å in the free ligand to 6.5 Å in the complex. The separation between the top and bottom macrocycles is less affected. The X-ray structure of the complex is illustrated in Fig. 24.

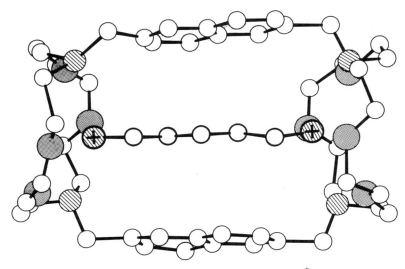

Fig. 24. X-ray structure of the complex [$\mathbf{64} \subset H_3\overset{+}{N}-CH_2-CH_2-CH_2-CH_2-CH_2-\overset{+}{N}H_3$].

3.2.2. Nitrogen and sulphur containing macropolycycles. Transition metal binuclear cryptates

Interest in ligands able to bind two metal ions in close proximity is growing rapidly. Compartmental ligands[148] and dinuclear cryptates[149] have been reviewed recently.

Due to the nature of the ligands (the large number of oxygen donor atoms) most of the dinuclear complexes described above are formed with AC or AEC. Introduction of other binding sites, nitrogen or sulphur, will lead to ligands displaying abilities to complex transition metal cations.

Although we are concerned here with tricyclic ligands some bicyclic compounds will be mentioned if this is justified by their complexation behaviour.

The ligands we will discuss here are represented in Fig. 25. Compound **69** was formed by the linking of two β, β', β''-triaminoethylamine (Tren)

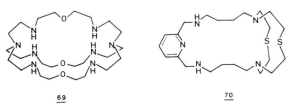

Fig. 25. Ligands designed for dinuclear complexation.

units by three identical bridges; this ligand was nicknamed Bis-Tren.[150]
Dinuclear complexes can be formed with Ag^+, Zn^{2+}, Cu^{2+}, and Co^{2+} cations.
A schematic structure of these complexes is indicated in Fig. 26a. The
intercationic distances being in the range 4.5 ± 0.5 Å, it may be possible to
insert small substrates, CN^-, N_3^- and others, as shown in Fig. 26b. The
driving force of this process is provided by the free coordination sites left
on the two complexed cations. For this type of association the name "cascade
complex" was proposed.[149] Some preliminary evidence for this kind of
process has been obtained.[150] More detailed results have been published
recently.[151] Possible applications, such as the activation of fixed substrates
can easily be imagined.

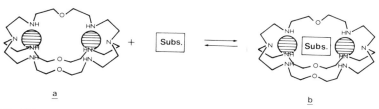

a   b

Fig. 26. Dinuclear complex (a) and cascade complex (b).

Ligand **70** has two very different binding units.[129] This is an important
approach to hetero-dinuclear complexes. In this respect the binding properties of this ligand are of interest. In the bis-Cu(II) complex the redox
properties of the two copper ions are very different: one monoelectronic
reduction takes place at $+550$ mV (vs. NHE), the second at $+70$ mV.[152] The
mixed valence Cu(I)–Cu(II) complex can then easily be obtained.

Syntheses of the sulphur containing ligands **71**, **72**, **73** (see Fig. 27) have
been reported.[153] Intensive studies have been pursued on the bis-copper
complexes of cryptand **71**. The X-ray structure[154,155] shows that a large
cavity exists between the two metal ions which are separated by 5.6 Å. The
redox properties of $[2Cu^{2+} \subset \mathbf{71}]$ are of primary importance. The electrochemical reduction takes place at a positive potential ($E_{1/2} = +200 \pm 5$ mV
vs. SCE) and the process is dielectronic, i.e. the two copper ions are reduced
at the same potential.[156] In other words the two redox centres, Cu(II), are
equivalent. This result has considerable application possibilities, for
example, in catalytic reactions which require multielectronic transfer.
Ligand **72**, having two different macrocyclic units, shows the same behaviour
as compound **70**, i.e. a mixed Cu(II)–Cu(I) complex can be formed.[149]

Compound **74** containing six sulphur atoms forms dinuclear complexes
with both Cu(I) and Cu(II) salts.[157]

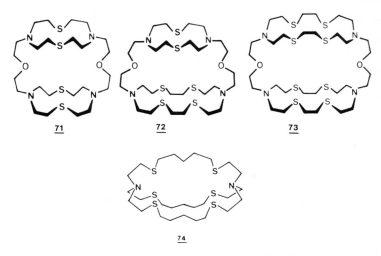

Fig. 27. Sulphur containing ligands designed to be dinuclear receptors.

All the above copper complexing agents display, in many respects, analogies with copper proteins[158] and they may represent informative models for them.

Recently published reviews should be consulted for various aspects of macropolycyclic compounds and their complexes.[159,160]

4. Functionalized macrocycles

The advantages of incorporating side-chains, or arms, in macrocycles, were recognized quite early.[8,161] Very rapidly evidence was obtained that these crown derivatives fulfil many requirements for the design of a synthetic enzyme. A large number of investigations of this type of complexing agent have brought the state of the art to a high level; enzyme-like receptors are by now almost in hand.[162]

A question arises; why monomacrocycles and not the usually more efficient macropolycycles? The latter compounds are in many respects strictly receptors: the substrates, ions or molecules, are deeply enclosed in a cavity and in most of the cases they have no interaction with the outside medium (solvent, ions, molecules). Consequently, functionalized side chains cannot be expected to have any kind of interaction (stabilization of the complex, chemical reaction and others) with the bound substrate. In addition the rate of exchange is usually slow, an unacceptable situation for enzymatic reactions. By contrast in monomacrocycles all the necessary features (binding, chirality, functionality) can be introduced into the receptor structure. Expressed in another way, in most of the macropolycycles all the space surrounding the substrate is occupied by the ligand and no space is left for additional functions. This is not the case in monocycles in which room is still freely available for functionalization above and below the cycle. In general, the monocycle brings the chirality and the capacity to bind the substrate. Side-chains can give additional binding and bear reactive functions able to interact with the bound substrate. Finally the side-chains protect the substrate quite well from the outer environment leading to a situation quite similar to that which is observed in cryptates. This aspect was one reason which made us decide to mention these compounds briefly even though they are not strictly cryptates. The second, and decisive, reason was that one of the final goals of the research which we will present here, is the synthesis of a cation channel which has all the cryptate characteristics.

Only the tartaro-crowns will be presented here; for the numerous other examples, detailed information can be found in several reviews.[163-166]

4.1. Chiral macrocycle derived from L-(+)tartaric acid

The chiral crown-ether **75** based on tartaric acid was synthesized in 1975 by Girodeau, Lehn and Sauvage[167,168] (see Fig. 28). This ligand possesses the requirements we discussed above: chirality, complexation ability and four functions for structural modification. Several studies have been carried out in recent years on this bis-tartaro-crown unit.[169] The nature of the ligand side-chain X (see Fig. 28) has a large effect on the complexation ability. The general trends are indicated under the following headings.

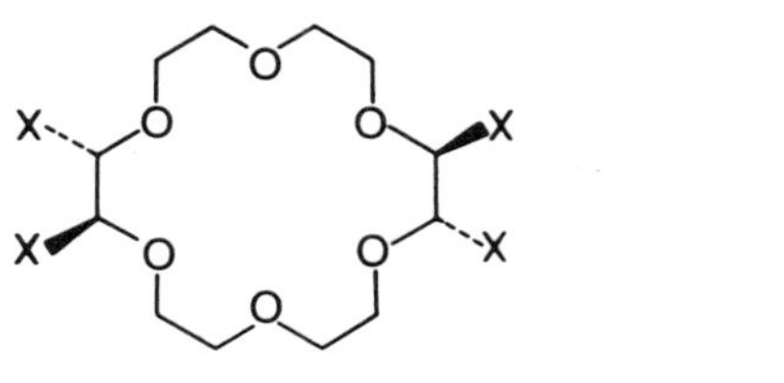

75	X CONMe$_2$	77	X = CONH-CH-CH$_2$-(3-indole)
			CO$_2^-$NMe$_4^+$
76	X - COOH	78	X = CONH-CH$_2$-CO$_2^-$NMe$_4^+$

Fig. 28. Chiral macrocycle derived from L-tartaric acid.

Table 11. Stability constants[a] (solvent, water)[169]

	K^+	NH_4^+	$Me-NH_3^+$	$Ph-(CH_2)_2\,NH_3^+$	$^+H_3N-(CH_2)_2-NH_3^+$	$^+H_3N-(CH_2)_3-NH_3^+$	$^+H_3N-(CH_2)_4-NH_3^+$
75	70	<5	—	—	—	—	—
76	3×10^5	3.2×10^3	750	260	—	—	—
77	5.5×10^4	340	95	100	3000	800	570
78	180	10	<10	—	—	—	—

[a] Determined by selective electrodes (by competition with NH_4^+ for the $R-NH_3^+$ salts).

(i) Carboxylate groups strongly increase the stabilities of the complexes, with the largest effect being observed when the carboxylates are close to the macrocycle as in compound **76** (compare with **75** and **78** in Table 11).

(ii) Hydrophobic effects are apparent in comparing the stability constant of $MeNH_3^+$ and $Ph–CH_2–CH_2–NH_3^+$ with ligands **76** and **77**. Ligand **76** binds $MeNH_3^+$ more strongly than $Ph–CH_2–CH_2–NH_3^+$ whereas **77** shows the same stability with both substrates.

(iii) Electrostatic interactions between X and R are illustrated by the complexes of compound **77**: the diammonium salts give much higher stabilities than the monoammonium ones. In addition, the length of the aliphatic chain is very important. The optimal length is $n = 2$; when $n = 3$ or 4 the stabilities decrease markedly. A crystal structure of complex $[NH_3^+–(CH_2)_2–NH_3^+ \subset \mathbf{76}]$ has been published.[170] It shows that the macrocycle is almost planar and that the carboxy groups of each tartaric acid residue are diaxial and extend above and below the plane.

(iv) Lateral interactions have also been established for the case where R = pyridinium and X = indole, by observation of a charge transfer band in the U.V. spectrum[171] (see Fig. 29). More complete information on several of these aspects can be found in ref. 172.

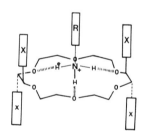

Fig. 29. Lateral interactions between R and X side-chains.

A crucial step towards enzyme-like systems was taken by the incorporation in the receptor of a reactive site which can react with the bound substrate. In Fig. 30, two schematic representations are presented. In complex **79** it has been established that the rate of H-transfer from 1,4-dihydropyridine to pyridinium is accelerated.[173] In the case of complex **80** the rate of intramolecular thiolysis of the bound substrate (glycyl–glycine *p*-nitrophenyl ester) is accelerated by a factor of 140. Systematic studies of a large series of substrates show that the molecular catalyst displays in addition to the rate enhancement: (i) structural selectivity for dipeptide esters and (ii) high chiral recognition for the (L) antipode of glycyl-phenyl-alanine esters.[174]

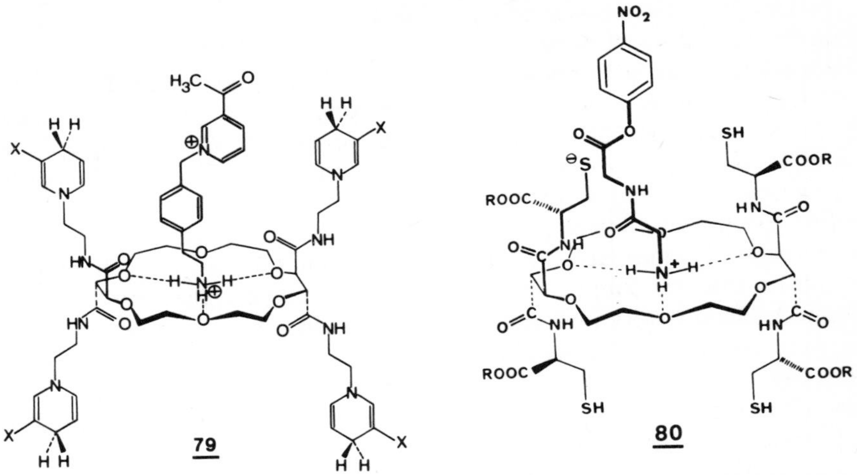

Fig. 30. Reaction between the receptor side-chain and the bound substrate.

4.2. Face-discriminated and side-discriminated macrocycles

As shown above, the tetrasubstituted macrocycles possess many interesting features. Even more interesting would be the independent handling of the four side-chains. After long and painstaking efforts this was realized by Behr and Lehn.[175] The results obtained are summarized in Fig. 31. The dianhydride treated with an aromatic amine gives a mixture of *syn* (face discriminated) and *anti* (side discriminated) diamide–diacid isomers. These isomers can be separated by chromatography. If the reaction is performed in the presence of triethylamine only the important *syn* isomer is obtained. X-ray structures of complexes allowed the assignments of the *syn* and *anti* isomers. In Fig. 32 the structures of [Ca^{2+} ⊂ anti **81**] and [Sr^{2+} ⊂ syn **82**] are represented.[175] The face discriminated *syn* compound, **82**, is the ultimate precursor for the development of a macropolycyclic system. By coupling of several macrocycles a channel may be constructed. The number of macrocycles should be large enough to give an overall length of the macropolycycle in the range of 40–50 Å. This size is required in order to act as a transmembrane cation channel. This synthetic compound would constitute a model for the naturally occurring channels, gramicidin A[176–179] and alamethicin[180] which are still largely under investigation. Encouraging X-ray results have been obtained with the potassium complex formed with compound **75**.[181] In Fig. 33 it can be seen that a channel-like packing is obtained and, equally important, potassium is either inside or outside the

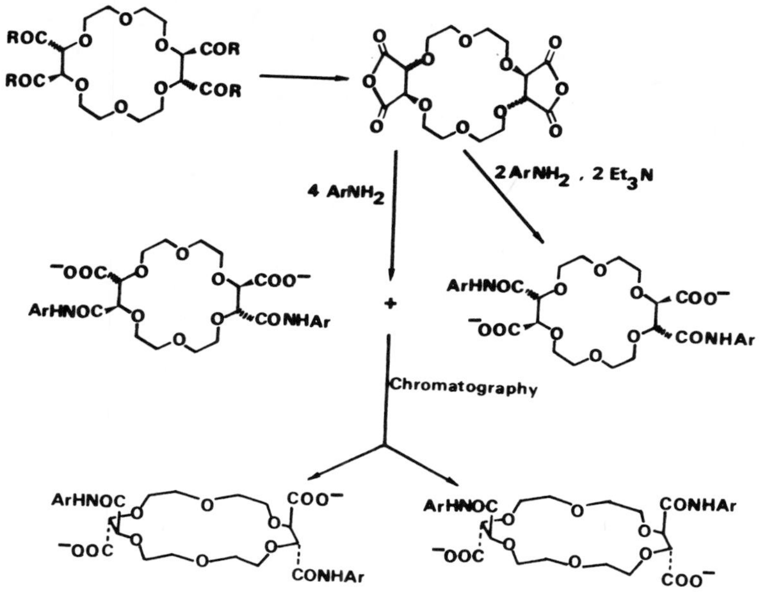

Fig. 31. Synthetic pathways to side-discriminated (**81**) and face-discriminated (**82**) macrocycles.

macrocycles. This structure can be considered as a solid-state model of the channel.[181] Intensive synthetic efforts are being pursued to achieve the actual channel.

5. Anion complexation

It is quite surprising to note that compared to cation complexation, anion complexation has received very little attention. Hundreds of papers have appeared on complexes of cations of groups I and II, thousands on complexes with the other cations and only a few tens specifically on synthetic anion complexing agents. This hiatus has many origins which we will explain later but the reason is certainly not that anion complexation has no interest.

In biology, anions are the necessary partners of all positive centres (ammonium, guanidinium and other cations). Their functions, transfer across membranes, *etc.* are of the same importance as those of cations.

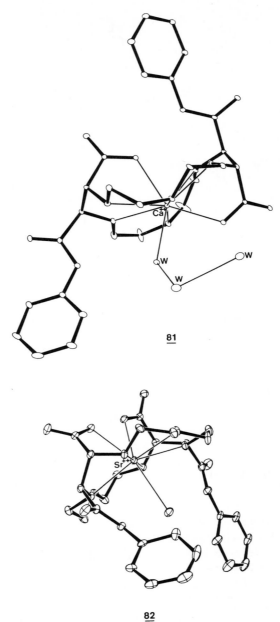

Fig. 32. X-ray structures of the side-discriminated (**81**) and face-discriminated (**82**) 18-crown-6 derivatives.

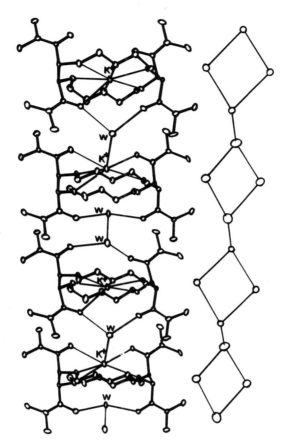

Fig. 33. Channel-like packing observed with the crown complex [K$^+$ $\subset$ **75**].

Enzyme substrates, as pointed out several years ago,[182] are more often anions than cations. Some details on anion occurrence in biology can be found in ref. 183.

In chemistry, anions play many roles, for example as nucleophiles, as bases, as redox agents, in phase-transfer catalysis and others. Complexation of anions can bring about changes in chemical reactivity as does cation complexation.

In this chapter, we will discuss the most important results in this field and we would be very pleased if this encourages others to explore the almost virgin territory of anion complexation.

5.1. Introduction

To point out the problems of anion complexation we will first mention the most important characteristics of anions, then discuss the design of potential anion complexing agents.

Compared to metal cations, anions are very large. The small anion F^- (1.36 Å) is about the same size as K^+ (1.33 Å). Some selected values[184] are indicated in Table 12. A large variety of geometries is encountered:[185] spherical (F^-, Cl^-, Br^-, I^-), linear (N_3^-, CN^-, SCN^- etc.), planar (NO_3^-, CO_3^{2-}, $R-CO_2^-$, etc.), tetrahedral (PO_4^{3-}, SO_4^{2-}, ClO_4^-, MnO_4^-, etc.) and octahedral ($Fe(CN)_6^{4-}$, $Co(CN)_6^{3-}$ etc.). Anions, compared to cations of the same size have larger free energies of solvation,[212] e.g. $\Delta G_{F^-} = 434.3$ kJ mol^{-1} $\Delta G_{K^+} = 337.2$ kJ·mol^{-1}; as complexation is a competition with solvation this makes anions more difficult to complex. Finally, most of the anions exist only in a limited range of the normal pH scale; for example, above pH 5–6 for the carboxylates, above 7 for CO_3H^-.

Table 12. Anion radii[184] (Å)

Anion	radius	Anion	radius	Anion	radius	Anion	radius
F^-	1.36	OH^-	1.40	NO_2^-	1.55	CO_3^{2-}	1.85
Cl^-	1.81	CN^-	1.82	NO_3^-	1.89	SO_4^{2-}	2.30
Br^-	1.95	IO_3^-	1.82	MnO_4^-	2.40	PO_4^{3-}	2.38
I^-	2.16						

The ligand then has to take account of all the various anion peculiarities: size, geometry, pH dependence. The largest handicap for anion complexing ligands orginates directly from the periodic table. Whereas quite a number of donor atoms (O, S, N, P etc.) are available for cation complexation, no acceptor displaying the required characteristics for a ligand (no major synthetic difficulties, high chemical stability) is present in the periodic table. The alternative is then to use either neutral ligands in which binding sites are –OH, –SH etc. or positively charged ligands having ammonium or guanidinium as binding sites. We considered in fact only this latter class of ligands. The charged ligands introduce a new difficulty; they are all more or less pH dependent. This is particularly critical for ammonium ions (except tetrasubstituted ones), but less so for the guanidinium ions. In short, anion complexation can occur when, over a certain range of pH the anion exists, and the ligand is in its protonated form. To fulfil these conditions the ligand must be meticulously designed. Guidance for this design was found in biological examples.

Biochemists have devoted a tremendous amount of work to anions in biological systems. Unfortunately we cannot give here an in-depth discussion of all the interesting results obtained. As guanidinium and ammonium are our chosen binding sites for anions we will simply point out some typical results. In enzymes it has been demonstrated by several techniques that guanidinium ions of arginine residues play a major role in the binding of anionic substrates or cofactors. This was shown by X-ray,[186–187] analysis, by chemical modification of the guanidinium[188] moiety and by ^{15}N NMR.[189] The ammonium ions of lysine residues are also anion binding sites.[190] The naturally occurring polyamines (putrescine, spermidine, spermine) interact strongly with nucleotides.[191–193] The binding of DNA by protamines and histones (these nucleoproteins contain a larger number of positively charged arginyl and lysyl residues) has also been widely investigated.[194,195] With these facts in mind we decided to synthesize systems containing guanidinium or ammonium as binding sites.

5.2. Guanidinium containing anion receptors

The most attractive feature of the guanidinium ion is its high pK (13.5 for guanidinium); it remains then protonated over a wide range of pH. Its anion complexing ability has been demonstrated by X-ray analysis[196] which showed the existence of a bidentate ionic hydrogen bond as pictured in Fig. 34. The macrocycles shown in Fig. 35 were synthesized, as well as the

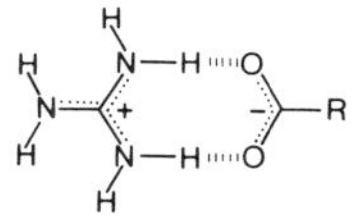

Fig. 34. Bidentate ionic hydrogen bonding.

three open-chain analogues.[197] Anion complexation was investigated by pH-metric titration measurements. The computer analysis of these curves[198] gave the following results with the phosphate PO_4^{3-} anion: log K_s = 1.7, 2.2, 2.4 respectively for **83, 84** and **85**. The stabilities of the complexes formed with the open-chain analogues are slightly lower (~0.5 in log K_s). These results show that the complexes formed with **83, 84** and **85** are of low stability, that almost no macrocyclic effect is observed. Since the synthesis of the guanidinium containing macrocycles proved to be rather difficult, we decided to carry out no further investigations on this kind of system. In addition, comparative studies of open-chain anion complexones containing

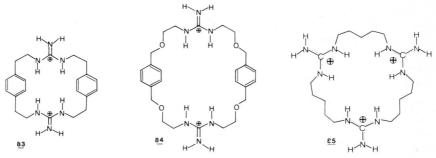

Fig. 35. Guanidinium containing macrocycles.

either guanidinium or ammonium as binding sites (see examples in Fig. 36) demonstrated that ammonium is a more efficient binding site than guanidinium.[199] This can be explained by the higher charge density displayed by the ammonium ion compared to the guanidinium ion.

Another type of guanidinium based anionic complexing agent was described recently by Schmidtchen.[200]

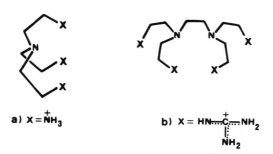

Fig. 36. Open-chain anion receptors containing either ammonium or guanidinium groups.

5.3. Ammonium containing receptors

5.3.1. Macropolycycles
The first synthetic anion complexing agent was described by Simmons and Park in 1968[201] (Fig. 37). Depending on the size of the macrobicycle the

$\underline{86}\ (n = 7-10)$

Fig. 37. Diaza-macrobicycles as halide anion receptors.

various halide ions (Cl⁻, Br⁻, I⁻) are complexed to varying degrees. X-ray analysis later demonstrated that the ion is held inside the cavity.[202] The name of *katapinate* (G. καταπινω = swallow up, engulf) was proposed for this type of complex).

In their tetraprotonated forms the spheroidal macrotricycles mentioned in Section 3.1 form very stable complexes with Cl⁻ and Br⁻.[203] The stability constants, obtained by selective electrode measurements, appear in Table 13.

Table 13. Stability constants[a] *in log K_S[203] (solvent, water)*

	$41-H_4^{4+}$	$42-H_4^{4+}$	$44-H_4^{4+}$	$86-H_2^{2+}$ $n = 10$
Cl⁻	>4	>4.5	1.7	~0.7
Br⁻	<1.0	1.55	<1.0	<1.0

[a] Determined by Cl⁻ and Br⁻ selective electrodes. (See formulae **41, 42, 44**, on p. 358.)

The stabilities of the chloride complexes are very high with $41-H_4^{4+}$ and $42-H_4^{4+}$, at least three orders of magnitude more stable than the katapinate–Cl⁻ complex. They also display very high Cl⁻/Br⁻ selectivity (>1000). Comparing $41-H_4^{4+}$ and $44-H_4^{4+}$, a large macrotricyclic cryptate effect is revealed. An X-ray structure of the complex [Cl⁻ ⊂ $41-H_4^{4+}$] confirmed the anion inclusion in the cavity.[124] In the schematic representation, the tetrahedral array of $\overset{+}{N}$–H···X⁻ hydrogen bonds is shown (see Fig. 38).

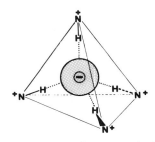

Fig. 38. Schematic view of a **41** anion complex. Note the four ionic-hydrogen bonds.

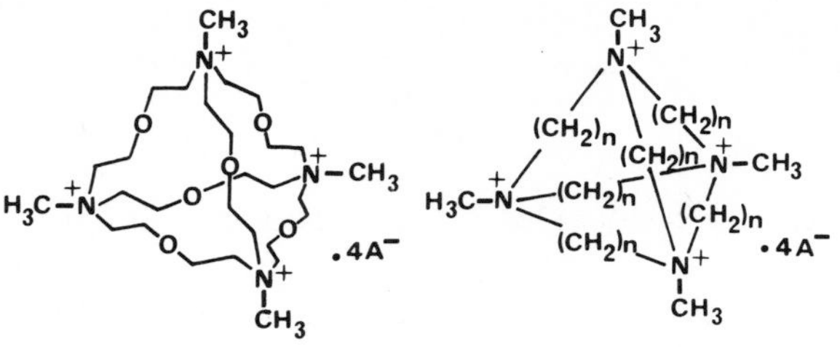

Fig. 39. Anion receptors containing quaternary ammonium salts.

Schmidtchen later synthesized a series of macrotricycles in which the binding sites are quaternary ammonium salts[204-206] (Fig. 39). The stability constants are given in Table 14. It is noteworthy that: (i) the stabilities are much lower than those observed in Table 13, and (ii) Br^- is better complexed than Cl^-. These results demonstrated that the quaternary ammonium salt is a less efficient binding site than the tertiary one. This can be explained by the lack of ionic hydrogen bonds in the quaternary ion; the electrostatic interaction is also expected to be lower in this case due to steric hindrance. The reverse Cl^-/Br^- selectivity can be explained by the larger cavity in the quaternary ammonium compounds.

Table 14. Stability constants[a] in log K_s^{204} (solvent, water)

	87	88	89
Cl^-	1.0	1.3	<0.5
Br^-	1.8	2.45	2.45
I^-	—	2.2	2.4

[a] Determined by selective electrodes.

All the above ligands are receptors for spherical anions. To enlarge the field of anion complexation the shape of the ligand has to be changed in order to accommodate the various anion geometries encountered.

The hexaprotonated Bis-Tren 69 already described in Section 3.2.2, having a more elliptical cavity, forms complexes with many anions.[207] This ligand was specially designed for linear anions and ^{13}C NMR experiments showed that the azide anion forms a 1:1 complex with [69–6H$^+$].[207] The inclusion of this anion has recently been completely demonstrated by X-ray structure analysis.[208] This structure is represented in Fig. 40. Structures with halide anions, Cl^- and Br^-, have also been obtained recently.[209] In both cases a halide anion is included in the central cavity. Other linear anions like the hydrogen dihalide anions HX_2^- are still under investigation.

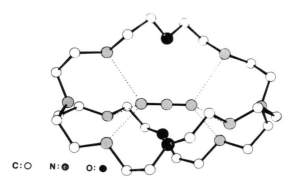

C:O   N:◉   O:●

Fig. 40. X-ray structure of the [azide ⊂ bis-tren-6H$^+$] anion complex.

Vögtle has synthesized the macrobicycle **90** shown in Fig. 41. NMR experiments have shown Br^- and I^- complexation.[210]

The syntheses of very large macrobicycles, (Fig. 41) compounds **91** and **92**, have been performed.[211] The successive nitrogen atoms are separated by three carbon atoms (for an explanation of this feature see below). In their hexa- and nona-protonated forms these ligands will certainly display high stability constants with large anions. These studies are in progress.

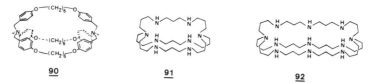

       90   91   92

Fig. 41. Macrobicyclic anion receptors.

5.3.2. Macrocycles

The previously described anion complexones are well designed for many anions: halogenides, azide, sulphate etc., but not for most of the important classes of carboxylate and adenosine phosphate anions. The major problem

in dealing with these anions is the high pH corresponding to anion formation: maleate ($pK_2 = 6.0$), AMP ($pK_2 = 6.1$), ADP ($pK_3 = 6.4$), ATP ($pK_4 =$
6.9). At these high pH's the ammonium containing ligand must still be fully
(or at least highly) protonated. Moreover it is well known that in the polyaza
macrocycles, in which two successive nitrogen atoms are separated by an
ethylene unit, full protonation occurs at very low pH. For example in
[12]–ane–N_4, $pK_1 = 10.70$, $pK_2 = 9.70$, $pK_3 = 1.7$, $pK_4 = 0.9$[213] (for compilation and discussion of this important aspect see ref. 213–216). It is clear
that at the pH at which full protonation will be realized (pH < 1), the anions
will all be fully protonated too. Ligands based on this pattern are then
totally excluded. In contrast, in the case of [17]–ane–N_5, containing three
ethylene units and two propylene ones, the pH of full protonation rises:
$pK_1 = 10.3$, $pK_2 = 9.6$, $pK_3 = 7.4$, $pK_4 = 4.10$, $pK_5 = 2.4$. The propylene units,
in separating the nitrogen atoms, decrease the positive charge density i.e.
the electrostatic repulsion an approaching proton has to overcome is diminished. Suitable ligands can then be obtained if the macrocycles are totally
based on propylene units. A further validation of this pattern is given by
the natural polyamines[191] in which two successive nitrogen atoms are separated by three or four carbon atoms. Nevertheless the ethylene diamine unit
still has interesting features, in its diprotonated form it displays a higher
local charge density than the propylene diammonium species. To solve the
protonation problem we separated two successive ethylene diammonium
units by five atoms.

Three macrocycles having these characteristics have been synthesized[217]
(see Fig. 42). The key step of cyclization was achieved by the Richman and

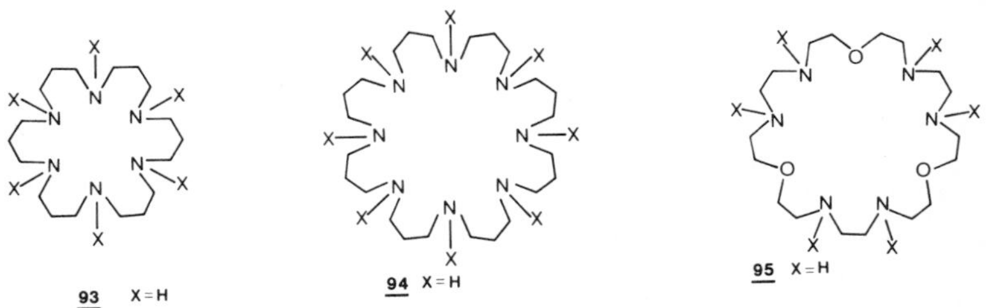

Fig. 42. Polyaza macrocycles designed for anion complexation.

Atkins method.[218] The last pKs of protonation are: $pK_6 = 6.60$ for **93**,
$pK_8 = 6.45$ for **94**, $pK_6 = 5.70$ for **95**; which are reasonable values for anion
complexation. The stability constants obtained by the pH-metric titration
method are shown in Table 15. It should be noted that the macrocycle **95**,
based on ethylene diamine units forms the most stable complexes with the
highly charged anions: oxalate, sulphate and to a lesser extent AMP, ADP,

Table 15. Stability constants[a] *in log K_S*[217] *(solvent, water)*

	sulphate[2-]	oxalate[2-]	citrate[3-]	trimesate[3-]
93–6H[+]	4.0	3.8	4.7	3.5
94–8H[+]	4.0	3.7	7.6	6.1
95–6H[+]	4.5	4.7	5.8	3.8
	$Co(CN)_6^{3-}$	$Fe(CN)_6^{4-}$	AMP[2-]	ADP[3-]	ATP[4-]
93–6H[+]	3.9	6.9	3.4	6.5	8.9
94–8H[+]	6.0	8.9	4.1	7.5	8.5
95–6H[+]	3.3	6.3	4.7	7.7	9.1

[a] Determined by pH-metric titrations.

ATP. The largest ligand **94** has the best binding constants with large anions: citrate, trimesate, $Co(CN)_6^{3-}$, $Fe(CN)_6^{4-}$. Complexation has been studied by other methods as well. For example, for the nucleotide phosphate polyanions observed by ^{31}P NMR; changes in chemical shifts are observed on addition of ligands, and in some cases 1:2 ligand:substrate species are detected (for example with **94**). Electrochemical studies on the complexes formed by the ligands **93** and **94** with $Fe(CN)_6^{4-}$ and $Ru(CN)_6^{4-}$ have been made.[219] They indicate that: (i) the stoichiometries are always 1:1 and (ii) the anions $Fe(CN)_6^{4-}$ and $Ru(CN)_6^{4-}$ are more difficult to oxidize. By an appropriate selection of the receptor the redox potential of anionic species may then be shifted and controlled. This can have important applications in several fields.

Receptors for dicarboxylate substrates: $^-O_2C-(CH_2)_m-CO_2^-$ have been synthesized[220] (see Fig. 43). In their hexaprotonated forms these two compounds form very stable and selective complexes with dicarboxylates.

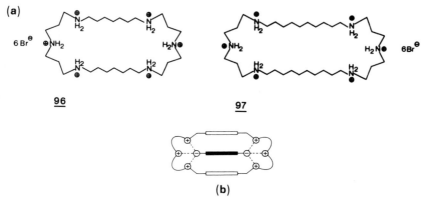

Fig. 43. Ditopic receptors (a) designed for dicarboxylate anion complexation (b).

Ligand **96**–6H$^+$ complexes preferentially the succinate ($m = 2$; log $K_s = 4.3$) and glutarate ($m = 3$; log $K_s = 4.4$) dicarboxylates. Ligand **97**–6H$^+$ has a preference for pimelate ($m = 5$; log $K_s = 4.4$) and suberate ($m = 6$; log $K_2 = 4.25$). The chain length selectivities observed originate from the structural complementarity between the receptor and the diammonium substrates (see Fig. 43b). Each receptor has one or two optimal substrates. Those which are too large or too short form less stable complexes.

A series of polyaza macrocycles containing two distinct sets of donor atoms, as in **96** and **97**, has been published[221,222] (see Fig. 44). These compounds certainly display interesting features for both anion and cation complexation.

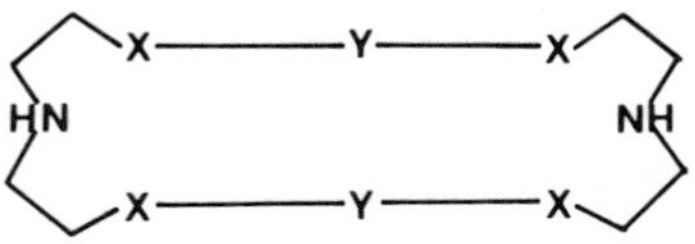

98   X = NH : Y = (CH$_2$)$_5$
99   X = NH : Y = (CH$_2$CH$_2$)$_2$O
100 X = S : Y = (CH$_2$)$_5$

Fig. 44. Polyaza macrocycles.

Recently, it has been shown by a polarographic method, that penta and hexa–aza-macrocycles interact with several polycarboxylates at neutral pH.[223] The ligands are in their triprotonated forms at this pH. The most stable complex is formed with the citrate anion (log $K_s = 3.0$).

5.3.3. Lipophilic diammonium salts

Using a very simple receptor (see Fig. 45), Tabushi *et al.* were able to extract adenosine phosphates from an aqueous into an organic phase.[224–227] Discriminative extractions were obtained. For example, the selectivities for ADP/AMP and ATP/AMP are respectively 45 and 7500 at pH = 3.[227] Active transport of ADP was also described.[226]

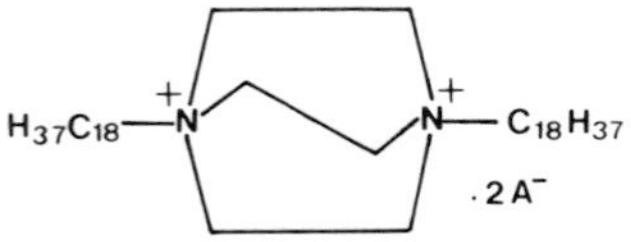

101

Fig. 45. Lipophilic diammonium salt for anion extraction.

6. Cryptate applications

As mentioned at the beginning of this chapter, crown-ethers and cryptates have found very wide applications in numerous fields. We will indicate in brief the most important contributions. This part will be subdivided into three main aspects: (i) applications of cation complexation, (ii) applications of the cryptate counteranions, both to inorganic and organic chemistry and (iii) applications of anion complexation.

6.1. Applications of cation complexation

Cryptates can be formed with the cations obtained from most of the elements of the periodic table. Direct applications of cation complexation include: separation and analysis of cations, enrichment of precious metals, isotopic separations etc. Several reviews have appeared on these aspects.[228-230] Most of these applications can be carried out either in a two phase water/organic solvent system or on resins. Quite a large number of polymer-bound crown-ethers or cryptates have been synthesized.[230]

6.1.1. Cryptates of heavy metals

In Section 2.1.2 attention was focused mainly on AC and AEC but cryptates of a larger number of cations can be obtained[231] (Cu^{2+}, Ni^{2+}, Co^{2+}, Zn^{2+}, Pb^{2+}, Ag^{+}, Cd^{2+} etc.) with the diaza-polyoxa-macrobicycles **1**, **5** and **6**. The stabilities and selectivities of these complexes have been studied.[232] Solvent dependencies have also been investigated.[233-235] Complexes of heavy metals with nitrogen and sulphur rich cryptands have already been mentioned (see Sections 2.2.2 and 2.2.3).

The important class of lanthanides has been studied by several groups.[236-238] Stability constants for the complexes of [2.1.1], [2.2.1], [2.2.2] with eleven members of this class have been measured.[237] The range of values for log K_s is too small (between 5.9–6.9) to allow their application to separation techniques. X-ray studies of $[La^{3+} \subset 2.2.2]$,[239] $[Eu^{3+} \subset 2.2.2]$[240] and $[Sm^{3+} \subset 2.2.2]$[241] have been reported. Electrochemical studies have also been realized.[236,242]

More interesting are the results obtained with the macrocycle **75** based on tartaric acid (see Section 4.1). The order of stability is $La^{3+} > Pr^{3+} > Nd^{3+} \cdots > Yb^{3+} > Lu^{3+}$. Between La^{3+} and Lu^{3+} the selectivity is higher than 10^3.[243]

Finally actinides have also been investigated. Compound **1** [2.2.2] forms complexes with U^{4+}, Th^{4+}, UO_2^{2+},[244] which are soluble in chloroform. Some controversy appeared recently as to the existence of the $[UO_2^{2+} \subset 2.2.2]$ complex. It seems that there is no direct bonding between UO_2^{2+} and the ligand but rather an addition compound.[245]

6.1.2. Isotopic separation

The possibility of isotopic separation was first investigated with crown-ethers (separation of $^{40}Ca^{2+}/^{44}Ca^{2+}$).[246] An exchange resin having a cryptand anchor group (compound **15**, 222$_B$) was used for $^{40}Ca^{2+}/^{48}Ca^{2+}$ separation.[247] It seems that this constitutes a promising method: calcium enriched by 3.3% of ^{48}Ca was obtained apparently with no major difficulty (natural abundance of $^{48}Ca = 0.18\%$). Knöchel *et al.* have published an extensive study on $^{24}Na^+$ and $^{22}Na^+$ isotopic separation.[248] A screening of eighteen ligands (crown-ethers, cryptands) was realized using a two-phase water/chloroform system. The separation factor:

$$SF = \frac{^{24}Na^+/^{22}Na^+ \text{ in chloroform}}{^{24}Na^+/^{22}Na^+ \text{ in water}}$$

varies largely depending on the ligand: [2.2.1] = 1.052, [2.2.2$_B$] = 1.030, [2.2$_B$.2$_B$] = 1.0025, [2.2.2] = 0.946. The isotopic shifts are then either in the direction of the heavier $^{24}Na^+$ or the lighter $^{22}Na^+$ cations. An attempt to rationalize the results based on vibrational degrees of freedom was proposed.[248] Further examples of the use of inclusion compounds in isotopic separation can be found in Chapter 8, Volume 3.

6.1.3. Detoxication—cryptatotherapy

In addition to the possibilities of detoxication discussed in Section 2.2.2 various other investigations have been made. On lead poisoning, for example.[249]

The increasing involvement of radioactive materials in life today (energy production by nuclear reactors, medical applications, etc.) increases the probability of poisoning with these substances. Müller has devoted intensive studies to this aspect. Removal of radioactive elements: $^{85}Sr^{2+}$,[250] $^{224}Ra^{2+}$,[251,252] and $^{140}Ba^{2+}$[253] was performed *in vivo* (on rats). Enhanced excretion of the dangerous material was observed in all cases in the presence of the cryptand [2.2.2].

As already mentioned the high stability constants found in many cryptates are associated with high selectivities making these compounds candidates for medical applications. Detoxication is one possibility; another aspect is to use the cryptand as a carrier. The lithium cation is widely used as a antipsychotic for the manic-depressive condition. A suitable carrier facilitating the flow of Li^+ through membranes would allow a reduction of the quite toxic lithium salt absorption by the patient.[254]

6.1.4. Transport of cations

The crossing of membranes by ions or molecules is of fundamental importance in biological systems.[255-258] This justifies the impressive amount of work dedicated to this area in the past decades. The transport of ions across membranes mediated by ionophores is one of the numerous mechanisms used in biology.[259,260] Crown-ethers and cryptates have been widely used in order to elucidate the underlying factors of transport. The major advantage of this type of compound, compared to the natural ones, is that there are innumerable possibilities to carry out structural modifications by chemical synthesis.

Cryptands have been investigated as cation-carriers through liquid membranes (chloroform).[261] Efficient rates of transport were observed with several ligands: $[2.2.1]$–Li^+, $[2.2.C_5]$–Na^+; $[2.2.C_8]$–K^+, $[3.2.2]$–Na^+; $[3.3.3]$–K^+; $[3.3.3]$–Cs^+. For these various systems the initial flux of cations is about $0.1 \ \mu mol/h^{-1}$ per mol of carrier cm^{-2}. Some interesting selectivities of transport were obtained: with $[2.2.C_5]$, $Na^+/Li^+ > 100$; with $[2.2.2]$, $Na^+/K^+ = 20$, $Cs^+/K^+ = 100$. Many factors affecting the transport of cations have been investigated: nature of the membrane, carrier concentration, concentration gradient, pH, lipophilicity of the anion, lipophilicity of the carrier etc. It has been established that efficient transport is obtained when the stability of the complex has a value of about 10^5 in methanol. Weaker complexes ($<10^3$) or very stable complexes ($>10^8$) lead to poor transport. This rule, $K_s \sim 10^5$ in methanol, explains some odd results described above: $[2.2.2]$ forms a very stable complex with K^+, but Na^+ and Cs^+ which form less stable complexes are transported more efficiently.[261] Similar rules have been found by another group.[262] Recent reviews on cation transport should be consulted for more general information.[263-265] See also Chapter 16, Volume 3.

6.2. Cryptate associated anions

The complexation of the cation by a cryptand transforms a small inorganic cation into a very large organic one. For example in the case of K^+ ($r = 1.33$ Å) the complexation by $[2.2.2]$ leads to a cation size of about $r = 5$ Å. In addition, the cation is quite well shielded from the environment, the solvent and the counteranion (see detailed discussion of this problem in ref. 266). The large cation size and the shielding of the complexed cation affect the cation–anion interaction which becomes very weak. Also important is the fact that the cryptate salts are soluble in many low polarity solvents (chloroform, tetrahydrofuran, benzene, etc.). All these combined factors lead to an unusual situation for anions and modify all their properties. Applications of this type of anion are numerous spanning inorganic and organic chemistry.

6.2.1. Inorganic chemistry

6.2.1.1. *Metal anions.* One of the most exciting results in cryptate chemistry was obtained by Dye's group. People working with metal amine solutions had for several years gathered evidence that metal anions, M^- (Na^-, K^- etc.) exist. This offensive -1 oxidation state for alkali metals was fully demonstrated in 1974 by X-ray crystal analysis of $[Na^+ \subset 2.2.2]Na^-$ obtained by elaborate experiments.[267] This gold coloured complex has a metallic appearance. Its process of formation can be described by the equation:

$$2Na_{(sol.)} + [2.2.2] \rightleftharpoons [Na^+ \subset 2.2.2]Na^-$$

Further research has shown that other metal anions can be obtained: K^-, Rb^-, Cs^-. Characterization of these species, *alkalides*, was based on optical and NMR spectroscopy of the various nuclei.[268–270] Using a 1 : 1 metal-ligand stoichiometry and by rapid evaporation of the solvent (methylamine) a thin blue film is obtained. Properties of this film seem to indicate the existence of the species: $[M^+ \subset 2.2.2]e^-$ which was called *electride*.[271] Applications of this new class of compounds will certainly be found even if at first glance their very high reactivity constitutes a serious handicap.

6.2.1.2. *Polyatomic anions.* Sodium and potassium salts of anions such as Pb_9^{4-}; Sn_9^{4-}, Sb_7^{3-} etc. in ammonium solution have been known for a long time. But in almost no case was it possible to isolate a solid and characterize the quite unstable salts. It has been demonstrated that complexation of the cation by a cryptand such as [2.2.2] has a stabilizing effect on the ionic state.[272] By this procedure many homo-polyatomic anions have been isolated and analysed by X-ray crystallography. They are Sb_7^{3-},[272] Te_3^{3-},[273] Pb_5^{2-} and Sn_5^{2-},[274] Sn_9^{4-},[275] Ge_9^{2-} and Ge_9^{4-},[276] Sn_4^{2-} and Ge_4^{2-},[277] As_{11}^{3-}.[278] Heteroatomic polyanions have also been obtained recently: Tl_2Te^{2-}[279] $HgTe_2^{2-}$.[280]

Complexation of cations by cryptands has also been applied in the study of metal carbonyl anions. The metal carbonylate ion geometry is strongly influenced by its environment. This dependence has been investigated in the $[Fe(CO)_4]^{2-}$ anion by comparing the X-ray structures of the potassium salt to the sodium-cryptate salt.[281] The potassium cryptate salt of $[Cr_2(CO)_{10}(\mu\text{-H})]^-$ has also been analysed by X-ray crystallography.[282]

6.2.2. *Organic chemistry—chemical reactivity*

As mentioned above cryptates have interesting features: (i) the cation is very large, i.e. interaction with the anion is weak, and (ii) they are often soluble in organic media. The associated anion can then be considered almost naked i.e. very reactive. Large numbers of applications can then be imagined, theoretically all reactions in which anions are involved (and there are many) are candidates for anionic activation by cryptands. A limited number of examples will be indicated.

6.2.2.1. *Anionic activation. Strongly basic systems*: A solution of potassium *tertio*amylate (*ter*-AmOK) in benzene does not ionize triphenylmethane, (pK = 31.4) or diphenylmethane (pK = 33.4) (pK from Streitwieser scale).[283] By addition of [2.2.2] the red coloured anions are formed immediately and in high yield.[284] But $[K^+ \subset 2.2.2]$*ter*-AmO$^-$ in benzene does not ionize paramethylbiphenyl (pK = 38.7). This can be obtained only when the very strongly basic solution of $[K^+ \subset 2.2.2, C_{10}]$*ter*-AmO$^-$ in heptane is used[285,286] (see formula of $[2.2.2, C_{10}]$ in Fig. 46).

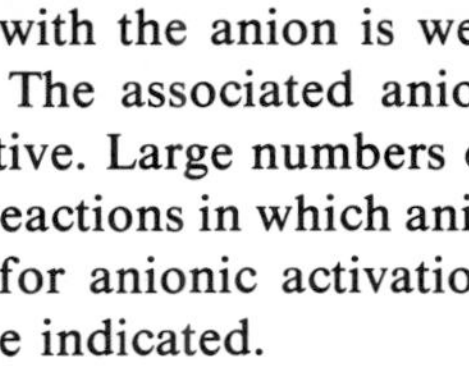

102

Fig. 46. Lipophilic cryptand.

Hydrolysis of a hindered ester: Methylmesitoate, a highly hindered ester, is hydrolysed very rapidly by a 0.8M solution of potassium hydroxide cryptate in DMSO (half-time reaction ~1 min).[284] In benzene, the heterogeneous reaction carried out with powdered KOH +[2.2.2], is much slower (70% in 12 h at 25°) but still faster than the hydrolysis with crown-ether +powdered KOH (58% in 31 h at 74°) in toluene.[3] Note that the potassium hydroxide is not dissolved in benzene or toluene by the ligands, the reactions take place on the solid-liquid interface of the suspended KOH.

Ester and ether synthesis: It has been demonstrated that a suspension of potassium alkanoates in benzene in the presence of [2.2.2], reacts with organic bromides to give esters.[287] A catalytic effect is observed. The driving force for this solid–liquid phase transfer catalysis is the lower solubility of the potassium bromide compared to potassium alkanoates. This can be represented by

$$[K^+ \subset 2.2.2]R\text{–}CO_2^- + R'\text{–}Br \rightarrow R\text{–}CO_2 - R' + [K^+ \subset 2.2.2]Br^-$$

$$[K^+ \subset 2.2.2]Br^- + R\text{–}CO_2K \rightarrow [K^+ \subset 2.2.2]RCO_2^- + \underline{BrK}$$
$$\text{solid} \qquad\qquad\qquad\qquad\qquad\qquad\qquad\qquad \downarrow$$

Knöchel studied similar reactions: acetate ion + cryptand + benzyl chloride in both homogeneous[288] and heterogeneous[289] conditions. Solid–liquid phase transfer catalysis was also observed and discussed. By a similar solid–liquid phase transfer mechanism the Williamson synthesis of ethers has been improved.[290] Alkylation of phenolates has also been performed in the presence of cryptands.[291]

Alkali hydrides have been activated. Sodium hydride and potassium hydride in the presence respectively of [2.2.1] and [2.2.2] (solvents being THF or benzene) react with many weak acids: phenol, alcohols, phthalimide, triphenylmethane and diphenylmethane.[292] No solubilization of the reagent seems to occur. Alkylation of the formed phenoxides, alkoxides, phthalimidate etc. are very rapid and were performed at room temperature.[293] In most of these reactions only catalytic amounts of cryptand are necessary. This important result can be explained by a solid–liquid phase transfer catalysis. The driving force as mentioned above is the precipitation of the alkali halide. This method gives good results for the Williamson reaction and for the Gabriel synthesis.[293] Potassium hydride, activated by cryptand, can abstract chlorine, bromine or iodine from halo-substituted benzenes. Thus, monohalobenzene derivatives are simply transformed into benzene.[294]

Reactivity and regioselectivity of alkylation of enolates have been intensively studied.[295] Alkylation of the sodium enolate of ethyl acetoacetate in THF gives only C alkylation. Addition of [2.2.2] increases the rate of reaction by a factor of 100 and 21% O-alkylation is obtained.[296,297]

Phase transfer catalysis (PTC). We have already mentioned examples of solid–liquid PTC. Cryptands have also been used as liquid–liquid phase transfer catalysts. It has been demonstrated that lipophilic cryptands (see Fig. 47) are more effective than the simple [2.2.2] which is too soluble in the aqueous phase.[298] Comparative studies have shown that compared to onium salts and crown-ethers they are the best phase transfer catalysts.[299–301] But as pointed out by one author they are also much more expensive than crown-ethers or the classical onium salts.[301] An even more lipophilic cryp-

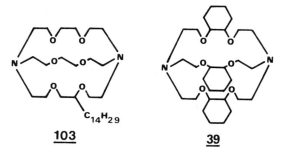

103

39

Fig. 47. Lipophilic cryptand.

tand, compound **39** has been synthesized.[302] In PTC this compound, expected to be more efficient, shows about the same trends as [2.2.2, C_{14}], compound **103**. A search for more lipophilic ligands is then not necessary, as ligand **103** seems to have all the requirements for a phase transfer catalyst. One must note that **39** forms less stable complexes than **103**; this constitutes an unfavourable point in regard to PTC.

Polymer supported cryptands have been synthesized by several groups.[230] The obvious advantage of this fixed cryptand is the possibility of recycling the catalyst after the reaction. They may be used in analytical chemistry, i.e. in cation separation by liquid chromatography, or in organic synthesis, as phase transfer catalysts.[303-305] In PTC, compared to non-immobilized phase transfer catalysts, a decrease in reactivity is observed for the polymer-supported catalyst. The mechanical properties of these polymers seem to be a problem in some cases.[305] Several types of polymer have been reported: (i) reaction of an amino-side-chain-macrobicycle derivative on chloromethylated polystyrene,[303,305] (ii) radical polymerization of a vinyl-group containing cryptand[304] and (iii), condensation of diepoxides with the cryptand [2.2.2$_{NN}$] compound **20** (Fig. 9) followed by grafting on a Merrifield polymer.[306,307] Many valuable details can be found in the papers by Blasius and Janzen[230] and Montanari.[334]

Anionic polymerization. Cryptands have found a large field of application in polymer chemistry. Many very active initiators have been obtained. Solutions of alkali metals have been obtained in non-polar solvents (benzene, dioxan, THF, etc.) when the metals are complexed by cryptands.[308] These solutions react on several monomers: isoprene, styrene, methyl methacrylate etc.[308] Most of these polymerizations are instantaneous. Several bases or salts in the presence of cryptands induce efficient polymerization: [$K^+ \subset$ 2.2.2]OH^- in benzene on hexamethylcyclotrisiloxane; [$Na^+ \subset$ 2.2.2]*ter*-AmO^- in benzene on styrene; [$K^+ \subset$ 2.2.2]SCN^- in THF on propylene sulphide etc.[309] The polymerization of this last monomer has been closely studied. It is quite surprising that cryptated ion pairs are more reactive than free ions.[310] Other anionic polymerizations in the presence of cryptates have been described: lactones,[311] ethylene oxide,[312,313] 2-pyrrolidone and 2-piperidone.[314] Two reviews on crown-ether and cryptates in polymer chemistry appeared recently.[315,316]

6.2.2.2. Effect on cation participation. As shown in the above examples cation complexation usually leads to an activated anion i.e. to an increase in the rate of reaction. By contrast, for reactions in which the cation is directly implicated in the chemical mechanism, a decrease of the reaction rate is often observed. This has been largely demonstrated by Pierre and Handel on very simple examples. Cyclohexanone, as is well known, is readily reduced by $LiAlH_4$; on the contrary if the reaction is performed in the presence of a stoichiometric amount (based on $LiAlH_4$) of [2.1.1] no reaction occurs; if this last reaction is carried out in the presence of a large excess of LiI the normal reduction takes place.[317] These three experiments clearly show the cation assistance displayed by Li^+. This cation coordinates to the carbonyl group which is more polarized and then more susceptible to hydride attack.

The same trends were demonstrated using many other substrates: aldehydes, carboxylic acids, esters, amides, etc.[318]

It has also been demonstrated that the rate of acetone reduction by potassium hydrido (phosphine) ruthenate complex decreases significantly on addition of cryptand [2.2.2].[319] The same conclusion, cation assistance, was suggested by the authors. In the addition of organolithiums to carbonyls, lithium also displays an assisting role.[320]

Reactions in which stereoselectivity originates from a cyclic chelate (see Fig. 48) are very sensitive to cation complexation. By using the appropriate conditions predominant quantities of S*R* or R*R* can be obtained.[321]

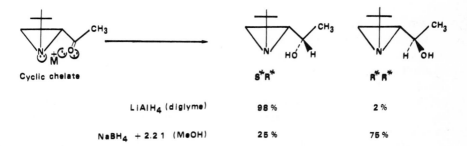

Fig. 48. Effect of cation complexation on the stereoselectivity.

Internal chelation also plays an important role in the stereochemical properties of carbanions α to sulphoxide as demonstrated by complexation experiments.[322]

Regioselectivity of the reduction of 2-cyclohexenone is reversed in the presence of cryptands.[323] In ether the reduction of 2-cyclohexenone with $LiAlH_4$ gives 98% of cyclohexenol (1,2-addition) and 2% of 1,4-addition. If the reaction is performed in the presence of [2.1.1] in stoichiometric amounts based on $LiAlH_4$: (i) the rate is slower and (ii) 77% of 1,4-addition is obtained.[323,324] Similar studies have been made on various substituted 2-cyclohexenones. The general trends described above are observed.[325]

It has been demonstrated that the addition of complexing agents modifies the regioselectivity of the reaction of the allylic carbanion Ph–S–CHLi–CH=CMe$_2$ with various substrates.[326]

Finally both the contributions of anionic activation and cation participation have been nicely illustrated in the reaction of the acetonitrile anion with benzaldehyde. Results are indicated in Table 16. It is noteworthy that: (i) in the case of $M^+ = Li^+$ the addition of [2.1.1] leads to a large decrease in the reaction rate i.e. the cation participation is the dominant factor, (ii) when $M^+ = K^+$, addition of [2.2.2] brings about a rate increase i.e. the anion activation is the dominant factor and (iii) in the case of $M^+ = Na^+$ the rate of the reaction is not affected by addition of [2.2.1], the two factors neutralize each other.[327] A detailed discussion of this cation participation/anion activation balance can be found in ref. 266.

Note on cryptand recovery: Cryptand syntheses are long and time consuming; the commercially available ones are quite expensive. As has been shown both in our department and in Popov's laboratory cryptand recovery from solutions after reaction is possible.[328]

Table 16. Cation participation—Anion activation[327]
PhCHO + CH$_2$CN$^-$M$^+$ → Ph–CH–CH$_2$CN
                                      |
                                      OH

M^+	Cryptands	Time (mins)	Yield %
Li$^+$	—	5	98
Li$^+$	[2.1.1]	60	90
Na$^+$	—	30	70
Na$^+$	[2.2.1]	30	60
K$^+$	—	150	80
K$^+$	[2.2.2]	30	90

6.3. Applications of anion complexation

Only a few examples of chemical applications of anion complexation have appeared.

It has been demonstrated that the rate of the reaction (a) shown in Fig. 49 decreases when guanidinium GH$^+$ is added. The authors suggested that guanidinium interacts with the carboxylate and thus diminishes its availability for nucleophilic attack leading to the intermediate compound a'.[329]

Examples of rate enhancement in the anion attack on diaryl phosphate, upon addition of guanidinium, have been obtained by the same group[330] (see Fig. 49b).

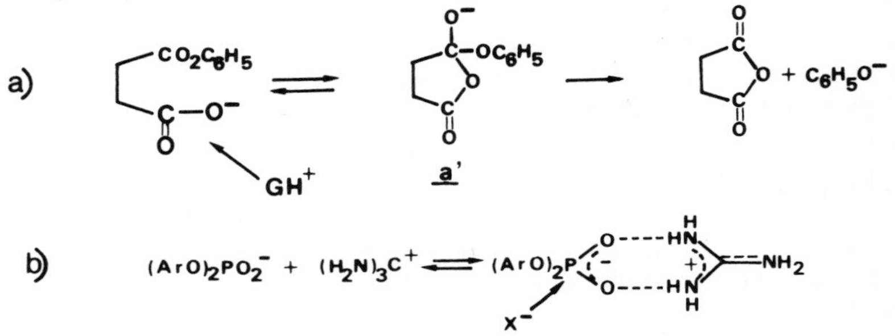

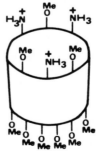

Fig. 49. Rate decrease (a) and rate increase (b) by anion complexation.

Schmidtchen using the large compound **89** (Fig. 39) obtained acceleration of three types of reactions. The author suggested that these rate enhancements originate from the lowering of the energy of the anionic transition state by inclusion in the macrotricycle cavity.[331]

Boger and Knowles have synthesized the triamino-per-*O*-methylcyclodextrin (Fig. 50). At pH = 7 the three amino groups are protonated and efficient binding of the dianion 4-nitrophenyl phosphate is observed.[332]

Fig. 50. Tri-ammonium cyclodextrin derivative showing complexation properties toward phosphate anions.

Conclusion

We have tried in this short review to summarize the major aspects of cryptate chemistry. We hope that the rapid treatment of each of the parts will not make the reading too cumbersome. Some aspects like the complexes of uncharged molecules (for this see ref. 333) have not been treated. Finally the reader can find more treatment of the subject in the additional reference list.

Acknowledgements

The author thanks Professor Jean-Marie Lehn for his valuable advice and his inspiration in the field of cryptate chemistry, Professor Jean-Pierre Sauvage for his comments, Dr. Cindy Burrows for her help in preparing the manuscript, Serge Wechsler for the photographic work and Jacqueline Claudon and Agnès Kintzinger for the typing, and the Centre National de la Recherche Scientifique for financial support.

References

1. A. Lüttringhaus, *Justus Liebigs Ann. Chem.*, 1937, **528**, 181.
2. J. L. Down, J. Lewis, B. Moore and G. Wilkinson, *J. Chem. Soc.*, 1959, 3767.
3. C. J. Pedersen, *J. Am. Chem. Soc.*, 1967, **89**, 2495, 7017.
4. C. Moore and B. Pressman, *Biochem. Biophys. Res. Commun.*, 1964, **15**, 562.
5. Yu. A. Ovchinnikov, V. Y. Ivanov and A. M. Shkrob, *Membrane-Active Complexones*, Elsevier Scientific Publishing Company, Amsterdam, 1974.
6. B. T. Kilbourn, J. D. Dunitz, L. A. R. Pioda and W. Simon, *J. Mol. Biol.*, 1967, **30**, 559.
7. M. Dobler, *Ionophores and Their Structures*, John Wiley & Sons, New York, 1981.
8. J. M. Lehn, *Struct. Bonding (Berlin)*, 1973, **16**, 1.
9. B. Dietrich, J. M. Lehn and J. P. Sauvage, *Tetrahedron Lett.*, 1969, 2885.
10. A. H. Haines and P. Karntiang, *J. Chem. Soc., Perkin Trans. 1*, 1979, 2577.
11. A. C. Coxon and J. F. Stoddart, *J. Chem. Soc., Perkin Trans. 1*, 1977, 767.
12. B. Metz, D. Moras and R. Weiss, *J. Chem. Soc., Chem. Commun.*, 1970, 217.
13. B. Dietrich, J. M. Lehn and J. P. Sauvage, *Tetrahedron Lett.*, 1969, 2889.
14. B. Dietrich, J. M. Lehn, J. P. Sauvage and J. Blanzat, *Tetrahedron*, 1973, **29**, 1629.
15 J. M. Lehn and J. P. Sauvage, *J. Am. Chem. Soc.*, 1975, **97**, 6700.
16. B. G. Cox, J. Garcia-Rosas and H. Schneider, *J. Am. Chem. Soc.*, 1981, **103**, 1384.
17. J. Cheney, J. P. Kintzinger and J. M. Lehn, *Nouv. J. Chim.*, 1978, **2**, 411.

18. P. B. Smith, J. L. Dye, J. Cheney and J. M. Lehn, *J. Am. Chem. Soc.*, 1981, **103**, 6044.
19. R. W. Alder and R. B. Sessions, *J. Am. Chem. Soc.*, 1979, **101**, 3651.
20. R. W. Alder, A. Casson and R. B. Sessions, *J. Am. Chem. Soc.*, 1979, **101**, 3652.
21. R. W. Alder, R. J. Arrowsmith, A. Casson, R. B. Sessions, E. Heilbronner, B. Kovač, H. Huber and M. Taagepera, *J. Am. Chem. Soc.*, 1981, **103**, 6137.
22. D. K. Cabbiness and D. W. Margerum, *J. Am. Chem. Soc.*, 1969, **91**, 6540.
23. F. P. Hinz and D. W. Margerum, *Inorg. Chem.*, 1974, **13**, 2941.
24. H. K. Frensdorff, *J. Am. Chem. Soc.*, 1971, **93**, 600.
25. G. W. Gokel, D. M. Dishong and C. J. Diamond, *J. Chem. Soc., Chem. Commun.*, 1980, 1053.
26. D. J. Cram and K. N. Trueblood, in *Host–Guest Complex Chemistry I*, (ed. F. Vögtle) Springer-Verlag, Berlin, 1981, p. 43.
27. D. J. Cram, G. M. Lein, T. Kaneda, R. C. Helgeson, C. B. Knobler, E. Maverick and K. N. Trueblood, *J. Am. Chem. Soc.*, 1981, **103**, 6228.
28. G. Anderegg, *Helv. Chim. Acta*, 1975, **58**, 1218.
29. E. Kauffmann, J. M. Lehn and J. P. Sauvage, *Helv. Chim. Acta*, 1976, **59**, 1099.
30. J. D. Lamb, R. M. Izatt, J. J. Christensen and D. J. Eatough, in *Coordination Chemistry of Macrocyclic Compounds*, (ed. G. A. Melson) Plenum Press, New York, 1979, p. 145.
31. A. I. Popov and J. M. Lehn, in *Coordination Chemistry of Macrocyclic Compounds*, (ed. G. A. Melson) Plenum Press, New York, London, 1979, p. 537.
32. G. W. Liesegang and E. M. Eyring, in *Synthetic Multidentate Macrocyclic Compounds*, (eds. R. M. Izatt and J. J. Christensen) Academic Press, New York, London, 1978, p. 245.
33. R. Gresser, D. W. Boyd, A. M. Albrecht-Gary and J. P. Schwing, *J. Am. Chem. Soc.*, 1980, **102**, 651.
34. B. G. Cox, J. Garcia-Rosas and H. Schneider, *J. Am. Chem. Soc.*, 1981, **103**, 1054.
35. B. G. Cox, W. Jedral, P. Firman and H. Schneider, *J. Chem. Soc., Perkin Trans. 2*, 1981, 1486.
36. G. W. Liesegang, *J. Am. Chem. Soc.*, 1981, **103**, 953.
37. J. M. Lehn, J. P. Sauvage and B. Dietrich, *J. Am. Chem. Soc.*, 1970, **92**, 2916.
38. C. Cambillau and M. Ourevitch, *J. Chem. Soc., Chem. Commun.*, 1981, 996.
39. Y. M. Cahen, J. L. Dye and A. I. Popov, *J. Phys. Chem.*, 1975, **79**, 1292.
40. P. Laszlo, *Angew. Chem., Int. Ed. Engl.*, 1978, **17**, 254.
41. A. Knöchel and R. D. Wilken, *J. Am. Chem. Soc.*, 1981, **103**, 5707.
42. J. M. Ceraso, P. B. Smith, J. S. Landers and J. L. Dye, *J. Phys. Chem.*, 1977, **81**, 760.
43. J. P. Kintzinger and J. M. Lehn, *J. Am. Chem. Soc.*, 1974, **96**, 3313.
44. E. Mei, J. L. Dye and A. I. Popov, *J. Am. Chem. Soc.*, 1976, **98**, 1619.
45. E. Kauffmann, J. L. Dye, J. M. Lehn and A. I. Popov, *J. Am. Chem. Soc.*, 1980, **102**, 2274.
46. A. I. Popov, A. J. Smetana, J. P. Kintzinger and T. T-T. Nguyên, *Helv. Chim. Acta*, 1980, **63**, 668.
47. H. G. Förster and J. D. Roberts, *J. Am. Chem. Soc.*, 1980, **102**, 6984.
48. S. Boileau, P. Hemery and J. C. Justice, *J. Solution Chem.*, 1975, **4**, 873.
49. J. Koryta and M. Gross, *Topics in Bioelectrochemistry and Bioenergetics*, Vol. 3, Wiley & Sons, 1980, p. 93.
50. U. Takaki and J. Smid, *J. Am. Chem. Soc.*, 1974, **96**, 2588.

51. B. Tümmler, G. Maass, E. Weber, W. Wehner and F. Vögtle, *J. Am. Chem. Soc.*, 1977, **99**, 4683.
52. P. Gans, J. B. Gill and J. N. Towning, *J. Chem. Soc., Dalton Trans.*, 1977, 2202.
53. O. Bohman, P. Ahlberg, R. Nyholm, N. Mårtensson, K. Siegbahn and J. M. Lehn, *Chemica Scripta*, 1981, **18**, 44.
54. W. E. Morf and W. Simon, *Helv. Chim. Acta*, 1971, **54**, 2683.
55. B. Dietrich, J. M. Lehn and J. P. Sauvage, *J. Chem. Soc., Chem. Commun.*, 1973, 15.
56. C. J. Pedersen and M. H. Bromels, U.S. Patent, 3, 847, 949, 1974.
57. N. Wester and F. Vögtle, *J. Chem. Res. (S)*, 1978, 400.
58. J. M. Lehn and F. Montavon, *Helv. Chim. Acta*, 1976, **59**, 1566.
59. J. M. Lehn and F. Montavon, *Helv. Chim. Acta*, 1978, **61**, 67.
60. T. L. Blundell and J. A. Jenkins, *Chem. Soc. Rev.*, 1977, **6**, 139.
61. B. Dietrich, J. M. Lehn and J. P. Sauvage, *J. Chem. Soc., Chem. Commun.*, 1970, 1055.
62. B. Dietrich, Thèse de Doctorat d'Etat, Strasbourg, 1973.
63. B. Dietrich and J. M. Lehn, unpublished results.
64. D. J. Cram and J. M. Cram, *Acc. Chem. Res.*, 1978, **11**, 8.
65. J. F. Stoddart, *Chem. Soc. Rev.*, 1979, **8**, 85.
66. F. Vögtle and E. Weber, *Angew. Chem., Int. Ed. Engl.*, 1979, **18**, 753.
67. D. G. Parsons, *J. Chem. Soc., Perkin Trans. 1*, 1978, 451.
68. I. R. Hanson, D. G. Parsons and M. R. Truter, *J. Chem. Soc., Chem. Commun.*, 1979, 486.
69. J. A. Bandy, D. G. Parsons and M. R. Truter, *J. Chem. Soc., Chem. Commun.*, 1981, 729.
70. I. R. Hanson and M. R. Truter, *J. Chem. Soc., Perkin Trans. 2*, 1981, 1.
71. J. D. Owen, *J. Chem. Soc., Perkin Trans. 2*, 1981, 12.
72. I. R. Hanson, J. D. Owen and M. R. Truter, *J. Chem. Soc., Perkin Trans. 2*, 1981, 1606.
73. W. Wehner and F. Vögtle, *Tetrahedron Lett.*, 1976, 2603.
74. G. R. Newkome, V. Majestic, F. Fronczek and J. L. Atwood, *J. Am. Chem. Soc.*, 1979, **101**, 1047.
75. G. R. Newkome, V. K. Majestic and F. R. Fronczek, *Tetrahedron Lett.*, 1981, 3035.
76. G. R. Newkome, V. K. Majestic and F. R. Fronczek, *Tetrahedron Lett.*, 1981, 3039.
77. E. Buhleier, W. Wehner and F. Vögtle, *Chem. Ber.*, 1978, **111**, 200.
78. S. Shinkai, T. Ogawa, T. Nakaji, Y. Kusano and O. Manabe, *Tetrahedron Lett.*, 1979, 4569.
79. S. Shinkai, T. Nakaji, Y. Nishida, T. Ogawa and O. Manabe, *J. Am. Chem. Soc.*, 1980, **102**, 5860.
80. S. Shinkai, T. Nakaji, T. Ogawa, K. Shigematsu and O. Manabe, *J. Am. Chem. Soc.*, 1981, **103**, 111.
81. T. Asano, T. Okada, S. Shinkai, K. Shigematsu, Y. Kusano and O. Manabe, *J. Am. Chem. Soc.*, 1981, **103**, 5161.
82. J. S. Bradshaw and P. E. Stott, *Tetrahedron*, 1980, **36**, 461.
83. D. Landini, F. Montanari and F. Rolla, *Synthesis*, 1978, 223.
84. I. I. Creaser, J. MacB. Harrowfield, A. J. Herlt, A. M. Sargeson, J. Springborg, R. J. Geue and M. R. Snow, *J. Am. Chem. Soc.*, 1977, **99**, 3181.
85. A. M. Sargeson, *Pure Appl. Chem.*, 1978, **50**, 905.
86. G. J. Gainsford, R. J. Geue and A. M. Sargeson, *J. Chem. Soc., Chem. Commun.*, 1982, 233.

87. G. Oepen and F. Vögtle, *Justus Liebigs Ann. Chem.*, 1979, 1094.
88. A. P. Bell and C. D. Hall, *J. Chem. Soc., Chem. Commun.*, 1980, 163.
89. M. Sato, M. Kubo, S. Ebine and S. Akabori, *Tetrahedron Lett.*, 1982, **23**, 185.
90. E. Buhleier, K. Frensch, F. Luppertz and F. Vögtle, *Justus Liebigs Ann. Chem.*, 1978, 1586.
91. L. Rossa and F. Vögtle, *Justus Liebigs Ann. Chem.*, 1981, 459.
92. G. Weber, *Acta Crystallogr.*, 1981, **B37**, 1832.
93. B. Dietrich, J. M. Lehn and J. Simon, *Angew. Chem., Int. Ed. Engl.*, 1974, **13**, 406.
94. I. Tabushi, K. Fujita and L. C. Yuan, *Tetrahedron Lett.*, 1977, 2503.
95. A. Ueno, H. Yoshimura, R. Saka and T. Osa, *J. Am. Chem. Soc.*, 1979, **101**, 2779.
96. I. Tabushi, Y. Kuroda and K. Shimokawa, *J. Am. Chem. Soc.*, 1979, **101**, 1614.
97. I. Tabushi, Y. Kuroda and A. Mochizuki, *J. Am. Chem. Soc.*, 1980, **102**, 1152.
98. R. Breslow, M. F. Czarniecki, J. Emert and H. Hamaguchi, *J. Am. Chem. Soc.*, 1980, **102**, 762.
99. A. Ueno, K. Takahashi and T. Osa, *J. Chem. Soc., Chem. Commun.*, 1981, 94.
100. I. Tabushi and L. C. Yuan, *J. Am. Chem. Soc.*, 1981, **103**, 3574.
101. C. K. Chang, *J. Am. Chem. Soc.*, 1977, **99**, 2819.
102. J. P. Collman, J. I. Brauman, T. J. Collins, B. Iverson and J. L. Sessler, *J. Am. Chem. Soc.*, 1981, **103**, 2450.
103. J. P. Collman, C. M. Elliott, T. R. Halbert and B. S. Tovrog, *Proc. Natl. Acad. Sci. USA*, 1977, **74**, 18.
104. J. P. Collman, A. O. Chong, G. B. Jameson, R. T. Oakley, E. Rose, E. R. Schmittou and J. A. Ibers, *J. Am. Chem. Soc.*, 1981, **103**, 516.
105. J. P. Collman, P. Denisevich, Y. Konai, M. Marrocco, C. Koval and F. C. Anson, *J. Am. Chem. Soc.*, 1980, **102**, 6027.
106. B. Ward, C. B. Wang and C. K. Chang, *J. Am. Chem. Soc.*, 1981, **103**, 5236.
107. M. H. Hatada, A. Tulinsky and C. K. Chang, *J. Am. Chem. Soc.*, 1980, **102**, 7115.
108. R. R. Bucks and S. G. Boxer, *J. Am. Chem. Soc.*, 1982, **104**, 340.
109. J. Almog, J. E. Baldwin, R. L. Dyer and M. Peters, *J. Am. Chem. Soc.*, 1975, **97**, 226.
110. A. R. Battersby, D. G. Buckley, S. G. Hartley and M. D. Turnbull, *J. Chem. Soc., Chem. Commun.*, 1976, 879.
111. J. E. Baldwin and J. F. DeBernardis, *J. Org. Chem.*, 1977, **42**, 3986.
112. A. R. Battersby, S. G. Hartley and M. D. Turnbull, *Tetrahedron Lett.*, 1978, 3169.
113. M. Momenteau, B. Loock, J. Mispelter and E. Bisagni, *Nouv. J. Chim.*, 1979, **3**, 77.
114. C. K. Chang and M. S. Kuo, *J. Am. Chem. Soc.*, 1979, **101**, 3413.
115. J. R. Budge, P. E. Ellis, Jr., R. D. Jones, J. E. Linard, T. Szymanski, F. Basolo, J. E. Baldwin and R. L. Dyer, *J. Am. Chem. Soc.*, 1979, **101**, 4762.
116. A. R. Battersby and A. D. Hamilton, *J. Chem. Soc., Chem. Commun.*, 1980, 117.
117. K. Nagappa Ganesh and J. K. M. Sanders, *J. Chem. Soc., Chem. Commun.*, 1980, 1129.
118. M. J. Gunter, L. N. Mander, K. S. Murray and P. E. Clark, *J. Am. Chem. Soc.*, 1981, **103**, 6784.
119 T. G. Traylor, *Acc. Chem. Res.*, 1981, **14**, 102.
120. E. Graf and J. M. Lehn, *J. Am. Chem. Soc.*, 1975, **97**, 5022.
121. E. Graf and J. M. Lehn, *Helv. Chim. Acta*, 1981, **64**, 1040.
122. E. Graf, *Thèse de Doctorat d'Etat*, Strasbourg, 1979.
123. E. Graf, J. P. Kintzinger, J. M. Lehn and J. LeMoigne, *J. Am. Chem. Soc.*, 1982, **104**, 1672.

124. B. Metz, J. M. Rosalky and R. Weiss, *J. Chem. Soc., Chem. Commun.*, 1976, 533.
125. R. Annunziata, F. Montanari, S. Quici and M. T. Vitali, *J. Chem. Soc., Chem. Commun.*, 1981, 777.
126. J. Cheney, J. M. Lehn, J. P. Sauvage and M. E. Stubbs, *J. Chem. Soc., Chem. Commun.*, 1972, 1100.
127. J. M. Lehn and M. E. Stubbs, *J. Am. Chem. Soc.*, 1974, **96**, 4011.
128. R. Wiest and R. Weiss, *J. Chem. Soc., Chem. Commun.*, 1973, 678.
129. J. Comarmond, *Thèse de Docteur-Ingénieur, Strasbourg*, 1981.
130. M. J. Calverley and J. Dale, *J. Chem. Soc., Chem. Commun.*, 1981, 1084.
131. P. Groth, *Acta Chem. Scand., Ser. A* 1981, **35**, 717.
132. J. M. Lehn, J. Simon and J. Wagner, *Angew. Chem., Int. Ed. Engl.*, 1973, **12**, 579.
133. J. M. Lehn, J. Simon and J. Wagner, *Nouv. J. Chim.*, 1977, **1**, 77.
134. J. M. Lehn and J. Simon, *Helv. Chim. Acta*, 1977, **60**, 141.
135. M. Mellinger, J. Fischer and R. Weiss, *Angew. Chem. Int. Ed. Engl.*, 1973, **12**, 771.
136. G. M. Edelman and W. O. McClure, *Acc. Chem. Res.*, 1968, **1**, 65.
137. F. Vögtle and P. Dix, *Justus Liebigs Ann. Chem.*, 1977, 1698.
138. J. M. Lehn, J. Simon and A. Moradpour, *Helv. Chim. Acta*, 1978, **61**, 2407.
139. R. C. Hayward, *Nachrichten aus Chem. Tech. Lab.*, 1977, **25**, 15.
140. M. R. Johnson, I. O. Sutherland and R. F. Newton, *J. Chem. Soc., Chem. Commun.*, 1979, 309.
141. R. Mageswaran, S. Mageswaran and I. O. Sutherland, *J. Chem. Soc., Chem. Commun.*, 1979, 722.
142. N. F. Jones, A. Kumar and I. O. Sutherland, *J. Chem. Soc., Chem. Commun.*, 1981, 990.
143. F. Kotzyba-Hibert, J. M. Lehn and P. Vierling, *Tetrahedron Lett.*, 1980, **21**, 941.
144. J. P. Kintzinger, F. Kotzyba-Hibert, J. M. Lehn, A. Pagelot and K. Saigo, *J. Chem. Soc., Chem. Commun.*, 1981, 833.
145. J. M. Lehn and P. Vierling, *Tetrahedron Lett.*, 1980, **21**, 1323.
146. F. Kotzyba-Hibert, J. M. Lehn and K. Saigo, *J. Am. Chem. Soc.*, 1981, **103**, 4266.
147. C. Pascard, C. Riche, M. Cesario, F. Kotzyba-Hibert and J. M. Lehn, *J. Chem. Soc., Chem. Commun.*, 1982, 557.
148. U. Casellato, P. A. Vigato, D. E. Fenton and M. Vidali, *Chem. Soc. Rev.*, 1979, **8**, 199.
149. J. M. Lehn, *Pure Appl. Chem.*, 1980, **52**, 2441.
150. J. M. Lehn, S. H. Pine, E. Watanabe and A. K. Willard, *J. Am. Chem. Soc.*, 1977, **99**, 6766.
151. R. J. Motekaitis, A. E. Martell, J. M. Lehn and E. I. Watanabe, *Inorg. Chem.*, 1982, **21**, 4253.
152. J. P. Gisselbrecht and M. Gross, unpublished results.
153. A. H. Alberts, R. Annunziata and J. M. Lehn, *J. Am. Chem. Soc.*, 1977, **99**, 8502.
154. R. Louis, Y. Agnus and R. Weiss, *J. Am. Chem. Soc.*, 1978, **100**, 3604.
155. Y. Agnus and R. Louis, *Nouv. J. Chim.*, 1981, **5**, 305.
156. J. P. Gisselbrecht, M. Gross, A. H. Alberts and J. M. Lehn, *Inorg. Chem.*, 1980, **19**, 1386.
157. Y. Agnus, R. Louis and R. Weiss, *J. Am. Chem. Soc.*, 1979, **101**, 3381.
158. J. A. Fee, *Struct. Bonding (Berlin)*, 1975, **23**, 1.
159. D. H. Busch, *Pure Appl. Chem.*, 1980, **52**, 2477.
160. J. M. Lehn, in "Frontiers of Chemistry" (IUPAC Symposium Series), ed. K. J. Laidler, Pergamon Press, Oxford, 1982, p. 265.
161. D. J. Cram and J. M. Cram, *Science*, 1974, **183**, 803.

162. J. P. Behr and J. M. Lehn, in *Structural and Functional Aspects of Enzyme Catalysis*, (eds. H. Eggerer and R. Huber) Springer-Verlag, Berlin, Heidelberg, New York, 1981, p. 24.
163. Y. Pocker, *Ann. Rev. Phys. Chem.*, 1979, **30**, 579.
164. J. F. Stoddart, in *Bioenergetics and Thermodynamics: Model Systems* (ed. A. Braibanti) D. Reidel Publishing Company, Dordrecht, Boston, London, 1980, p. 43.
165. F. de Jong and D. N. Reinhoudt, *Adv. Phys. Org. Chem.*, 1980, **17**, 279.
166. R. M. Kellogg, in *Host Guest Complex Chemistry II* (ed. F. Vögtle) Springer-Verlag, Berlin, Heidelberg, New York, 1982, p. 111.
167. J. M. Girodeau, J. M. Lehn and J. P. Sauvage, *Angew. Chem., Int. Edit. Engl.*, 1975, **14**, 764.
168. J. P. Behr, J. M. Girodeau, R. C. Hayward, J. M. Lehn and J. P. Sauvage, *Helv. Chim. Acta*, 1980, **63**, 2096.
169. J. P. Behr, J. M. Lehn and P. Vierling, *J. Chem. Soc., Chem. Comun.*, 1976, 621.
170. J. J. Daly, P. Schönholzer, J. P. Behr and J. M. Lehn, *Helv. Chim. Acta*, 1981, **64**, 1444.
171. J. P. Behr and J. M. Lehn, *Helv. Chim. Acta*, 1980, **63**, 2112.
172. J. P. Behr, in *Bioenergetics and Thermodynamics: Model Systems* (ed. A. Braibanti) D. Reidel Publishing Company, Dordrecht, Boston, London, 1980, p. 425.
173. J. P. Behr and J. M. Lehn, *J. Chem. Soc., Chem. Commun.*, 1978, 143.
174. J. M. Lehn and C. Sirlin, *J. Chem. Soc., Chem. Commun.*, 1978, 949.
175. J. P. Behr, J. M. Lehn, D. Moras and J. C. Thierry, *J. Am. Chem. Soc.*, 1981, **103**, 701.
176. G. Szabo and D. W. Urry, *Science*, 1979, **203**, 55.
177. R. E. Koeppe, J. M. Berg, K. O. Hodgson and L. Stryer, *Nature (London)*, 1979, **279**, 723.
178. H. G. Khorana, *Bioorg. Chem.*, 1980, **9**, 363.
179. S. V. Sychev, N. A. Nevskaya, St. Jordanov, E. N. Shepel, A. I. Miroshnikov and V. T. Ivanov, *Bioorg. Chem.*, 1980, **9**, 121.
180. R. Nagaraj and P. Balaram, *Acc. Chem. Res.*, 1981, **14**, 356.
181. J. P. Behr, J. M. Lehn, A. C. Dock and D. Moras, *Nature (London)*, 1982, **295**, 526.
182. L. G. Lange, III, J. F. Riordan and B. L. Vallèe, *Biochem.*, 1974, **13**, 4361.
183. J. J. R. Fraústo da Silva and R. J. P. Williams, *Struct. Bonding (Berlin)*, 1976, **29**, 67.
184. C. S. G. Phillips and R. J. P. Williams, *Inorganic Chemistry*, Oxford University Press, 1965, p. 159.
185. L. Radom, *Aust. J. Chem.*, 1976, **29**, 1635.
186. F. A. Cotton, E. E. Hazen, Jr and M. J. Legg, *Proc. Natl. Acad. Sci. USA*, 1979, **76**, 2551.
187. P. W. Tucker, E. E. Hazen, Jr and F. A. Cotton, *Mol. Cell. Biochem.*, 1978, **22**, 67; 1979, **23**, 3, 67 and 131.
188. J. F. Riordan, K. D. McElvany and C. L. Borders, Jr, *Science*, 1977, **195**, 884.
189. K. Kanamori, A. H. Cain and J. D. Roberts, *J. Am. Chem. Soc.*, 1978, **100**, 4979.
190. G. Taborsky and K. McCollum, *J. Biol. Chem.*, 1979, **254**, 7069.
191. C. W. Tabor and H. Tabor, *Annu. Rev. Biochem.*, 1976, **45**, 285.
192. C. Nakai and W. Glinsmann, *Biochem.*, 1977, **16**, 5636.
193. S. Bunce and E. S. W. Kong, *Biophys. Chem.*, 1978, **8**, 357.

194. T. T. Herskovits and J. Brahms, *Biopolymers*, 1976, **15**, 687.
195. G. Voordouw and H. Eisenberg, *Nature (London)*, 1978, **273**, 446.
196. F. A. Cotton, V. W. Day, E. E. Hazen, Jr and S. Larsen, *J. Am. Chem. Soc.*, 1973, **95**, 4834.
197. B. Dietrich, T. M. Fyles, J. M. Lehn, L. G. Pease and D. L. Fyles, *J. Chem. Soc., Chem. Commun.*, 1978, 934.
198. I. G. Sayce, *Talanta*, 1968, **15**, 1397; and 1971, **18**, 653.
199. B. Dietrich, D. L. Fyles, T. M. Fyles and J. M. Lehn, *Helv. Chim. Acta*, 1979, **62**, 2763.
200. F. P. Schmidtchen, *Chem. Ber.*, 1980, **113**, 2175.
201. C. H. Park and H. E. Simmons, *J. Am. Chem. Soc.*, 1968, **90**, 2431.
202. R. A. Bell, G. G. Christoph, F. R. Fronczek and R. E. Marsh, *Science*, 1975, **190**, 151.
203. E. Graf and J. M. Lehn, *J. Am. Chem. Soc.*, 1976, **98**, 6403.
204. F. P. Schmidtchen, *Angew. Chem. Int. Ed. Engl.*, 1977, **16**, 720.
205. F. P. Schmidtchen, *Chem. Ber.*, 1980, **113**, 864.
206. F. P. Schmidtchen, *Chem. Ber.*, 1981, **114**, 597.
207. J. M. Lehn, E. Sonveaux and A. K. Willard, *J. Am. Chem. Soc.*, 1978, **100**, 4914.
208. C. Pascard, J. Guilhem, B. Dietrich, J. M. Lehn and A. K. Willard, Colloque International du C.N.R.S., "Composés Macrocycliques", Strasbourg, 1982.
209. B. Dietrich, J. Guilhem, J. M. Lehn, C. Pascard and E. Sonveaux, *Helv. Chim. Acta*, in press.
210. N. Wester and F. Vögtle, *Chem. Ber.*, 1980, **113**, 1487.
211. B. Dietrich, M. W. Hosseini, J. M. Lehn and R. B. Sessions, *Helv. Chim. Acta*, 1983, **66**, 1262.
212. S. Goldman and R. G. Bates, *J. Am. Chem. Soc.*, 1972, **94**, 1476.
213. A. P. Leugger, L. Hertli and T. A. Kaden, *Helv. Chim. Acta*, 1978, **61**, 2296.
214. E. Kimura and M. Kodama, *Yuki Gosei Kagaku Kyokaishi*, 1977, **35**, 632.
215. M. Micheloni, A. Sabatini and P. Paoletti, *J. Chem. Soc., Perkin Trans. 2*, 1978, 828.
216. M. Micheloni, P. Paoletti and A. Vacca, *J. Chem. Soc., Perkin Trans. 2*, 1978, 945.
217. B. Dietrich, M. W. Hosseini, J. M. Lehn and R. B. Sessions, *J. Am. Chem. Soc.*, 1981, **193**, 1282.
218. J. E. Richman and T. J. Atkins, *J. Am. Chem. Soc.*, 1974, **96**, 2268.
219. F. Peter, M. Gross, M. W. Hosseini, J. M. Lehn and R. B. Sessions, *J. Chem. Soc., Chem. Commun.*, 1981, 1067.
220. M. W. Hosseini and J. M. Lehn, *J. Am. Chem. Soc.*, 1982, **104**, 3525.
221. A. E. Martin, T. M. Ford and J. E. Bulkowski, *J. Org. Chem.*, 1982, **47**, 412.
222. A. E. Martin and J. E. Bulkowski, *J. Org. Chem.*, 1982, **47**, 415.
223. E. Kimura, A. Sakonaka, T. Yatsunami and M. Kodama, *J. Am. Chem. Soc.*, 1981, **103**, 3041.
224. I. Tabushi, J. Imuta, N. Seko and Y. Kobuke, *J. Am. Chem. Soc.*, 1978, **100**, 6287.
225. I. Tabushi, Y. Kobuke and J. Imuta, *Nucleic Acids Research. Symposium Series* 1979, **6**, 175.
226. I. Tabushi, Y. Kobuke and J. Imuta, *J. Am. Chem. Soc.*, 1980, **102**, 1744.
227. I. Tabushi, Y. Kobuke and J. Imuta, *J. Am. Chem. Soc.*, 1981, **103**, 6152.
228. H. M. N. H. Irving, *Pure Appl. Chem.*, 1978, **50**, 1129.
229. I. M. Kolthoff, *Anal. Chem.*, 1979, **51**, 1r.
230. E. Blasius and K. P. Janzen, in *Host Guest Complex Chemistry* I. (ed. F. Vögtle) Springer-Verlag, Berlin, 1981, 163.

231. B. Dietrich, J. M. Lehn and J. P. Sauvage, *Tetrahedron*, 1973, **29**, 1647.

232. F. Arnaud-Neu, B. Spiess and M. J. Schwing-Weill, *Helv. Chim. Acta*, 1977, **60**, 2633.

233. B. Spiess, F. Arnaud-Neu and M. J. Schwing-Weill, *Helv. Chim. Acta*, 1979, **62**, 1531.

234. B. Spiess, F. Arnaud-Neu and M. J. Schwing-Weill, *Helv. Chim. Acta*, 1980, **63**, 2287.

235. B. Spiess, D. Martin-Faber, F. Arnaud-Neu and M. J. Schwing-Weill, *Inorg. Chim. Acta*, 1981, **54**, L91.

236. O. A. Gansow, A. R. Kausar, K. M. Triplett, M. J. Weaver and E. L. Yee, *J. Am. Chem. Soc.*, 1977, **99**, 7087.

237. J. H. Burns and C. F. Baes, Jr, *Inorg. Chem.*, 1981, **20**, 616.

238. G. Anderegg, *Helv. Chim. Acta*, 1981, **64**, 1790.

239. F. A. Hart, M. B. Hursthouse, K. M. A. Malik and S. Moorhouse, *J. Chem. Soc., Chem. Commun.*, 1978, 549.

240. M. Ciampolini, P. Dapporto and N. Nardi, *J. Chem. Soc., Chem. Commun.*, 1978, 788.

241. J. H. Burns, *Inorg. Chem.*, 1979, **18**, 3044.

242. E. L. Yee, O. A. Gansow and M. J. Weaver, *J. Am. Chem. Soc.*, 1980, **102**, 2278.

243. J. M. Girodeau, Thèse de Docteur-Ingénieur 1977 (Strasbourg).

244. R. M. Costes, G. Folcher, P. Plurien and P. Rigny, *Inorg. Nucl. Chem. Lett.*, 1976, **12**, 491.

245. B. Spiess, F. Arnaud-Neu and M. J. Schwing-Weill, *Inorg. Nucl. Chem. Lett.*, 1979, **15**, 13.

246. B. E. Jepson and R. DeWitt, *J. Inorg. Nucl. Chem.*, 1976, **38**, 1175.

247. K. G. Heumann and H. P. Schiefer, *Angew. Chem., Int. Ed. Engl.*, 1980, **19**, 406.

248. A. Knöchel and R. D. Wilken, *J. Am. Chem. Soc.*, 1981, **103**, 5707.

249. P. Baudot, M. Jacque and M. Robin, *Toxicology and Applied Pharmacology*, 1977, **41**, 113.

250. W. H. Müller, *Naturwissenshaften*, 1970, **57**, 248.

251. W. H. Müller and W. A. Müller, *Naturwissenshaften*, 1974, **61**, 455.

252. W. H. Müller, *Strahlentherapie*, 1977, 570.

253. W. H. Müller, W. A. Müller and U. Linzner, *Naturwissenshaften*, 1977, **64**, 96.

254. J. M. Lehn, *Neurosciences Res. Prog. Bull.*, 1976, **14**, 133.

255. J. T. Edward, in *Intestinal Absorption of Metal Ions, Trace Elements and Radionuclides*, (eds. S. C. Skoryna and D. Waldron-Edward) Pergamon Press, Oxford, 1970, p. 1.

256. R. J. P. Williams, *Chem. Soc. Rev.*, 1980, **9**, 281.

257. R. D. Keynes, *Scientific American* 1979, **240**, 98.

258. P. C. Hinkle and R. E. McCarty, *Scientific American*, 1978, **238**, 104.

259. G. M. Shean and K. Sollner, *Ann. N.Y. Acad. Sci.*, 1966, **137**, 759.

260. B. C. Pressman, *Annu. Rev. Biochem.*, 1978, **45**, 501.

261. M. Kirch, Thèse de Doctorat d'Etat, 1980 (Strasbourg).

262. J. D. Lamb, J. J. Christensen, J. L. Oscarson, B. L. Nielsen, B. W. Asay and R. M. Izatt, *J. Am. Chem. Soc.*, 1980, **102**, 6820.

263. G. R. Painter and B. C. Pressman, in *Host–Guest Complex Chemistry II.* (ed. F. Vögtle) Springer-Verlag, Berlin, 1982, p. 83.

264. W. E. Morf, D. Ammann, R. Bissig, E. Pretsch and W. Simon, in *Progress in Macrocyclic Chemistry*, (eds. R. M. Izatt and J. J. Christensen) John Wiley and Sons, New York, 1979, p. 1.

265. S. Lindenbaum, J. H. Rytting and L. A. Sternson, see ref. in 264, p. 220.
266. J. M. Lehn, *Pure Appl. Chem.*, 1980, **52**, 2303.
267. F. J. Tehan, B. L. Barnett and J. L. Dye, *J. Am. Chem. Soc.*, 1974, **96**, 7203.
268. J. L. Dye, *J. Chem. Educ.*, 1977, **54**, 332.
269. J. L. Dye, *Angew. Chem., Int. Ed. Engl.*, 1979, **18**, 587.
270. J. L. Dye, see ref. in 264, p. 63.
271. J. L. Dye, M. R. Yemen and M. G. DaGue, *J. Chem. Phys.*, 1978, **68**, 1665; D. Issa and J. L. Dye, *J. Am. Chem. Soc.*, 1982, **104**, 3781.
272. D. G. Adolphson, J. D. Corbett and D. J. Merryman, *J. Am. Chem. Soc.*, 1976, **98**, 7234.
273. A. Cisar and J. D. Corbett, *Inorg. Chem.*, 1977, **16**, 632.
274. P. A. Edwards and J. D. Corbett, *Inorg. Chem.*, 1977, **16**, 903.
275. J. D. Corbett and P. A. Edwards, *J. Am. Chem. Soc.*, 1977, **99**, 3313.
276. C. H. E. Belin, J. D. Corbett and A. Cisar, *J. Am. Chem. Soc.*, 1977, **99**, 7163.
277. S. C. Critchlow and J. D. Corbett, *J. Chem. Soc., Chem. Commun.*, 1981, 236.
278. C. H. E. Belin, *J. Am. Chem. Soc.*, 1980, **102**, 6036.
279. R. C. Burns and J. D. Corbett, *J. Am. Chem. Soc.*, 1981, **103**, 2627.
280. R. C. Burns and J. D. Corbett, *Inorg. Chem.*, 1981, **20**, 4433.
281. R. G. Teller, R. G. Finke, J. P. Collman, H. B. Chin and R. Bau, *J. Am. Chem. Soc.*, 1977, **99**, 1104.
282. J. L. Petersen, R. K. Brown and J. M. Williams, *Inorg. Chem.*, 1981, **20**, 158.
283. A. Streitwieser, J. H. Hammons, E. Ciuffarin and J. I. Brauman, *J. Am. Chem. Soc.*, 1967, **89**, 59.
284. B. Dietrich and J. M. Lehn, *Tetrahedron Lett.*, 1973, 1225.
285. D. Clément, F. Damm and J. M. Lehn, *Heterocycles*, 1976, **5**, 477.
286. D. Clément, Thèse de Doctorat de 3e Cycle 1976 (Strasbourg).
287. S. Akabori and M. Ohtomi, *Bull. Chem. Soc. Jpn.*, 1975, **48**, 2991.
288. W. L. Dorn, A. Knöchel, J. Oehler and G. Rudolph, *Z. Naturforsch., Teil B*, 1977, **32**, 776.
289. A. Knöchel, J. Oehler and G. Rudolph, *Z. Naturforsch, Teil B*, 1977, **32**, 783.
290. J. F. Biellmann, H. D'Orchymont and M. P. Goeldner, *Tetrahedron Lett.*, 1979, 4209.
291. S. Akabori and H. Tuji, *Bull. Chem. Soc. Jpn.*, 1978, **51**, 1197.
292. R. Le Goaller, M. A. Pasquini and J. L. Pierre, *Tetrahedron*, 1980, **36**, 237.
293. M. A. Pasquini, R. Le Goaller and J. L. Pierre, *Tetrahedron*, 1980, **36**, 1223.
294. H. Handel, M. A. Pasquini and J. L. Pierre, *Tetrahedron*, 1980, **36**, 3205.
295. F. Guibe and G. Bram, *Bull. Soc. Chim. Fr.*, 1975, 933.
296. C. Cambillau, P. Sarthou and G. Bram, *Tetrahedron Lett.*, 1976, 281.
297. C. Cambillau, G. Bram, J. Corset, C. Riche and C. Pascard-Billy, *Tetrahedron*, 1978, **34**, 2675.
298. M. Cinquini, F. Montanari and P. Tundo, *J. Chem. Soc., Chem. Commun.*, 1975, 393.
299. D. Landini, A. Maia, F. Montanari and P. Tundo, *J. Am. Chem. Soc.*, 1979, **101**, 2526.
300. D. Landini, A. Maia, F. Montanari and F. M. Pirisi, *J. Chem. Soc., Perkin Trans. 2*, 1980, 46.
301. P. E. Stott, J. S. Bradshaw and W. W. Parish, *J. Am. Chem. Soc.*, 1980, **102**, 4810.
302. D. Landini, A. Maia, F. Montanari and F. Rolla, *J. Chem. Soc., Perkin Trans. 2*, 1981, 821.

303. M. Cinquini, S. Colonna, H. Molinari, F. Montanari and P. Tundo, *J. Chem. Soc., Chem. Commun.*, 1976, 394.
304. P. Reuter, A. Krämer and G. Manecke, IUPAC Mainz 1979, 562.
305. F. Montanari and P. Tundo, *J. Org. Chem.*, 1981, **46**, 2125.
306. P. Gramain and Y. Frère, *Ind. Eng. Chem. Prod. Res. Dev.*, 1981, **20**, 524.
307. Y. Frère, Thèse de Doctorat d'Etat 1981 (Strasbourg).
308. S. Boileau, B. Kaempf, S. Raynal, J. Lacoste and F. Schué, *J. Polym. Sci., Polym. Lett. Ed.*, 1974, **12**, 211.
309. S. Boileau, P. Hemery, B. Kaempf, F. Schué and M. Viguier, *J. Polym. Sci., Polym. Lett. Ed.*, 1974, **12**, 217.
310. P. Hemery, S. Boileau and P. Sigwalt, *J. Polym. Sci., Polym. Symp.*, 1975, **52**, 189.
311. A. Deffieux and S. Boileau, *Macromolecules*, 1976, **9**, 369.
312. S. Boileau, 1st European Discussion Meeting on Polymer Science. Strasbourg 1978, p. 10.
313. K. S. Kazanskij, see ref. in 312, p. 15.
314. B. Coutin and H. Sekiguchi, see ref. in 312, p. 88.
315. L. J. Mathias and K. B. Al-Jumah, *Polymer News*, 1979, **6**, 9.
316. L. J. Mathias, *J. Macromol. Sci., Chem.*, 1981, 853.
317. J. L. Pierre and H. Handel, *Tetrahedron Lett.*, 1974, 2317.
318. J. L. Pierre, H. Handel and R. Perraud, *Tetrahedron*, 1975, **31**, 2795.
319. R. A. Grey, G. P. Pez and A. Wallo, *J. Am. Chem. Soc.*, 1981, **103**, 7536.
320. J. L. Pierre, H. Handel and R. Perraud, *Tetrahedron Lett.*, 1977, 2013.
321. H. Handel and J. L. Pierre, *Tetrahedron*, 1975, **31**, 997.
322. J. F. Biellmann and J. J. Vicens, *Tetrahedron Lett.*, 1974, 2915.
323. H. Handel and J. L. Pierre, *Tetrahedron*, 1975, **31**, 2799.
324. A. Loupy and J. Seyden-Penne, *Tetrahedron Lett.*, 1978, 2571.
325. A. Loupy and J. Seyden-Penne, *Tetrahedron*, 1980, **36**, 1937.
326. P. M. Atlani, J. F. Biellmann, S. Dubé and J. J. Vicens, *Tetrahedron Lett.*, 1974, 2665.
327. A. Loupy, M. C. Roux-Schmitt and J. Seyden-Penne, *Tetrahedron Lett.*, 1981, 1685.
328. J. S. Shih, L. Liu and A. I. Popov, *J. Inorg. Nucl. Chem.*, 1977, **39**, 552.
329. B. L. Knier and P. Haake, *Tetrahedron Lett.*, 1977, 3219.
330. B. Springs and P. Haake, *Tetrahedron Lett.*, 1977, 3223.
331. F. P. Schmidtchen, *Angew. Chem., Int. Ed. Engl.*, 1981, **20**, 466.
332. J. Boger and J. R. Knowles, in *Molecular Interactions and Activity in Proteins*, Ciba Foundation Symposium **60**, 1977, p. 225.
333. F. Vögtle, H. Sieger and W. M. Müller, in *Host Guest Complex Chemistry I* (ed. F. Vögtle) Springer-Verlag, Berlin, Heidelberg, New York 1981, p. 107.
334. F. Montanari, D. Landini and F. Rolla, in *Host Guest Complex Chemistry II* (ed. F. Vögtle) Springer-Verlag, Berlin, Heidelberg, New York 1982, p. 147.

Further reading

J. J. Christensen, D. J. Eatough and R. M. Izatt, *Chem. Rev.*, 1974, **74**, 351.
N. S. Poonia and A. V. Bajaj, *Chem. Rev.*, 1979, **79**, 389.

J. M. Lehn, *Pure Appl. Chem.*, 1977, **49**, 857; and 1978, **50**, 871.

J. M. Lehn, *Acc. Chem. Res.*, 1978, **11**, 49.

*Structure and Bonding, 1973, **16**, Springer-Verlag, Berlin, Heidelberg, New York.

*Topics in Current Chemistry, 1977, **69**; 1981, **98**; and 1982, **101**.

*Coordination Chemistry of Macrocyclic Compounds, (ed. G. A. Melson) Plenum Press, New York, London 1979.

*Synthetic Multidentate Macrocyclic Compounds, (ed. R. M. Izatt and J. J. Christensen) Academic Press, New York 1978.

*Bioenergetics and Thermodynamics: *Model Systems*, (ed. A. Braibanti) D. Reidel Publishing Company, Dordrecht, Boston, London 1980.

*Progress in Macrocyclic Chemistry, (eds. R. M. Izatt and J. J. Christensen) John Wiley and Sons, New York, Chichester, Brisbane, Toronto 1979.

G. W. Gokel and S. H. Korzeniowski, *Macrocyclic Polyether Synthesis*, Springer-Verlag, Berlin, 1982.

* Parts have been mentioned in the references but these books contain many other valuable chapters which should be consulted.

11 · INCLUSION COMPOUNDS FORMED BY OTHER HOST LATTICES

J. E. D. DAVIES
University of Lancaster, Lancaster, U.K.

P. FINOCCHIARO
University of Catania, Catania, Italy

and

F. H. HERBSTEIN*
Technion-Israel Institute of Technology, Haifa, Israel

1. Introduction

This chapter deals mainly with inclusion compounds formed by host lattices which are not covered in the previous chapters. Many of the inclusion compounds reported have not been characterized crystallographically, and indeed many of them have been reported once only in the literature. Our hope is that the present chapter may stimulate further work on these inclusion compounds so that many of them may form the basis of complete chapters in further editions of this book. Most of the inclusion compounds reported here have been described as *inclusion compounds* or *clathrates* in

* In collaboration with R. E. Marsh (Caltech), M. Kapon and G. M. Reisner (Technion).

INCLUSION COMPOUNDS 2
ISBN 0-12-067101-8

the literature. There are in addition many more compounds which are described as *solvates* and many of these must surely be inclusion compounds.[190]

The purpose of the present chapter is therefore threefold: to illustrate the disparate nature of the compounds which can act as host lattices; to stimulate further work on the inclusion compounds reported here, and to try and ensure that the coverage of the present volume is as complete and as up to date as possible.

2. Organic host lattices

2.1. Trimesic acid

2.1.1. Introduction

Trimesic acid (benzene-1,3,5-tricarboxylic acid; TMA) is an attractive molecule for study as a potential former of molecular complexes. Firstly it is an acid ($pK_1 = 2.12$) and thus can form quasi-salts with basic compounds such as amino acids; the crystal structures of L- (or DL-) histidinium trimesate.1/3 acetone and glycine-trimesic acid monohydrate have been reported.[1,2] Of far greater concern in the present context is, however, the ability of trimesic acid to form, either on its own or together with other molecules (such as H_2O), extended hydrogen-bonded networks, which act as host lattices for clathrate and channel inclusion compounds. The composition of these networks provides a convenient basis for classification. In the first group of inclusion compounds the framework consists of TMA molecules alone, in the second group the framework has the composition $TMA.H_2O$ while in the third group the framework includes both components of the compound; only one example is known of the third group, $TMA.DMSO^*$. The TMA inclusion compounds show many unusual structural features.

2.1.2. Networks based on TMA alone

The hexagonal network shown in Fig. 1 has TMA molecules hydrogen-bonded through their carboxylic acid groups to form an infinite two-dimensional array. The holes in the network have diameters of 12–13 Å (after taking the van der Waals radii of the atoms into account); the matrix

---

* Dimethylsulphoxide.

formed by stacking such networks one above the other would have cylindrical channels analogous to those found in urea and thiourea channel inclusion compounds (see Chapter 2 Volume 2) but with cross-sectional areas perhaps ten times are large. Thus intriguing possibilities beckon for the inclusion of rather large molecules in TMA frameworks. However the *empty* "chicken-wire" network of Fig. 1 has not yet been found in any crystal, and inclusion compounds with large guest molecules in the postulated cylindrical channels have not yet been prepared. The first crystal

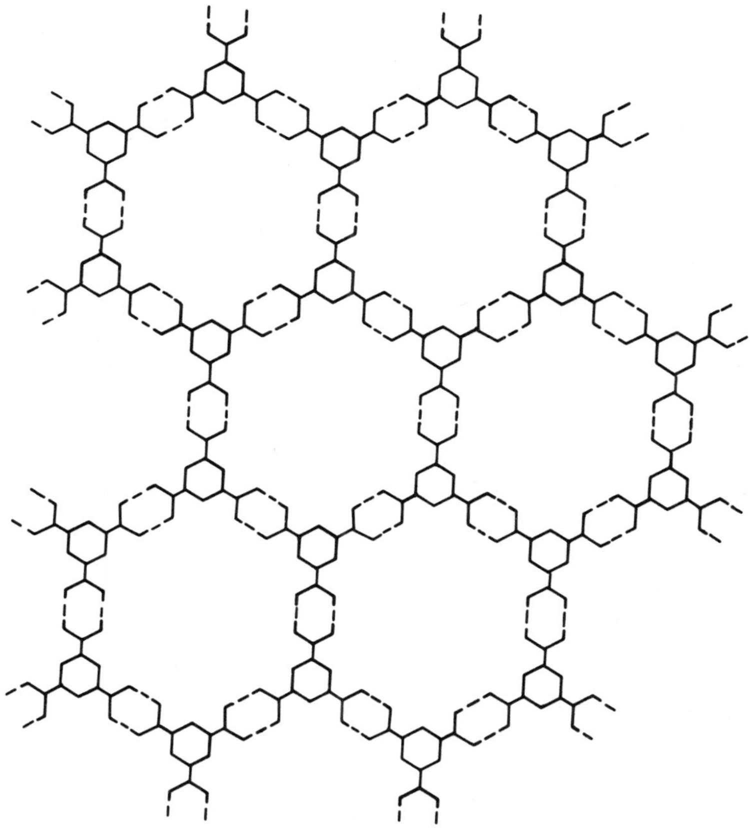

Fig. 1. A representation of the hexagonal network formed by six TMA molecules hydrogen-bonded through their carboxylic acid groups. The network extends indefinitely in the layer and the arrangement is essentially planar. Both ordered and disordered acid-dimer groups occur. Such hexagonal nets are found in α-TMA and in the TMA pentahalides. Diagram reproduced with permission from *Acta Crystallogr.*, 1969, **B25**, 5.

structure found to be based on an *interlaced* arrangement of "chicken-wire" networks was that of α-TMA[3], the polymorph of TMA stable at room temperature and a simpler version of this arrangement was later found[4] in TMA.0.7H$_2$O.0.09HI$_5$ which will be referred to as "TMA.I$_5$".

Inverting the chronology, the structure of "TMA.I$_5$" is first described. This compound was obtained[5] by crystallizing TMA from water containing HI$_3$ or KI$_3$ (no potassium is incorporated in the crystal despite its presence in the solution). Isomorphous crystals of TMA.0.7H$_2$O.0.167HIBr$_2$ are obtained from solutions containing I$^-$:Br$_2$ in equimolar ratio, and "TMA.Br$_5$" from solutions containing HBr$_3$. Each hexagonal network (in the form shown in Fig. 1) is interlaced (or catenated) by three other similar networks threaded through its large central hexagonal holes (Fig. 2), and

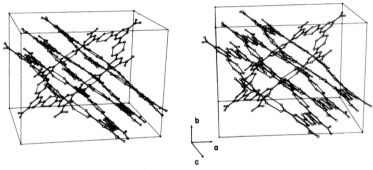

Fig. 2. Interlaced hexagonal networks in "TMA.I$_5$". The ORTEP stereodiagram shows a type-C ring through which a set of three rings (AB, C, AB in that order) have been interlaced. The two additional AB rings that flank the single C ring have been omitted for clarity. The enclosing box has the dimensions and orientation of the unit cell but its origin has been shifted to $0, \frac{1}{2}, 0$ for aesthetic reasons. Reproduced with permission from *Proc. R. Soc. London, Ser. A*, 1981, **376**, 301.

each of these networks in turn is similarly triply-catenated. In "TMA.I$_5$" and its analogues the individual networks maintain their planarity. Thus the interlaced triplets can be divided into two sets of mutually-parallel networks, the two sets intersecting at an angle of ~70° to one another, with channels of rhombic cross-section left in the interstices of the catenated arrangement (Fig. 3). The polyhalide ions (I$_5^-$, Br$_5^-$, IBr$_2^-$) are contained within these channels. The first two ions are disordered with respect to the matrix, the z coordinates of the halogen atoms in one channel bearing no fixed relationship to those of the ions in other channels; however the IBr$_2^-$ ions are ordered with respect to the matrix, the reason for the difference in

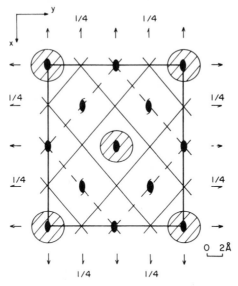

Fig. 3. The catenated hexagonal networks of TMA molecules viewed edge-on in the crystal structure of "TMA.I_5". The symmetry elements of space group I222 are shown in projection down [001]. The differences between the networks represented by full (AB-type) and broken (C-type) lines are small. The hatched circles represent the polyhalide chains viewed down their axes. Reproduced with permission from *Proc. R. Soc. London, Ser. A,* 1981, **376**, 301.

behaviour not being understood. The counter-ion to the polyhalide ion is a proton, presumed attached to the residual water molecules which are located in regions around $\frac{1}{2}0\frac{1}{2}$, etc., between the TMA networks. The TMA polyhalides are a new type of channel inclusion complex in the sense that the included species is an anion rather than a neutral molecule, the cation being remote from the channel. Perhaps the closest analogy is to the hydrated cyclodextrin polyiodide channel inclusion complexes[6] where the I_5^- ions are located within the central channels of the cyclodextrin doughnuts and the metal cations are coordinated to water molecules which lie between the cyclodextrin molecules (see Chapter 8, Volume 2).

We now return to the structure of α-TMA where there are the same hexagonal sheets of TMA molecules and these catenate in groups of three in a *local* manner remarkably similar to that found in "TMA.I_5". Indeed the angle between the two sets of triplets in a catenated region is $\sim 65°$, compared to the value of $\sim 70°$ found in "TMA.I_5". However the sheets as a whole are not planar in α-TMA but are bent so as to approach close packing. These features of α-TMA are shown schematically in Fig. 4

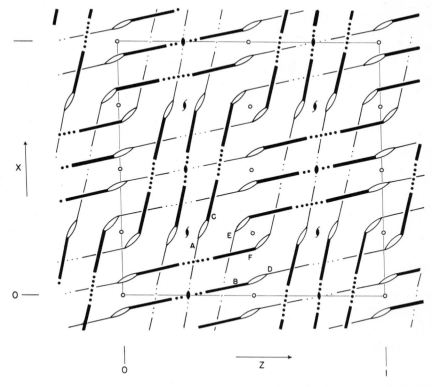

Fig. 4. Schematic diagram of the α-TMA structure in projection down [010] (monoclinic, space group C2/c, $Z = 48$, $a = 26.522$ (2), $b = 16.420$ (1), $c = 26.551$ (2) Å, $\beta = 91.53$ (1)°). Individual TMA molecules are represented by straight solid lines. Dotted lines represent hydrogen bonds between molecules having similar y coordinates while the pairs of curved lines represent hydrogen bonds between molecules lying at different heights along [010]. The TMA molecules A–F are crystallographically independent. Reproduced with permission from *Acta Crystallogr.*, 1969, **B25**, 5.

and, much more vividly in the original paper, by a series of colour stereophotographs of space-filling models.

If TMA is crystallized from water, colourless rhombs of the α-polymorph are obtained together with needles of TMA.3H$_2$O and lozenges of TMA.5/6H$_2$O (see below); if the water used is saturated with molecular I$_2$ or Br$_2$ then the rhombs are strikingly coloured (purple for I$_2$, red for Br$_2$), with (analysed) compositions of TMA.1/12I$_2$ and TMA.1/6Br$_2$. The crystals are isomorphous with those of α-TMA with negligible changes in cell-dimensions; the crystal densities match those calculated from the compositions.[5] These compounds can be called clathrates of TMA, or by analogy with metals, interstitial solid solutions. Unfortunately all the crystals were

twinned so the halogen molecules were not located*. Analogous organic crystals do not appear to be known; one could perhaps find similarities with the zeolites.

The catenation of the TMA framework in the TMA polyhalides and in α-TMA is essentially a solid-state phenomenon. A similar situation is found in the clathrate compounds of hydroquinone (quinol) with small molecules. These crystals are built up of two interpenetrating three-dimensional frameworks of hydrogen-bonded quinol molecules arranged so as to enclose the guest molecules, but without any chemical link between the frameworks—or with the enclathrated guest molecules.[7,8] (See Chapter 1, Volume 2.)

2.1.2. Networks based on TMA.H_2O
As noted above, TMA can crystallize from water in the form of two hydrates TMA.$3H_2O$ and TMA.$5/6H_2O$. The first of these loses water on exposure to the atmosphere but can be stabilized by adding picric acid [PA] to the solution, the resulting crystal composition being TMA.H_2O.[2/9PA]. This compound and TMA.$3H_2O$ are isostructural, the principal difference in the diffraction patterns being additional diffuse scattering from TMA.H_2O.[2/9PA]. The diffuse scattering was not taken into account in the solution of the structure[9] except to determine the picric acid content.

The dimensions of the unit cells of TMA.H_2O.[2/9PA] and TMA.$5/6H_2O$ are closely related (Table 1), suggesting a relationship between the structures

Table 1. Unit cell dimensions for TMA.H_2O.[2/9PA] and TMA.$5/6H_2O$

	I TMA.H_2O.[2/9PA]	II TMA.$5/6H_2O$
a (Å)	18.269 (8)	16.640 (1)
b	8.852 (5)	18.548 (1)
c	3.642 (4)	9.512 (1)
α (deg)	90.43 (8)	95.81 (1)
β	92.59 (6)	91.06 (4)
γ	99.56 (6)	94.35 (1)
Vol. (Å^3)	580.1 (7)	2911.3 (6)
Z	2	12

Note: The net of cell I with edges [101] (=18.47 Å), [011] (=9.55 Å), inter-edge angle = 95.5° is very similar to the bc net of cell II. These two planes, (111) in cell I and (100) in cell II, contain the TMA.H_2O repeating network. The stacking of these nets along c (in cell I) and a (in cell II) differs in detail. The volume of cell II is 5.02 times as large as that of cell I, showing that the packing density of the layers is essentially the same in both compounds.

* Added in proof: this structure has now been solved.

414   *J. E. D. Davies, P. Finocchiaro and F. H. Herbstein*

despite the difference in compositions. The TMA.H$_2$O framework in
TMA.H$_2$O.[2/9PA] is shown in Fig. 5; the channels with rectangular cross
section are occupied by disordered picric acid molecules, not located in
detail. Presumably the additional two water molecules in TMA.3H$_2$O occupy

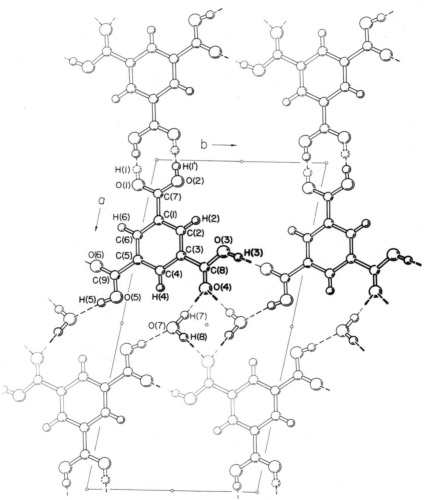

Fig. 5. The TMA.H$_2$O framework found in TMA.3H$_2$O and TMA.5/6H$_2$O. The open
regions show, in cross-section, the cylindrical channels which are occupied by picric
acid or water molecules in the two isomorphous TMA.H$_2$O[2/9PA] and
TMA.H$_2$O.[1/5TMA] structures. Reproduced with permission from *Acta Crystallogr.*, 1977, **B33**, 2358.

similar channels in the trihydrate but this has not been studied in detail. TMA.5/6H$_2$O has TMA.H$_2$O layers essentialy identical to those of TMA.H$_2$O.[2/9PA], with five such layers stacked in ordered arrangement one above the other. However the layers are offset in such a way that the channels of rectangular cross section have a zigzag rather than a straight profile. These channels accommodate a zigzag sequence of TMA molecules hydrogen-bonded through their carboxyl groups in the 1- and 3-positions (Fig. 6). Thus TMA.5/6H$_2$O is really a self complex and its formula is more

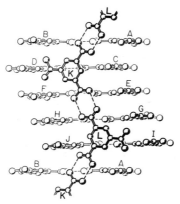

Fig. 6. The structure of TMA.5/6H$_2$O showing the chains of enclathrated molecules K and L and the zigzag stacking of framework molecules A–J. The view is along c^*, with a vertical. The layers containing the framework molecules A–J have the same structure as the TMA.H$_2$O network shown in Fig. 5. Reproduced with permission from *Acta Crystallogr.*, 1977, **33B**, 2358.

realistically written as TMA.H$_2$O.[1/5TMA]. The number of picric acid molecules in TMA.H$_2$O.[2/9PA] was inferred from the periodicity of the diffuse scattering pattern, supported by crystal density and chemical analysis; the composition of TMA.5/6H$_2$O is obtained directly from the crystal structure, following preliminary estimates being obtained from crystal density and chemical analysis. A somewhat similar situation is found in (TMTTF)$_{1.3}$(TCNQ)$_2$*,[10] the TMTTF molecules additional to the 1:2 composition being located in channels between the segregated stacks of TMTTF and TCNQ moieties, with their long axes along the channel axes.

The essential structural feature in both TMA compounds described above is a planar hydrogen-bonded layer of composition TMA.H$_2$O; the essential feature of the arrangement of TMA and H$_2$O molecules in such a layer is

* (TMTTF is 4,4′,5,5′-tetramethyl-$\Delta^{2,2'}$-bis-1,3-dithiole and TCNQ is 7,7,8,8-tetracyano-p- quinodimethane).

the gap left between molecules, large enough to allow insertion of a chain of guest molecules of the approximate size and shape of TMA. The bonding between such layers is appreciably weaker than within the layers, and consequently layers can be shifted laterally to accommodate guests of different types. Accomodation of the zigzag ribbon of guest TMA molecules requires appropriate shifts of adjacent TMA.H_2O layers. The picric acid

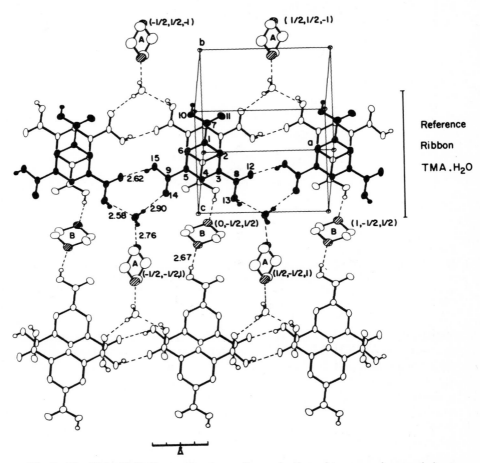

Fig. 7. The TMA.H_2O.dioxan structure—the projection of two superimposed sheets normal to the TMA planes is shown (this is approximately along $\lfloor 011 \rfloor$). The sheets are stepped, not flat. The reference ribbon TMA.H_2O shows the part of the structure similar to parts of the TMA.H_2O.[2/9PA] and TMA.5/6H_2O structures. Coordinates of the centres of some of the dioxane molecules are shown for reference (dioxans A and B are crystallographically independent). Reproduced with permission from *Acta Crystallogr.*, 1978, **B34**, 1608.

molecule, on the other hand, does not have regions capable of strong intermolecular bonding and thus fits without difficulty into a linear channel.

The compound TMA.H_2O.dioxan[11] is not of the host–guest type but there are clear resemblances between the arrangement of TMA and water molecules in ribbons along [100] (Fig. 7) and that found in appropriate parts of the TMA.H_2O framework in TMA.H_2O.2/9PA and TMA.5/6H_2O. This is another of the many examples of Nature's economy in design, where a particular feature is carried over from one structure to another.

2.1.3. Networks based on TMA and other molecules

There is only one example known of this situation—TMA.DMSO[12] (Fig. 8). In contrast to the usual behaviour of carboxylic acids, no dimerization of –CO_2H groups occurs in this compound; instead ribbons of TMA molecules linked by *single* hydrogen bonds between –CO_2H groups extend

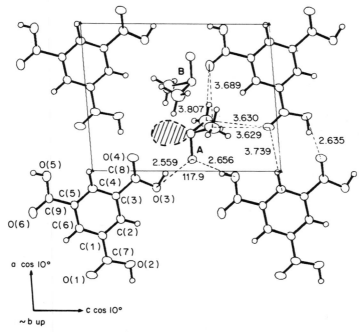

Fig. 8. TMA.DMSO in projection approximately down [010]. The presumed location of the sulphur lone pair is shown by the hatched lobe. The two DMSO molecules labelled A and B are related by the two-fold screw axis along [010]. Apart from the two CH_3 groups of DMSO all atoms shown are essentially co-planar. Reproduced with permission from *Acta Crystallogr.*, 1978, **B34**, 1613.

along [100], the ribbons being bridged through hydrogen-bonding of other −CO$_2$H groups to the oxygen of DMSO. This arrangement leaves a channel along the [010] direction, which is occupied by the sulphur and two methyl groups of the DMSO molecule. Furthermore there is an apparently vacant region in the channel in the direction of the sulphur lone pair, and it is inferrred that this lone pair is stereochemically active. Thus in TMA.DMSO the DMSO molecule both participates in the formation of the channel walls of the host lattice and fills the channel. Analogous situations do not appear to have been encountered previously.

2.2. Substituted methanes and ethanes

A common feature of the host lattices covered in previous chapters is their ability to assume trigonal symmetry in the solid state. It is thus conceivable that other classes of molecules able to assume a propeller-like conformation will display inclusion ability.

Since triphenylmethane and substituted triphenylmethanes adopt a three bladed propeller conformation in solution as well as in the solid state[13] they might be expected to act as host lattices. Triphenylmethane (TPM) itself has been reported to form a series of inclusion compounds with only a very limited range of guest molecules: benzene, thiophen, aniline and pyrrole with a 1 : 1 host : guest ratio.[14a] Substituted triphenylmethanes, provided they have two OH groups in *para*-positions, form inclusion compounds with a wider range of guest molecules[14b] with a host : guest ratio ranging from 2 : 1 to 1 : 2.

The results of competitive inclusion experiments[15] are shown in Table 2 where aniline seems to be the preferred guest. Whilst TPM does not form an inclusion compound with toluene, when recrystallized from an equimolar mixture of benzene and toluene the inclusion compound contains both guest species in a 80 : 20 ratio, suggesting that the toluene is incorporated as an auxiliary guest. The same effect does not occur using an equimolar benzene/xylenes mixture probably due to the greater size of the xylenes compared with benzene.

The crystal structure[16] of the TPM.C$_6$H$_6$ inclusion compound is illustrated in Fig. 9. The TPM benzene units are located on the C$_3$ axes and are related by centres of symmetry with TPM adopting a propeller conformation. The substituted triphenylmethanes form a layer-like structure whereby each hydroxyl group is hydrogen bonded to hydroxyl groups from two other molecules.[14c] The substituent on the third phenyl group is "sterically linked" to the corresponding group in another molecule forming an interlocking assembly.

Table 2. Guest selectivity properties of hosts TPM, 2 and 3

Host compound	Recrystallization solvent mixture[a]	Respective percentage of guest included[b]	Overall host/guest mole ratio
TPM	benzene/aniline	35/65	1:2
	thiophen/aniline	45/55	1:1
	pyrrole/aniline	28/72	1:3
	benzene/thiophen	72/28	3:1
	benzene/pyrrole	38/62	1:2
	benzene/toluene	80/20	5:1
	benzene/xylenes	100/0	1:0
2	benzene/toluene	80/20	4:1
	benzene/xylenes	100/0	1:0
	benzene/dioxan	84/16	5:1
	toluene/xylenes	100/0	1:0
	benzene/chloroform	43/57	1:1
	benzene/dichloroethane	100/0	1:0
3	o-xylene/toluene	75/25	3:1
	benzene/toluene	68/32	2:1
	o-xylene/p-xylene	80/20	4:1
	o-xylene/m-xylene	55/45	1:1
	m-xylene/p-xylene	55/45	1:1
	benzene/o-xylene	67/33	2:1
	o-xylene/mesitylene	74/26	3:1
	benzene/p-xylene	94/6	15:1

[a] Equimolar mixture.
[b] Analyses were performed with a C. Erba Fractovap C instrument using a 2.80 m column filled with Bentone 34 on Chromosorb.

In addition to the above layer structure which is found in the benzene, toluene, and *p*-xylene inclusion compounds, the substituted triphenyl-methanes can also form channel structures with guests such as *n*-alkanes, and *n*-alkenes, the channel diameter being at least 6.5 Å. γ-ray induced polymerization of guest isoprene did not result in a stereoregular polymer. Examples of hosts which do give rise to stereospecific polymerization are given in Chapter 10, Volume 3.

Other substituted methanes which have been reported as forming inclusion compounds are: triphenylchloromethane[17] (with acetone and CCl_4, as guests); triphenylcarbinol[17] (with acetone and CCl_4 as guests) and di(1-naphthyl) phenyl methanol[18] (with methanol as guest).

Tris-(5-acetyl-3-thienyl)methane (**1**, R = CH_3CO) has been reported as giving inclusion compounds with a wide variety of guest molecules, ranging

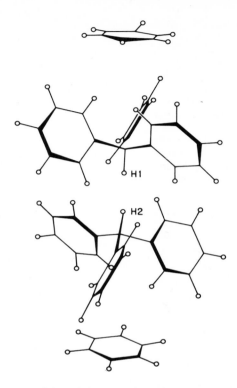

Fig. 9. Crystal structure of the triphenylmethane benzene (1 : 1) compound. Reproduced with permission from *Acta Crystallogr.*, 1975, **A31**, S130.

from aromatic hydrocarbons to aliphatic alcohols and ethers.[19] It also displays a selectivity for *o*-xylene when recrystallized from an equimolar mixture of *o*- and *m*-xylene. A variety of other derivatives were tested for inclusion ability but the methyl ester (**1**, R = CH_3O_2C) was the only other compound to form any inclusion compounds (with ethanol, cyclohexane and CCl_4 as guests).

(1)   **(2)**

The ability of tetraarylethanes (TAE) such as 1,2-dichloro-1,1,2,2-tetraphenylethane (**2**) to form inclusion compounds has been known for several decades,[17] but they have only been studied in detail in recent years. TAE **2** and **3** form inclusion compounds with a wide variety of guest molecules[15] and the results of some competitive inclusion experiments are summarized in Table 2.

It is well known that racemic compounds may be resolved by way of diastereomeric inclusion compounds in which the compound to be resolved functions as the *guest* and an enantiomerically pure compound as the *host* (See Chapter 9, Volume 3). TAE **3** has however been partially resolved using the converse procedure *viz* racemic **3** was recrystallized from (+)-α-pinene, *i.e.* the species to be resolved functioned as the *host* and an enantiomerically pure solvent as the *guest*.[20]

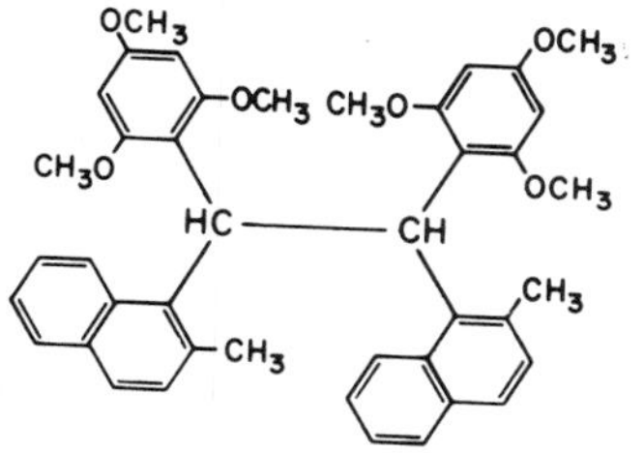

$$(C_6H_5)_2\overset{\parallel}{\underset{Se}{P}}CH_2CH_2\overset{\parallel}{\underset{Se}{P}}(C_6H_5)_2$$

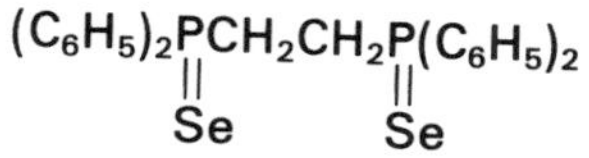

(4)

(3)

1,2-bis(diphenylphosphinoselenoyl)ethane (**4**) has been found to be a very versatile host towards a range of guest molecules.[21] It is also highly selective towards *p*-xylene rather than *o*- or *m*-xylene. In the *p*-xylene inclusion compound (Fig. 10) the host lattice structure is similar to that displayed by some solid tetraarylethanes.[22] The *p*-xylene guest molecule is however firmly held in the host lattice and the inclusion compound can rightly be regarded as a true clathrate compound. The sulphur analogue of **4** and the corresponding 1,3-di-substituted propane derivatives can also act as host lattices.[21]

2.3. Cyclophanes

In addition to cyclotriveratrylene (reviewed by A. Collet in Chapter 4, Volume 2) a number of other cyclic compounds have been shown to be capable of forming inclusion compounds as summarized in Table 3. The term "cyclophane" as used in this chapter encompasses cyclic compounds containing non-fused aromatic rings.

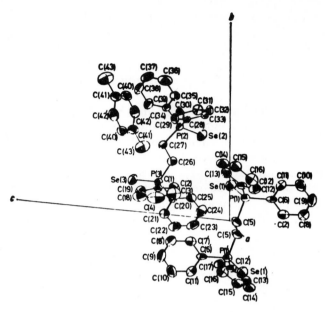

Fig. 10. Crystal structure of the $Ph_2P(Se)CH_2CH_2P(Se)Ph_2.p$-xylene (3:1) clathrate. Reproduced with permission from *J. Chem. Soc., Perkin Trans. 2,* 1980, 993.

Table 3. Inclusion compounds formed by cyclophanes

Host lattice	Typical guest molecules (host : guest stoichiometry)	Ref.
(5)	(a) $n = 3$ Benzene (1:1) Dioxane (1:1.25) (b) $n = 4$ Benzene (2:1) Dioxan (1:0.75)	23
(6)	$n = 4, m = 4$ 2-p-Toluidinylnaph- thalene-6-sulphonate (1:1) 2,7-Dihydroxynaph- thalene (1:1)	179

Table 3—continued

Host lattice	Typical guest molecules (host : guest stoichiometry)	Ref.
(7)	$X = NCH_3$ Benzene (1:1) Dioxan (1:1) $X = NCH_3$ and SCH_3^+ Sodium 1-anilino- 8-naphthalenesulphonate	24 25
(8)	*p*-xylene Terpenoids Cyclohexanol, Cyclohexanone Cinnamaldehyde Propylbenzene 1,4-Disubstituted benzenes	26 27 28 29 ·30 31
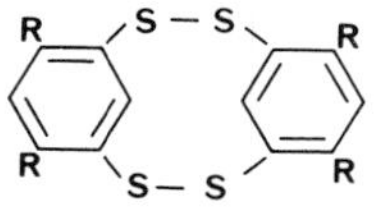 (9)	$R = OCH_3$ $CHCl_3$ (1:1) Dioxan (1:1) Ethyl Acetate (1:1) DMF (1:1) DMSO (1:1)	32
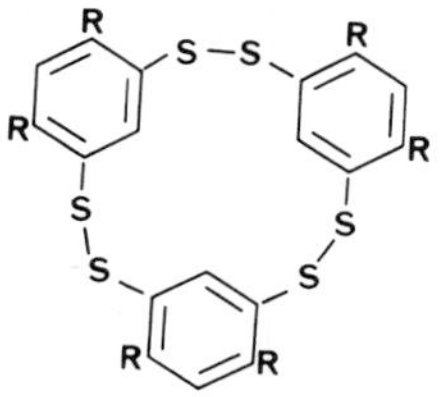 (10)	$R = OCH_3$ *ortho*-$C_6H_4Cl_2$(1:1)	33
(11)	(a) $m = 4, n = 4$ Durene (1:1) 2,7-Dihydroxynaphthalene (1:1) (b) $m = 5, n = 8$; $m = 4, n = 4$; $m = 5, n = 6$ Sodium 1-anilino-8- naphthalenesulphonate	34 35 36

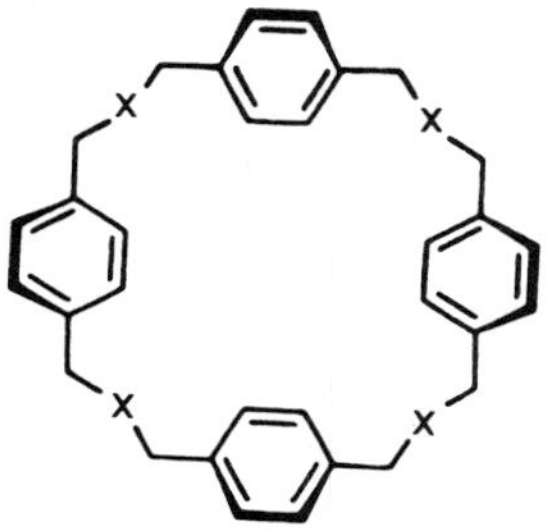

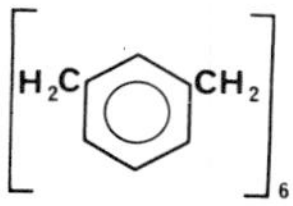

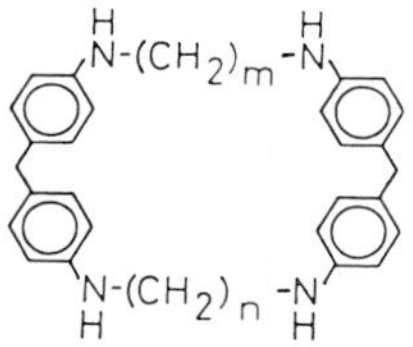

Table 3—continued

Host lattice	Typical guest molecules (host:guest stoichiometry)	Ref.

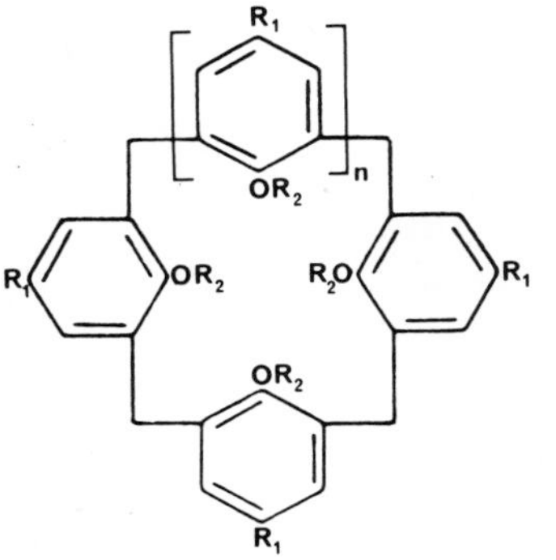 **(12)**	(a) $n = 1$, $R_1 = Bu^t$, $R_2 = H$
	      Toluene (1:1)
	      $CHCl_3$
	      Benzene
	(b) $n = 3$, $R_1 = Bu^t$, $R_2 = H$
	      $CH_3OH/CHCl_3$ (1:2:1)
	(c) $n = 1$, $R_1 = Bu^t$, $R_2 = CH_3CO$
	      Ethanoic Acid (1:1)
	(d) $n = 2$, $R_1 = R_2 = H$
	      Acetone (1:2)
		37a
		38
		38
		
		38
		
		37b
		
		37c

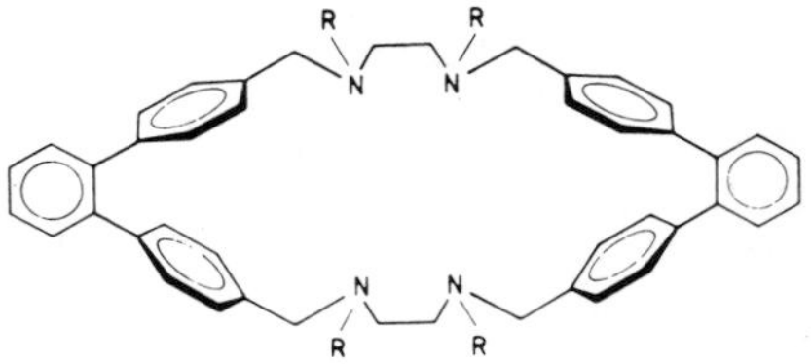 **(13)**	$R = C_6H_5CH_2$
	      Cyclohexane (1:1)
	      Dioxane
	      Morpholine
		39

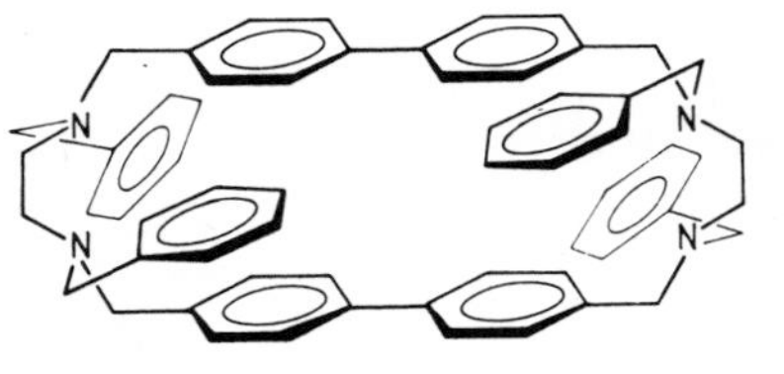 **(14)**	C_6F_5H C_6F_6	39
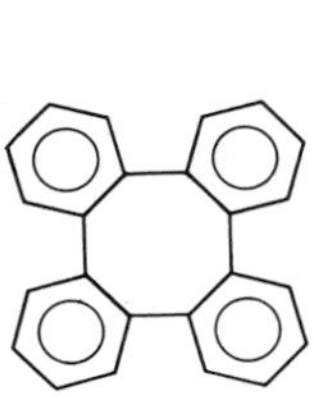 **(15)**	CH_2Cl_2, CH_2Br_2, CCl_4, $CHCl_3$ (2:1) i-PrBr, n-PrBr, CH_3COCH_3, THF, Dioxan, Benzene, Cyclohexane	40

Table 3—continued

Host lattice	Typical guest molecules (host: guest stoichiometry)	Ref.
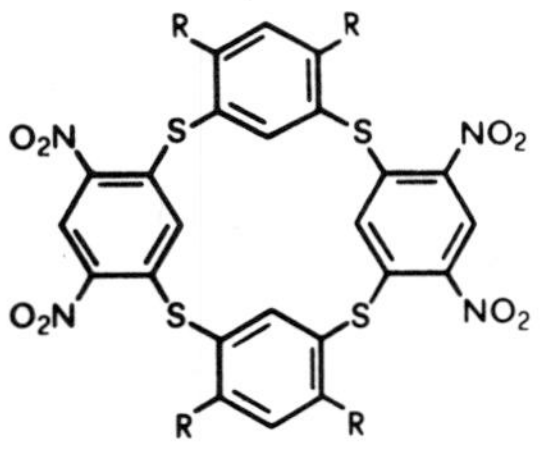 (16)	(a) R = H DMF, o-C$_6$H$_4$Cl$_2$ (1:2) DMSO (1:1) (b) R = Cl DMF (2:1) DMSO (3:1) o-C$_6$H$_4$Cl$_2$ (1:2) (c) R = CH$_3$ DMF (5:1) DMSO (4:1) o-C$_6$H$_4$Cl$_2$ (1:2) (d) R = OCH$_3$ o-C$_6$H$_4$Cl$_2$ (1:2)	41
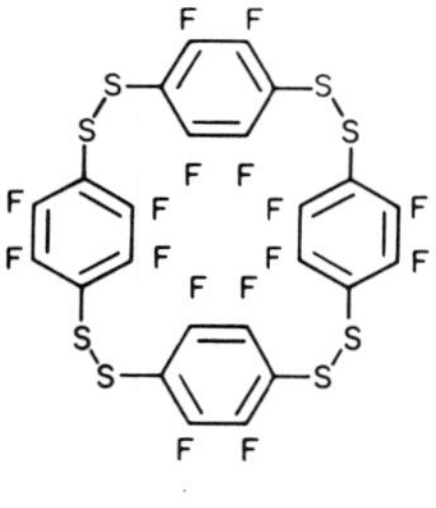 (17)	N,N'-Dimethyldihydro-phenazine (1:1) 1,3-Diphenylisobenzo-furan (1:4)	42
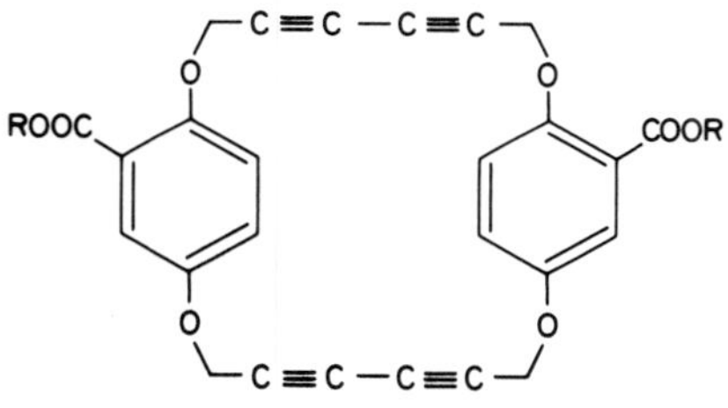 (18)	R = H 2-Naphthylmethyltriethyl-ammonium chloride (1:1)	43
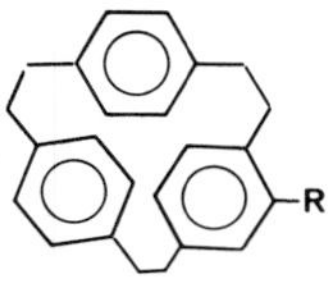 (19)	R = H and CH=CH$_2$ (H$_2$)[a]	180

Table 3—continued

Host lattice	Typical guest molecules (host : guest stoichiometry)	Ref.
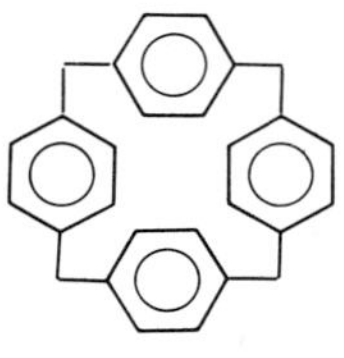 **(20)**	$(H_2)^{(a)}$	180
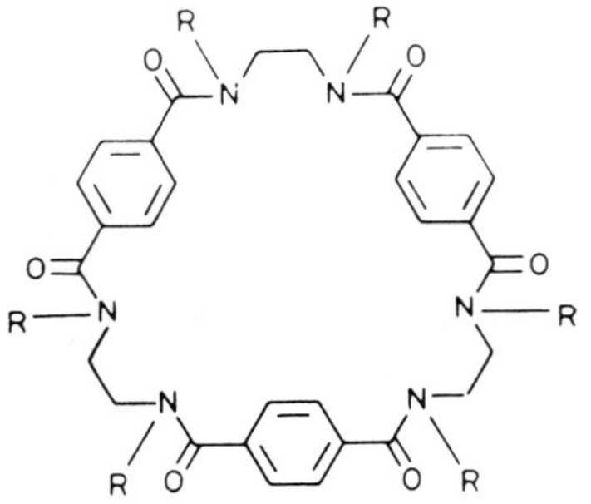 **(21)**	$R = C_6H_5CH_2$ $CHCl_3$ (1 : 1)	181

(a) **19** and **20** have been investigated as potential hydrogen-storage agents (ref. 180).

Crystallographically determined structures are available for only a few of the inclusion compounds,[23b,24b,34,37,40,181] some of which are illustrated in Figs. 11 and 12. It should be borne in mind that the preparation of a host : guest cyclophane compound does not necessarily mean that an inclusion compound has been formed, since hosts **17**[42] and **18**[43] have been found to form *charge transfer complexes* rather than *inclusion compounds*. (See Fig. 13).

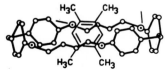

Fig. 11. Crystal structure of the inclusion compound formed between cyclophane **(11a)** and durene (1 : 1). (This adduct has complete composition **11a**.4HCl-durene.4H$_2$O.) Reproduced with permission from *J. Am. Chem. Soc.*, 1980, **102**, 2504.

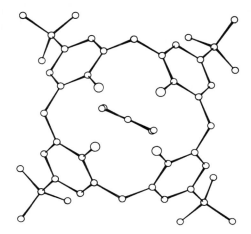

Fig. 12. Crystal structure of the inclusion compound formed between cyclophane (**12a**) and toluene (1:1). Reproduced with permission from *J. Chem. Soc., Chem. Commun.*, 1979, 1005.

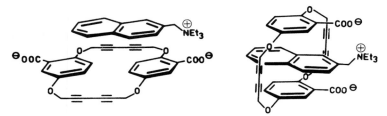

Fig. 13. The charge transfer (left) and inclusion (right) geometries of cyclophane (**18**), with the charge transfer geometry being the preferred geometry. Reproduced with permission from *J. Am. Chem. Soc.*, 1980, **102**, 657.

2.4. BSX and HMX

BSX, 1,7-diacetoxy-2,4,6-trinitro-2,4,6-triazaheptane (**22**) has been reported to form complexes with a variety of organic compounds.[44,45] Spectroscopic and thermal analyses[46] indicate two classes of complexes as shown in Table 4. The type I complexes seem to be inclusion compounds with very little

$$CH_3CO_2CH_2NCH_2NCH_2NCH_2O_2CCH_3$$
$$\quad\quad\quad\;\; NO_2 \;\; NO_2 \;\; NO_2$$

(22)

Table 4. Complexes formed by BSX

Type I		Type II	
Guest	Stoichiometry[a]	Guest	Stoichiometry[a]
Benzene	1:0.8	N,N-Dimethylfor-manide	1:1
Fluorobenzene	1:0.85		
Pentafluorobenzene	2:1.02	N,N-Dimethyl-acetamide	1:1
Hexafluorobenzene	2:1.04		
1,4-Dioxane	1:0.96	Acetylacetone	3:1
Dibromomethane	1:1.0	Cyclohexanone	2:1
3-Bromopyridine	2:0.9	Tetrachloroethane	2:1
Acetophenone	2:1.04	Pyridine	1:1
Acetonitrile	1:0.98	Nitrobenzene	1:0.97
		4-Hydroxybutanoic acid lactone	—
		N-Methyl-2-pyrrolidinone	1:1

[a] BSX:Guest.

interaction between the two components, whilst in the Type II complexes there seems to be strong specific interactions between nitro-nitrogens of the BSX molecules and the oxygen atoms of the organic molecule.

Crystallographic studies of the 1:1 complexes with N,N-dimethyl-formamide[47] and 1,4-dioxan[48] confirm the difference between these two complexes. The DMF molecules are grouped in channels between the columns of BSX dimers with $N \cdots O$ distances of 2.891 and 2.969 Å. The 1,4-dioxan molecules are also situated in channels between columns of BSX dimers but the $N \cdots O$ distances of 3.116 and 3.199 Å indicate much weaker host ... guest interaction than in the DMF complex. (Fig. 14).

The β-polymorph of HMX, 1,3,5,7-tetranitro-1,3,5,7-tetraazacyclo-octane (23) also forms complexes with a variety of organic compounds.[49] In the HMX–DMF (1:1) complex the interaction differs from that in the BSX–DMF complex, since the closest contacts are between the DMF carbonyl oxygen atom and CH_2 groups of the HMX molecule.[50]

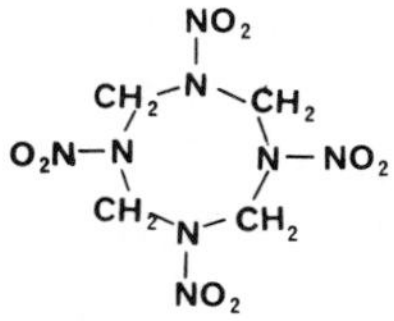

(23)

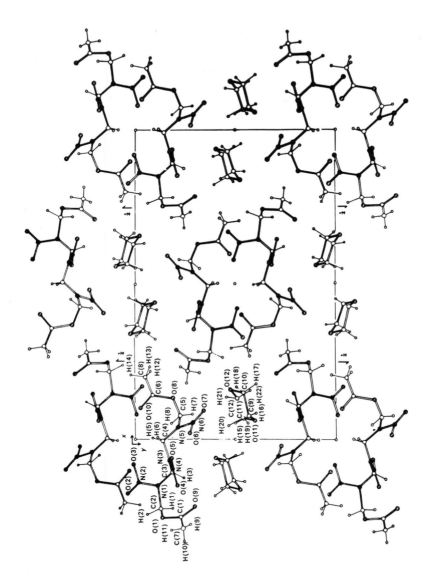

Fig. 14. The crystal structure of the BSX (**22**): dioxan (1 : 1) inclusion compound. Reproduced with permission from *Acta Crystallogr.*, 1973, **B29**, 2585.

2.5. 4,4′-Dinitrobiphenyl

4,4′-Dinitrobiphenyl (**24**) forms channel type inclusion compounds (Fig. 15) with a variety of aromatic guest molecules.[51-53] An interesting investigation has been the study of the effect on its reactivity of including cinnamyl

O_2N〈benzene〉—〈benzene〉NO_2

(24)

phenyl ether in this host lattice.[54] Whereas cinnamyl phenyl ether undergoes a facile Claisen rearrangement in isotropic solvents, the constraints imposed by the host lattice reduce its reactivity by at least a 1000 fold when it is included in the 4,4′-dinitrobiphenyl host lattice.

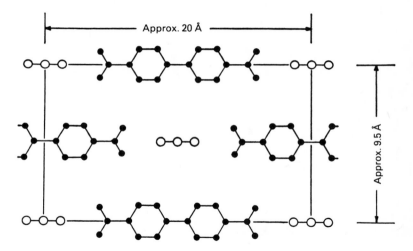

Fig. 15. The crystal structure of a 4,4′-dinitrobiphenyl inclusion compound. Reproduced with permission from *J. Chem. Soc.*, 1946, 1110.

2.6. *N*-(*p*-Tolyl)tetrachlorophthalimide

N-(*p*-Tolyl)tetrachlorophthalimide (**25**) forms channel-type inclusion compounds[55] (Fig. 16) with guest molecules satisfying the following criteria:
1. Some part of the guest molecule must be aromatic.
2. The thickness and width of the guest must not be too different from the thickness and width of a benzene ring.

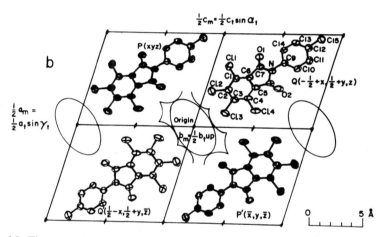

(25)

Some large aromatic molecules (perylene and fluoranthene) seem to form charge transfer complexes rather than inclusion compounds, whilst phenothiazine is unique in being able to form both inclusion compounds and charge transfer complexes.

Fig. 16. The crystal structure of the host lattice formed by *N*-(*p*-tolyl)tetrachlorophthalimide **(25)**. The channels where the guest molecules are accommodated are indicated by ellipses. Reproduced with permission from *Z. Kristallogr.*, 1981, **157**, 1.

2.7. Other organic host lattices

Information about the inclusion compounds formed by other organic host lattices is contained in Table 5, and the crystal structures of a few are illustrated in Figs. 17–19. Of particular interest is the inclusion compound formed between host 37 and *n*-hexane[69] since the *n*-hexane guest molecule does not adopt a completely planar zig-zag structure but takes up, rather, an extended zig-zag planar arrangement except for a terminal C atom torsional angle of 100° (Fig. 20). This is yet another example of the interesting situation which can be encountered in inclusion compounds where the

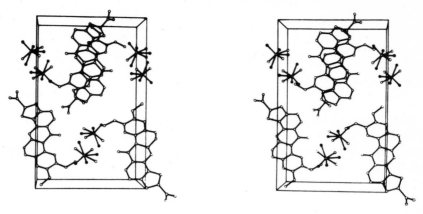

Fig. 17. Stereoscopic view of the unit cell of the host **32**: CCl_4 (1:1) inclusion compound. Reproduced with permission from *J. Am. Chem. Soc.*, 1975, **97**, 5752.

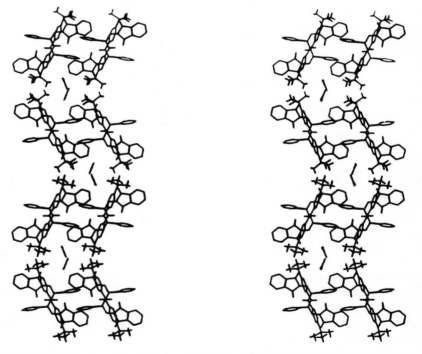

Fig. 18. Stereoscopic view of the crystal structure of the host **34**: acetone (1:1) inclusion compound. Reproduced with permission from *J. Chem. Soc., Perkin Trans. 2*, 1976, 1873.

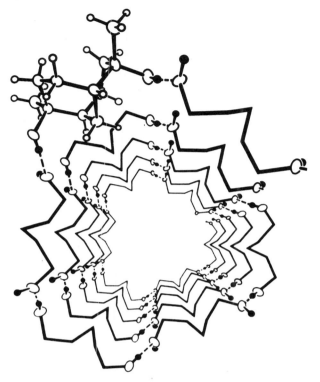

Fig. 19. Perspective view of the helical sequence of hydrogen bonded molecules of **38** forming a channel type host lattice. Reproduced with permission from *J. Chem. Soc., Chem. Commun.,* 1979, 992.

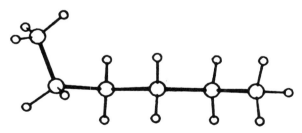

Fig. 20. Structure of the *n*-hexane guest molecule in the channel type host lattice formed by **37**. Reproduced with permission from *J. Chem. Soc., Chem. Commun.,* 1977, 928.

Table 5. Inclusion compounds formed by other organic host lattices

Host lattice	Typical guest molecules (host : guest stoichiometry)	Ref.
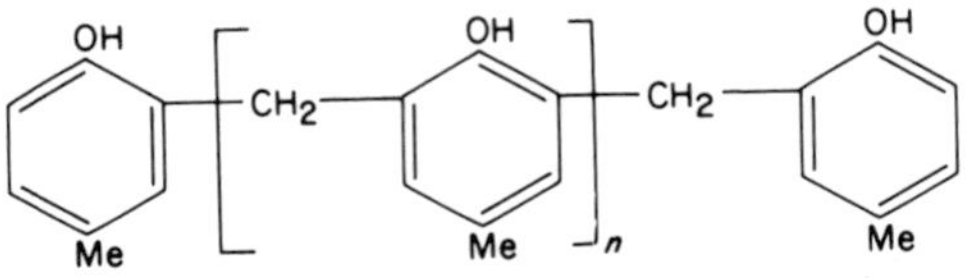 **(26)**	$n = 2$ $C_2H_4Cl_2$ (1 : 1)	56
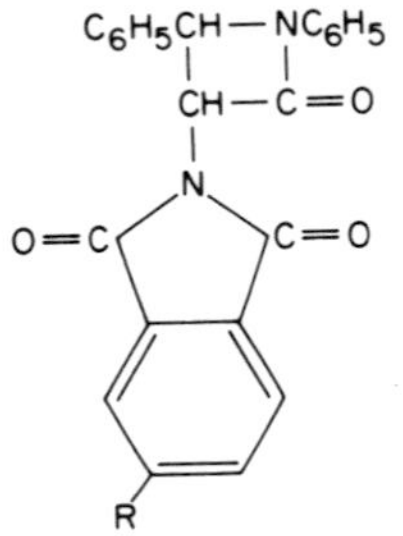 **(27)**	(a) R = H 1,4-Dioxan (2 : 1) (b) R = NO_2 C_6H_6 (1 : 1)	57
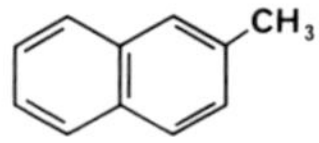 **(28)**	n-Heptane (1 : 1)	58
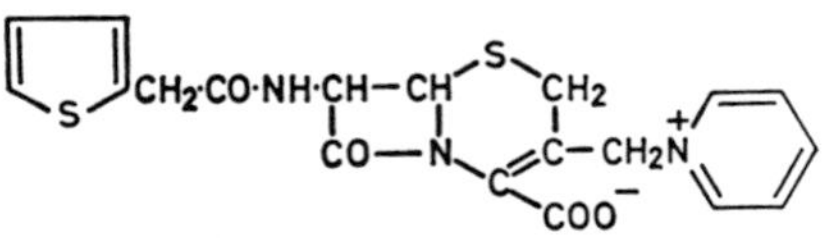 **(29)**	ζ-Polymorph DMSO (1 : 1) μ-Polymorph CH_3OH (1 : 1)	59
Benzidine : Tetracyano- p-quinodimethane (1 : 1) **(30)**	CH_2Br_2 CH_3COCH_3 CH_3NO_2 CH_3CN CH_2Cl_2 (1 : 1.8) C_6H_6 (1 : 1)	60 61 62

Table 5—continued

Host lattice	Typical guest molecules (host : guest stoichiometry)	Ref.
C_6H_5 $C_6H_5SO_2CCH_2CH_2CO_2H$ CH_3 **(31)**	CCl_4 (1 : 0.53) Et_2O (1 : 0.76) Toluene (1 : 0.49) C_6H_6 (1 : 0.47) C_6H_{12} (1 : 0.54) $CHCl_3$ (1 : 1.0)	63
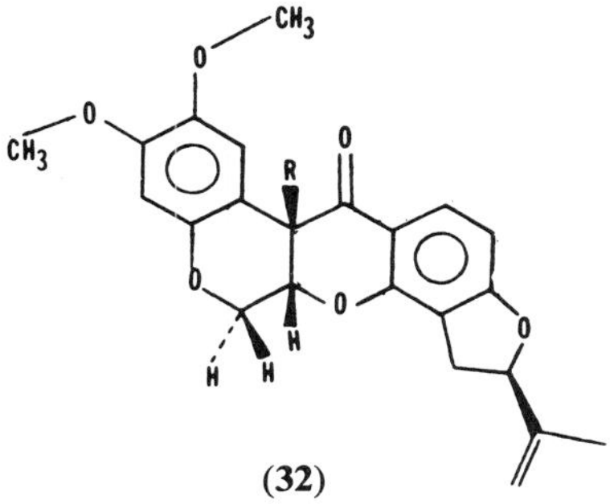 **(32)**	R = H CCl_4 (1 : 1)	64
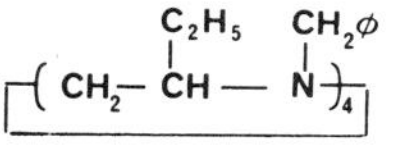 **(33)**	Substituted benzenes	65
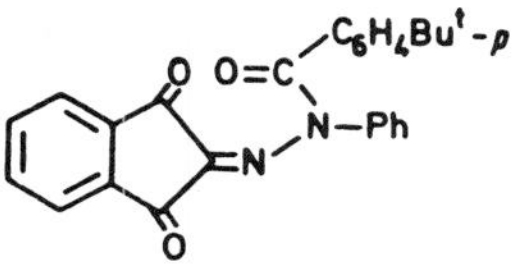 **(34)**	CH_3COCH_3 (1 : 1)	66
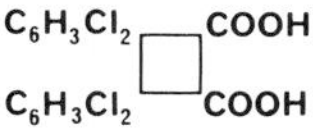 **(35)**	Pentane (2 : 1) Hexane (5 : 2) Heptane (4 : 1) Octane (4 : 1) Nonane (4 : 1) C_6H_6 (2 : 1) C_6H_{12} (2 : 1)	67
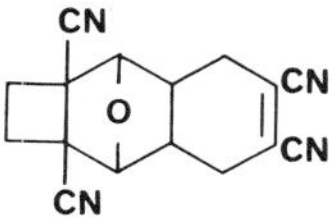 **(36)**	C_6H_6 (1 : 1) Pyridine (1 : 1) Furan (1 : 1) Thiophene (1 : 1) Pyrrole (1 : 1)	68

Table 5—continued

Host lattice	Typical guest molecules (host : guest stoichiometry)	Ref.
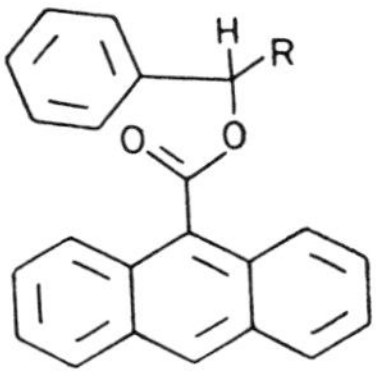 (37)	R = C_2H_5 and C_3H_7 n-Hexane (6:1)	69
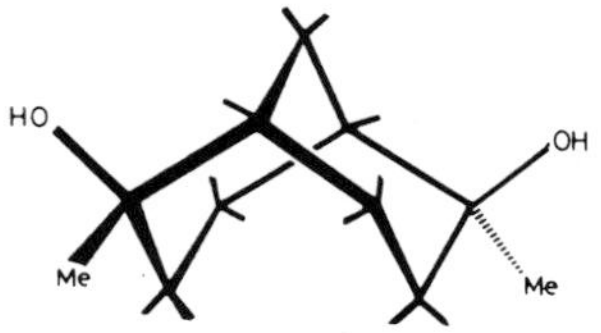 (38)	Ethyl acetate (3:1) $CHCl_3$ Toluene Dioxan Acetone	70
$CH_3CH_2NCH_2CH_2CO_2H$ (39) $\quad\quad\;\;\, CH_2CH_3$	C_6H_6 (1:1)	71
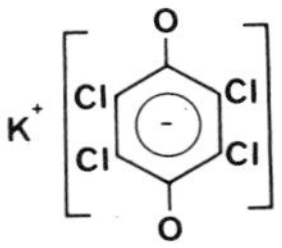 (40)	CH_3COCH_3 (1:1)	72
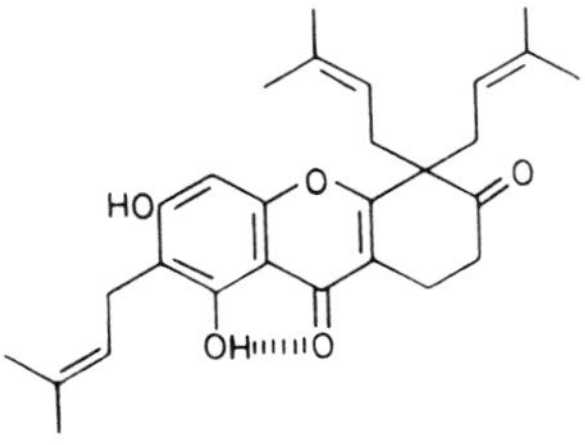 (41)	Palmitic acid (4:1)	73

Table 5—continued

Host lattice	Typical guest molecules (host : guest stoichiometry)	Ref.
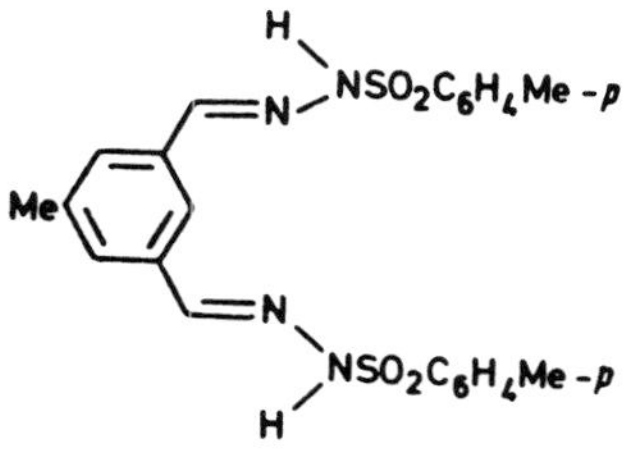 (42)	C_6H_6 (1 : 2)	74
1,2-$C_6F_4I_2$ (43)	*trans*-Ni$(CH_3OCH_2CH_2OCH_3)_2I_2$ (1 : 1)	75
$(C_6H_5)_2C-C{\equiv}C-C{\equiv}C-C(C_6H_5)_2$, OH, OH (44)	CH_3COCH_3 (1 : 2) Acetophenone (1 : 2) Cyclohexanone (1 : 2) DMF (1 : 2) DMSO (1 : 2)	76
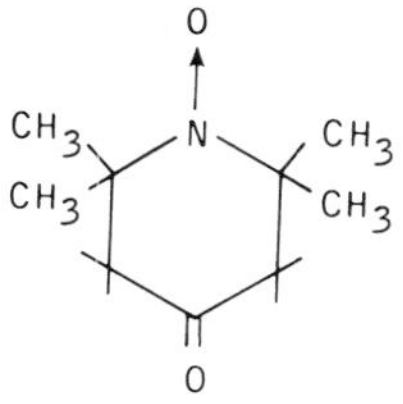 (45)	*n*-Alkanes	182
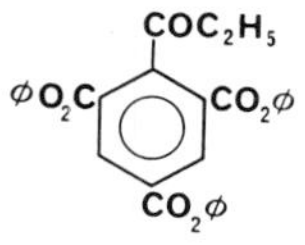 (46)	C_6H_6 (1 : 1) $C_6H_5CH_3$ (1 : 1) C_2H_5OH (1 : 1)	183
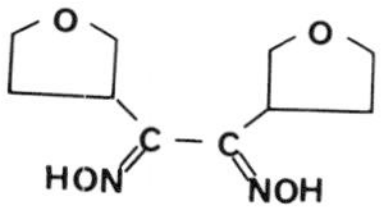 (47)	C_6H_6 (4 : 1)	184

Table 5—continued

Host lattice	Typical guest molecules (host : guest stoichiometry)	Ref.
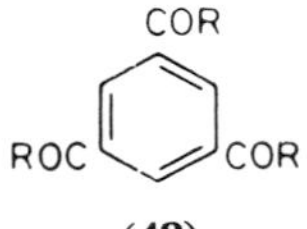 (48)	$R = O[CH_2]_{11}\overset{+}{N}C_5H_5$ C_6H_5OH	185
Sodium-*trans*-11-octadecen-9-ynoate (49)	$(CH_3)_2SO_4$	186
CCl_4 (50)	$N_3P_3(NC_2H_4)_6$ (3 : 1)	187
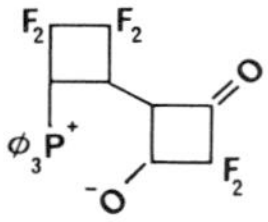 (51)	$CHCl_3$ (2 : 1)	188
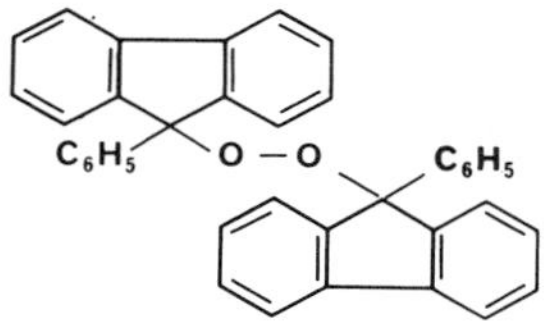 (52)	C_6H_6 (1 : 2)	189

constraints imposed by the host lattice can result in the guest molecule being forced to adopt an unusual conformation. Other examples are the stabilization of unusual conformations of squalene[77] and 3,3,6,6-tetramethyl-*s*-tetrathiane[78] by a hexa-host (See Chapter 5, Volume 2), and the stabilization of the axial conformers of monohalocyclohexanes by the thiourea host lattice.[79] (See Chapter 2, Volume 2).

The inclusion compound formed between host **41** and palmitic acid is a naturally occurring inclusion compound, and the elucidation of its structure involved a novel application of the use of inclusion compounds.[73] The removal of the palmitic acid proved to be difficult, but it was finally accomplished by dissolving the inclusion compound in warm methanol saturated with urea which, on cooling, preferentially crystallized the urea inclusion compound of palmitic acid and leaving guest free **41**.

It is also interesting to note that **26** is an open chain analogue of cyclophane **12a** and that both compounds form inclusion compounds.

3. Inorganic and organometallic host lattices

3.1. Cyanide and thiocyanate complexes

The cyanide ligand in inorganic complexes can be regarded as the equivalent of the hydroxyl group in organic compounds due to its ability to act as a bridging ligand and thus form an extended network which can act as a host lattice. The best known of the cyanide containing host lattices are the Hofmann complexes which have been reviewed in Chapter 2, Volume 1. A few other cyanide complexes have been reported to act as host lattices and are listed in Table 6.

Although the thiocyanate ligand is also capable of acting as a bridging ligand, it does actually act as a unidentate ligand in the Werner series of host lattices (see Chapters 3 and 4 in Volume 1). In the thiocyanate complexes listed in Table 6, the unidentate nature of the ligand is also preserved.

Table 6. *Inclusion compounds formed by inorganic and organometallic host lattices*

Host lattice	Typical guest molecules (host: guest stoichiometry)	Ref.
$Mg[Mg((CH_2)_6N_4)_2.Fe(CN)_6]_2$ (**53**)	H_2O, $(CH_3)_2O$, $C_6H_5NH_2$, $(CH_2OH)_2$	80a
$Fe^{II}(CN)_2(CH_3NC)_4$ (**54**)	$CHCl_3$ (1:4)	80b
$Mn_3[Co(CN)_6]_2$ (**55**)	H_2O, NH_3, CH_3OH, CH_3CHO, $(CH_3)_2O$, H_2S, PH_3, CH_3SH or $(CH_3)_2S$	81
$K_2Zn_3[Fe(CN)_6]_2$ (**56**)	H_2O (1:8), N_2, CO, CO_2, C_2H_4, C_2H_6, C_3H_8, C_4H_{10}	82
$K_3Mo(NCS)_6.H_2O$ (**57**)	CH_3CO_2H (1:1)	83, 84
$(NH_4)_3Mo(NCS)_6.H_2O$ (**58**)	HCl, C_2H_5OH (1:1)	83
$CoHg_2(SCN)_6$ (**59**)	C_6H_6 (1:1)	85, 86
$MHg_2(SCN)_6$ (**60**)	$M = Ni^{II}$, Co^{II}, Cd^{II}, Mn^{II} Benzene, Pyrrole, Furan, Thiophene (1:1)	87

Table 6—continued

Host lattice	Typical guest molecules (host : guest stoichiometry)	Ref.
trans-[Co(NH$_3$)$_4$(NCS$_2$)]O$_2$CCH$_3$ **(61)**	CH$_3$CO$_2$H (1:1)	88
[(CH$_3$)$_3$As.PdCl$_2$]$_2$ **(62)**	Dioxan N$_2$, H$_2$	89 90
C$_6$H$_5$SO$_3$OK **(63)**	C$_2$H$_4$ (1:0.22) C$_6$H$_6$ (1:0.16) CH$_3$OH (1:1) C$_5$H$_5$N (1:0.3)	91, 92

[PtM(S(CH$_2$)$_3$S(C$_2$H$_5$)X$_2$] **(64)**	(a) M = Pt^{II}
	   X = Cl C$_6$H$_6$ (1:0.5)
	   X = I C$_6$H$_6$ (1:1)
	   X = NH$_3$ C$_6$H$_6$ (1:0.67)
	(b) M = Pd^{II}
	   X = Cl C$_6$H$_6$ (1:0.5)
	   X = Br C$_6$H$_6$ (1:0.5),
	                  Acetone (1:1)
	   X = I C$_6$H$_6$ (1:0.5)
		93

M(Acetylacetonate)$_3$ **(65)**	M = Fe^{III}, Al^{III}, Cr^{III} CHCl$_3$ (1:2)	94
Pd$_2$dpt$_4$ **(66)** dpt = C$_6$H$_5$NNNC$_6$H$_5$	CH$_3$COCH$_3$ (1:1) C$_6$H$_6$ (1:0.5)	95
Pddpt$_2$py$_2$ **(67)**	Pyridine (1:1) C$_6$H$_6$ (1:1)	95
PdCl$_2$dpth$_2$ **(68)**	CH$_3$COCH$_3$ (1:0.5)	95
Be$_4$O(C$_6$H$_5$CO$_2$)$_6$ **(69)**	C$_6$H$_6$ (1:3) C$_6$H$_5$CH$_3$ (1:0.6) p-C$_6$H$_4$(CH$_3$)$_2$ (1:0.7)	96
Cu^{II}$\left(\begin{smallmatrix} C_6H_5CH-CHNHCH_3 \\ | \quad\quad | \\ OH \quad CH_3 \end{smallmatrix}\right)_2$ **(70)**	C$_6$H$_6$ (1:0.67)	97
[M(sexa)][ClO$_4$]$_2$ **(71)**	M = Co^{II}, Ni^{II}, Cu^{II} C$_2$H$_5$OH (1:1) (CH$_3$)$_2$CO (1:2)	98
(Ph$_3$PH)$_2$UO$_2$Cl$_4$ **(72)**	C$_6$H$_{11}$OH (1:1)	99
(Ph$_3$PH)$_2$U(OH)$_2$Cl$_4$ **(73)**	C$_6$H$_{11}$OH (1:1)	99
(Ph$_3$PH)$_2$Zr(OH)Cl$_5$ **(74)**	C$_6$H$_{11}$OH (1:1)	99
(Ph$_3$PH)$_2$U(OH)Cl$_5$ **(75)**	C$_6$H$_{11}$OH (1:1)	100
Cr(8-quinolinol)$_3$ **(76)**	CH$_3$OH (1:1)	101
Mn(8-quinolinol)$_3$ **(77)**	CH$_3$OH (1:1) C$_6$H$_{13}$OH (1:0.5)	102
Ni(ethylxanthate)$_2$ (4,4′-bipyridyl) **(78)**	C$_6$H$_6$ (1:1) (C$_2$H$_5$)$_2$O (1:1) CHCl$_3$ (1:1) (CH$_3$)$_2$CO (1:1)	103, 104

Table 6—continued

Host lattice	Typical guest molecules (host:guest stoichiometry)	Ref.
Ni(ethylxanthate)$_2$-(o-phenanthroline) **(79)**	C_6H_6 (1:1) $C_6H_5CH_3$ (1:1) $CHCl_3$ (1:1) C_6H_5Cl (1:1)	103
Pd(dimethyl-o-thiolo-phenylarsine)$_2$ **(80)**	C_5H_5N (1:1)	105
CuOSO$_2$CF$_3$ **(81)**	Aromatics (2:1)	106–108
[{FB(ONCHC$_5$H$_3$N)$_3$P}Fe^{II}][BF$_4$] **(82)**	CH_2Cl_2 (1:1)	109
[{FB(ONCHC$_5$H$_3$N)$_3$P}Co^{II}][BF$_4$] **(83)**	CH_3CN (1:1)	110
W(CO)$_4$[(C$_6$H$_5$)$_3$PCHCHCH$_2$] **(84)**	C_6H_6 (1:1.5)	111
Mo(CO)$_4$[(C$_6$H$_5$)$_3$PCHCHCH$_2$] **(85)**	C_6H_6 (1:1)	112
K$_2$CS$_4$ **(86)**	H_2O (1:0.5) CH_3OH (1:1) $(CH_3)_2NH$ (1:1)	113
Pt$_6$Cl$_{12}$ **(87)**	Br_2 (1:1) CCl_4 (1:0.75) CS_2 (1:1) $CHCl_3$ (1:1) C_6H_6 (1:1) CH_2Cl_2 (1:1)	114
4Na$_2$SO$_4$.NaCl **(88)**	H_2O_2 (1:2)	115
[Cr(urea)$_6$][Co(NH$_3$)$_2$(NO$_2$)$_4$]$_3$ **(89)**	$CHBr_3$ (1:1.5) CCl_4 (1:1.5)	116
Pt(1,10-phenanthroline)Cl$_2$I$_5$ **(90)**	CH_2Cl_2 (1:1) $CHCl_3$ (1:1)	117
Pd(C$_6$H$_5$C(S)CHCNHC$_6$H$_5$)$_2$ **(91)** ‖ S	Pyridine (1:2) DMSO (1:2)	118 119
Pd(C$_6$H$_5$C(S)CHCNHC$_6$H$_4$CH$_3$)$_2$ ‖ S **(92)**	Pyridine (1:2) DMSO (1:2)	120 119
Pd[C$_6$H$_5$C(S)CHC(S)NHC$_6$H$_3$-(CH$_3$)$_2$]$_2$ **(93)**	Pyridine (1:2) DMSO (1:2)	120 119
S$_4$(NCH$_2$OH)$_4$ **(94)**	Benzene	121
S$_5$(NCH$_2$OH)$_3$ **(95)**	Benzene	121
[C$_6$H$_5$CN.P(C$_6$H$_5$)$_3$.Ni]$_4$ **(96)**	Toluene:n-hexane:COD (1;2~1~1)	122
[(C$_6$H$_5$)$_3$PCH$_2$C$_6$H$_5$]$_2$[CdCl$_4$] **(97)**	$C_2H_4Cl_2$ (1:2)	123
(PPh$_3$)$_2$Ni(PhCOCHCHCO$_2$CH$_3$) **(98)**	C_6H_6 (1:1)	124
Mo$_2$(O$_2$CCH$_3$)$_4$.CH$_3$CO$_2$Na **(99)**	CH_3CO_2H (1:1)	125
N$_3$P$_3$(NC$_2$H$_4$)$_6$ **(100)**	CCl_4 (1:3) C_6H_6 (2:1)	126 127

Table 6—continued

Host lattice	Typical guest molecules (host: guest stoichiometry)	Ref.
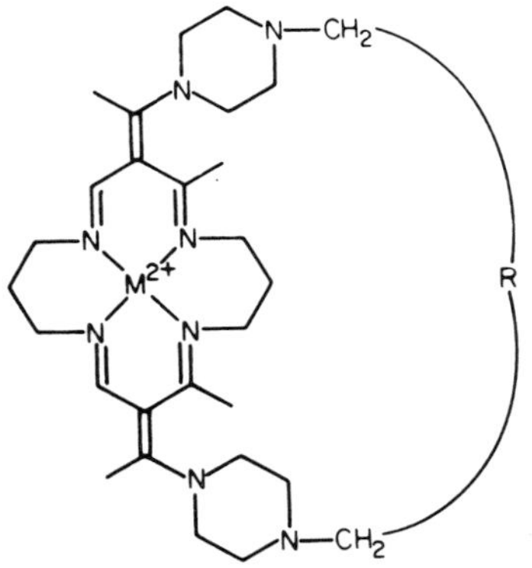 [PF$_6^-$]$_2$ **(101)**	R = Anthracene CH$_3$CN (1:2)	128
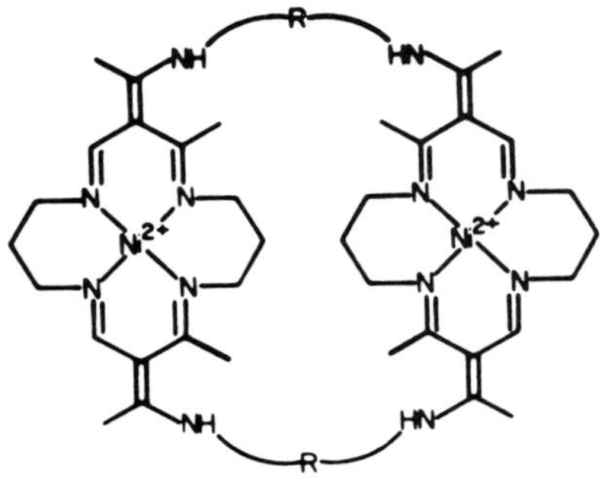 [PF$_6^-$]$_4$ **(102)**	R = m-xylyl CH$_3$COCH$_3$ (1:4)	129
trans-[Nipy$_4$(ONO)$_2$] **(103)**	Pyridine (1:2)	130
cis-[Ni(2-MeIm)$_4$ONO][NO$_3$] **(104)**	CH$_3$OH (1:0.5)	131
2-MeIm ≡ 2-Methylimidazole		
[Ni(4-Mepy)$_2$(NO$_2$)$_2$]$_2$ **(105)**	C$_6$H$_6$ (1:2)	132
[Ni(3-Mepy)$_2$(NO$_2$)$_2$]$_3$ **(106)**	C$_6$H$_6$ (1:1)	132
[M(H$_2$O)$_8$]Cl$_3$ **(107)**	M = Gd^{III}, Y^{III} 4,4′-Bipyridyl (1:2)	133

Table 6—continued

Host lattice	Typical guest molecules (host : guest stoichiometry)	Ref.
$Zn[2\text{-}NH_2.CH_2.py]_3Cl_2$ **(108)**	CH_3CH_2OH (1:1)	134
$InBr_3py_3$ **(109)**	Pyridine (1:1)	135
M(pyrrole-N-carbodithiamate)$_3$ **(110)**	$Me = Fe^{III}, Cr^{III}, Ir^{III}$	
	C_6H_6 (2:1)	136
	$M = Fe^{III}, Co^{III}$	137(Fe^{III})
	CH_2Cl_2 (2:1)	138(Co^{III})
$Sb(C_6H_5)_5$ **(111)**	Cyclohexane (2:1)	139
$Pt(C_5H_{10}NS_2)_2[P(C_6H_{11})_3]$ **(112)**	Cyclohexane (1:1)	140
$Ir(NO)[P(C_6H_5)_3]_3$ **(113)**	C_3H_6 (1:1)	141
$[Cu(C_{44}H_{60}N_4)Cl]Cl$ **(114)**	$CHCl_3$ (1:2)	142
$(\eta^5\text{-}C_5H_5)_2Fe_2(CO)_3[\mu\text{-}C{=}C(Ph)CH_2Ph]$ **(115)**	$(C_2H_5)_2O$ (1:1)	143
$[Ph_3C_3CO](C_{10}H_8N_2)(CO)_2MoBr$ **(116)**	C_6H_6 (1:1)	144
$Co[SC(NH_2)_2]_4SO_4$ **(117)**	CH_3OH (1:1)	145
SiO_2 (Dodecasils) **(118)**	CH_4 and $N(CH_3)_3$ (136:16:8)	146
SiO_2 (Melanophlogite) **(119)**	CH_4, N_2 or CO_2 (46:8)	147
$\{Zn[OP(CH_3)(C_6H_5)O]_2\}$ **(120)**	$C_4H_8O_2$ (n:n)	148
$K_2[Mn^{IV}(3,5\text{-}di\text{-}t\text{-}butylpyro\text{-}catecholate)_3]$ **(121)**	CH_3CN (6:1)	149
$Pb_2Ph_4(O_2CMe)_4.H_2O$ **(122)**	C_6H_6 (1:1)	150
$[Rh_2(\mu\text{-}H)(\mu\text{-}CO)(CO)_2\text{-}(PPh_2CH_2PPh_2)_2]\text{-}(p\text{-}CH_3C_6H_4SO_3)$ **(123)**	THF (1:2)	151
$Mo(NNPh_2)_2(S_2CNMe_2)_2$ **(124)**	Me_2CO (1:1)	152
$[MoCl(NNMe_2)_2(PPh_3)_2][BPh_4]$ **(125)**	CH_2Cl_2 (1:1)	152
$[Mo(NNMe_2)_2(bpy)_2][BPh_4]_2$ **(126)**	CH_2Cl_2 (1:1)	152
$Fe^{II}(N\text{-}MeIm)_2$(Diphenylboron-dimethylglyoximate)$_2$ **(127)**	CH_2Cl_2 (1:2)	153
$[\{(Me_4en)ClPt(CH_2)_2\text{-}\}_2NEt_2]\text{-}[ClO_4]$ **(128)**	$CHCl_3$ (1:1)	154
$Pt(CH_3)_4$ **(129)**	C_6H_6 (2:1)	155

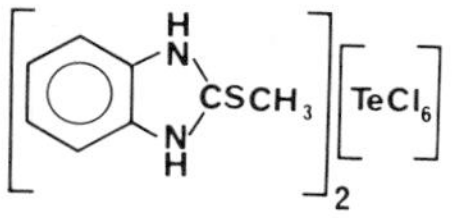

(130)

THF (1:2)   156

Table 6—continued

Host lattice	Typical guest molecules (host : guest stoichiometry)	Ref.
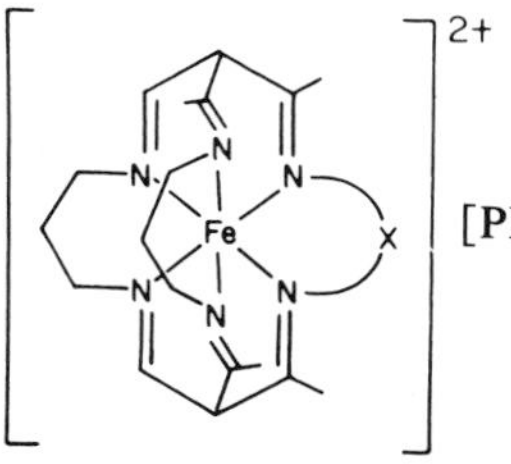 (131)	X = –(CH$_2$)$_5$– CH$_3$OH (2:1)	157
[Co(MeCN)(C$_2$H$_4$)(PMe$_3$)$_3$]BPh$_4$ (132)	CH$_3$CN (1:1)	158
[pipH]$_2$[Fe(o-catecholamide)$_2$-acac] (133)	THF (1:1)	159
[NMe$_4$]$_2$Co(NCS)$_4$ (134)	CH$_3$NO$_2$ (1:1)	160
[AsPh$_4$]$_2$[Mo$_2$Cl$_4$O$_4$] (135)	CH$_2$Cl$_2$.HCl (2:2:1)	161
[{MeC(CH$_2$PPh$_2$)$_3$}CoP$_2$S]BF$_4$ (136)	C$_6$H$_6$ (1:1)	162
[Pt$_3$(PPh$_3$)$_3$(μ-PPh$_2$)$_2$(μ-H)]BF$_4$ (137)	CH$_2$Cl$_2$ (1:2)	163
[PdCl(CH$_2$CN)(PPh$_3$)$_2$] (138)	(CH$_3$)$_2$CO, CH$_3$CN, or C$_6$H$_6$ (1:1)	164
[(PPh$_3$)$_2$N][Os$_{10}$C(CO)$_{23}$(NO)] (139)	CH$_2$Cl$_2$ (1:1)	165
[(dppe)Fe{S$_2$CHdppe}]BPh$_4$ dppe = 1,2-bis(diphenyl-phosphino)ethane (140)	CH$_2$Cl$_2$ (1:1)	166
[{Rh(C$_5$Me$_5$)}$_3$Cl$_5$np$_3$]PF$_6$ np$_3$ = tris(2-diphenyl-phosphinoethyl)amine (141)	*iso*-C$_3$H$_7$OH (2:1)	167
[Cu$_4$(OH)$_4$(SO$_3$CF$_3$)$_2${Npy$_3$}$_4$]-[SO$_3$CF$_3$]$_2$ (142)	(CH$_3$)$_2$CO (1:1)	168
[PPh$_3$]$_2$PtS$_4$ (143)	CHCl$_3$ (1:1)	169
[(η^5-C$_5$H$_5$)$_2$MO]$_3$ (144)	M = Zr　C$_6$H$_5$CH$_3$ (1:1) M = Hf　C$_6$H$_5$CH$_3$ (1:1)	170 171
{[*trans*-Ir(CO)(CH$_3$CN)-(PPh$_3$)$_2$]$_2^+$18C6}[PF$_6$]$_2^-$ (145)	CH$_2$Cl$_2$ (1:2)	172
PPh$_4$[Cl$_5$Os^v(NC(CCl$_3$)-NCCl(CCl$_3$))] (146)	CH$_2$Cl$_2$ (1:1)	173

Table 6—*continued*

Host lattice	Typical guest molecules (host : guest stoichiometry)	Ref.
[Cu(SALM edpt)Cu(hfa)$_2$]$_2$ **(147)** hfa = hexafluoraacetylacetonate SALM edpt = salicaldehyde + bis (3-aminopropyl)methylamine	CHCl$_3$ (1 : 0.6)	174
[PPh$_4$][Mo(NNMe$_2$)O(SC$_6$H$_5$)$_3$] **(148)**	Et$_2$O (1 : 1)	175
[η^5-C$_5$H$_5$]Ti(η^7-C$_7$H$_6$PPh$_2$)-Mo(CO)$_5$ **(149)**	C$_6$H$_5$CH$_3$ (1 : 1)	176
(ClHg)$_3$CCHO **(150)**	DMF (1 : 1)	177
(BrHg)$_3$CCHO **(151)**	DMSO (1 : 1)	177
Ru$_2$Cl$_3${p-CH$_3$C$_6$H$_4$-CH(CH$_3$)$_2$}$_2$BPh$_4$ **(152)**	MeOH (1 : 1)	178

3.2. Other inorganic host lattices

Information about the inclusion compounds formed by other inorganic host lattices is contained in Table 6, and the crystal structures of a few are illustrated in Figs. 21 and 22.

The selective inclusion ability of potassium benzenesulphonate **(63)** has been utilized in the gas chromatographic separation of alcohols, phenols,

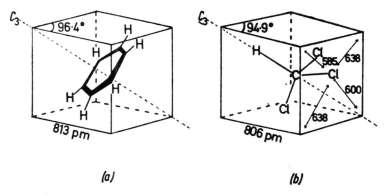

(a)   *(b)*

Fig. 21. Possible orientations of (a) benzene and (b) chloroform guest molecules in the rhombohedral unit cells formed by Pt$_6$Cl$_{12}$ **(87)**. Reproduced with permission from *J. Chem. Soc., Dalton Trans.*, 1975, 2432.

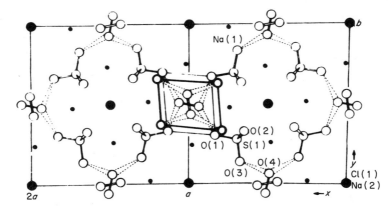

Fig. 22. Projection of the structure of $4Na_2SO_4.2H_2O_2.NaCl$ (**88**) showing the disordered hydrogen peroxide guest molecules. Reproduced with permission from *J. Chem. Soc., Chem. Commun.*, 1978, 288.

ethers and substituted pyridines.[92] The application of host lattice forming compounds as stationary phases in chromatography is reviewed in Chapter 6, Volume 3.

The metal complexes **71** contain the sexadentate ligand shown in Fig. 23. In the inclusion compound $[Fe(sexa)][FeCl_4]_2.(CH_3)_2CO$, a host structure containing four units per unit cell is compatible with the dimensions, density, and symmetry of the crystal lattice,[98] giving a host:guest ratio of 1:1.

The complexes formed between aromatic molecules and copper (I) tri-fluoromethanesulphonate (**81**) involve π bonding between the aromatic moiety and the copper atoms, (Fig. 24).[107] However the specific situation of the aromatic ring between the inorganic chains gives rise to an interesting shape selectivity by the host lattice, which will, for example, efficiently separate *p*-xylene from other eight carbon arenes.[108]

Hosts **82** and **83** have been classified as clathro-chelates[109,110] since the metal^{II} atom is effectively encapsulated by the ligand (Fig. 25) with the six nitrogen atoms arranged in a trigonal prismatic coordination around the metal^{II} atom.

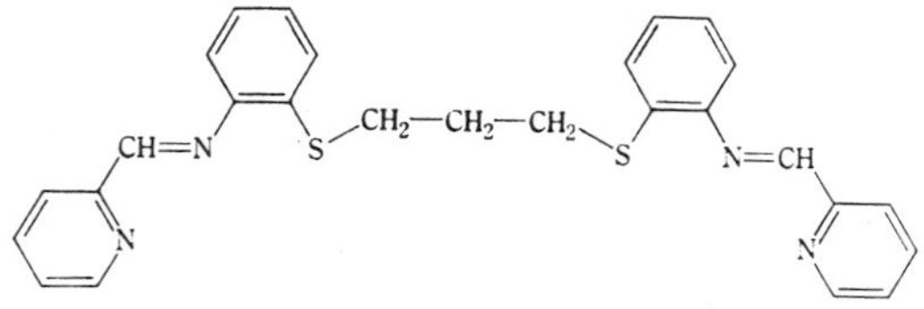

Fig. 23. The sexadentate ligand used in complexes **71**.

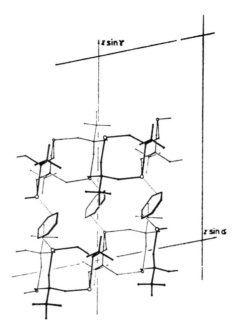

Fig. 24. The structure of the benzene: copper(I) trifluoromethanesulphonate (**81**) (1:2) complex. The copper atoms are shown as open circles. Reproduced with permission from *J. Chem. Soc., Chem. Commun.*, 1973, 12.

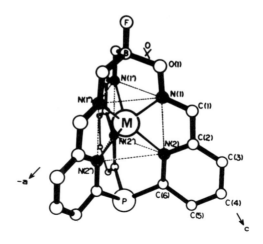

Fig. 25. Structure showing the complete encapsulation of the metal[II] atom in the hosts **82** and **83**. Reproduced with permission from *Inorg. Chem.*, 1972, **11**, 1811.

References

1. F. H. Herbstein and M. Kapon, *Acta Crystallogr.*, 1979, **B35**, 1614.
2. F. H. Herbstein, M. Kapon, I. Maor and G. M. Reisner, *Acta Crystallogr.*, 1981, **B37**, 136.
3. D. J. Duchamp and R. E. Marsh, *Acta Crystallogr.*, 1969, **B25**, 5.
4. F. H. Herbstein, M. Kapon and G. M. Reisner, *Proc. R. Soc., London, Ser. A,* 1981, **376**, 301.
5. F. H. Herbstein, *Isr. J. Chem.*, 1968, **6**, IVp.
6. M. Noltemeyer and W. Saenger, *J. Am. Chem. Soc.*, 1980, **102**, 2710.
7. D. E. Palin and H. M. Powell, *J. Chem. Soc.*, 1947, 208.
8. H. M. Powell, in *Non-Stoichiometric Compounds*, (ed. L. Mandelcorn) Academic Press, New York, 1964, Chapter 7, p. 438.
9. F. H. Herbstein and R. E. Marsh, *Acta Crystallogr.*, 1977, **B33**, 2358.
10. T. J. Kistenmacher, T. E. Phillips, D. O. Cowan, J. P. Ferraris, A. N. Bloch and T. O. Poehler, *Acta Crystallogr.*, 1976, **B32**, 539.
11. F. H. Herbstein and M. Kapon, *Acta Crystallogr.*, 1978, **B34**, 1608.
12. F. H. Herbstein, M. Kapon and S. Wasserman, *Acta Crystallogr.*, 1978, **B34**, 1613.
13. K. Mislow, D. Gust, P. Finocchiaro and R. J. Boettcher, *Top. Curr. Chem.*, 1974, **47**, 1.
14. (a) H. Hartley and N. G. Thomas, *J. Chem. Soc.*, 1906, 1013. (b) J. E. Driver and S. F. Mok, *J. Chem. Soc.*, 1955, 3914; and J. E. Driver and T. F. Lai, *J. Chem. Soc.*, 1958, 3219. (c) G. B. Barlow and A. C. Clamp, *J. Chem. Soc.*, 1961, 393.
15. A. Recca, F. A. Bottino, E. Libertini, and P. Finocchiaro, *Gazz. Chim. Ital.,* 1979, **109**, 213.
16. A. Allemand and R. Gerdil, *Acta Crystallogr.*, 1975, **A31**, S130.
17. J. F. Norris, *J. Am. Chem. Soc.*, 1916, **38**, 702.
18. A. Rahman and D. van der Helm, *Cryst. Struct. Commun.*, 1981, **10**, 731.
19. L. Bin Din and O. Meth-Cohn, *J. Chem. Soc., Chem. Commun.*, 1977, 741.
20. K. S. Hayes, W. D. Hounshell, P. Finocchiaro and K. Mislow, *J. Am. Chem. Soc.*, 1977, **99**, 4152.
21. (a) D. H. Brown, R. J. Cross and D. D. MacNicol, *Chem. Ind. (London)*, 1977, 766. (b) D. H. Brown, R. J. Cross, P. R. Mallinson and D. D. MacNicol, *J. Chem. Soc., Perkin Trans. 2*, 1980, 993.
22. (a) P. Finocchiaro, D. Gust, W. D. Hounshell, J. P. Hummel, P. Maravigna and K. Mislow, *J. Am. Chem. Soc.*, 1976, **98**, 4945. (b) P. Finocchiaro, W. D. Hounshell, and K. Mislow, 1976, **98**, 4952.
23. (a) H. Stetter and E.-E. Roos, *Chem. Ber.*, 1955, **88**, 1390. (b) R. Hilgenfeld and W. Saenger, *Angew. Chem., Int. Ed. Engl.*, 1982, **21**, 787.
24. (a) Y. Urushigawa, T. Inazu and T. Yoshino, *Bull. Chem. Soc. Jpn.*, 1971 **44**, 2546. (b) S. J. Abbott, A. G. M. Barrett, C. R. A. Godfrey, S. B. Kalindjian, G. W. Simpson and D. J. Williams, *J. Chem. Soc., Chem. Commun.*, 1982, 796.
25. (a) I. Tabushi, H. Sasaki and Y. Kuroda, *J. Am. Chem. Soc.*, 1976, **98**, 5727. (b) I. Tabushi, Y. Kuroda and Y. Kimura, *Tetrahedron Lett.*, 1976, 3327.
26. Y. Ichikawa, H. Tsuruta, K. Kato, Y. Yamanaka and M. Yamamoto, *Chem. Abs.*, 1977, **87**, 134101p.
27. Y. Ichikawa, M. Yamamoto, H. Tsuruta, K. Kato, T. Yamaji, E. Yoshisato and T. Hiramatsu, *Chem. Abs.*, 1978, **88**, 121459g.

28. T. Yamaji, E. Yoshisato and T. Hiramatsu, *Chem. Abs.*, 1978, **89**, 6017v.
29. T. Yamaji, E. Yoshisato and T. Hiramatsu, *Chem. Abs.*, 1978, **89**, 42589s.
30. T. Yamaji, E. Yoshisato and T. Hiramatsu, *Chem. Abs.*, 1978, **89**, 42730f.
31. Y. Ichikawa, H. Tsuruta, K. Kato, Y. Yamanaka and A. Yamamoto, *Chem. Abs.*, 1978, **89**, 197137s.
32. F. Bottino, S. Foti, S. Pappalardo, P. Finocchiaro and M. Ferrugia, *J. Chem. Soc., Perkin Trans. 1*, 1979, 198.
33. F. Bottino, S. Foti and S. Pappalardo, *J. Chem. Soc., Perkin Trans. 1*, 1979, 1712.
34. K. Odashima, A. Itai, Y. Iitaka and K. Koga, *J. Am. Chem. Soc.*, 1980, **102**, 2504.
35. K. Odashima, A. Itai, Y. Iitaka, Y. Arata and K. Koga, *Tetrahedron Lett.*, 1980, **21**, 4347.
36. T. Soga, K. Odashima and K. Koga, *Tetrahedron Lett.*, 1980, **21**, 4351; and 1981, **22**, 5311.
37. (a) G. D. Andreetti, R. Ungaro and A. Pochini, *J. Chem. Soc., Chem. Commun.*, 1979, 1005. (b) C. Rizzoli, G. D. Andreetti, R. Ungaro and A. Pochini, *J. Mol. Struct.*, 1982, **82**, 133. (c) M. Coruzzi, G. D. Andreetti, V. Bocchi, A. Pochini and R. Ungaro, *J. Chem. Soc., Perkin Trans. 2*, 1982, 1133.
38. C. D. Gutsche, B. Dhawan, K. H. No and R. Muthukrishnan, *J. Am. Chem. Soc.*, 1981, **103**, 3782.
39. F. Vögtle and W. M. Müller, *Angew. Chem., Int. Ed. Engl.*, 1982, **21**, 147.
40. N. Z. Huang and T. C. W. Mak, *J. Chem. Soc., Chem. Commun.*, 1982, 543.
41. F. A. Bottino, S. Pappalardo and G. Ronsisvalle, *Gazz. Chim. Ital.*, 1981, **111**, 437.
42. M. S. Raasch, *J. Org. Chem.*, 1979, **44**, 2629.
43. E. T. Jarvi and H. W. Whitlock, *J. Am. Chem. Soc.*, 1980, **102**, 657.
44. G. F. Claringbull and R. W. H. Small, *Acta Crystallogr.*, 1971, **B27**, 863.
45. R. E. Cobbledick and R. W. H. Small, *Acta Crystallogr.*, 1973, **B29**, 640.
46. R. E. Cobbledick, J. E. D. Davies, R. W. H. Small and D. Stubley, *Thermochim. Acta*, 1973, **7**, 317.
47. R. E. Cobbledick and R. W. H. Small, *Acta Crystallogr.*, 1973, **B29**, 1659.
48. R. E. Cobbledick and R. W. H. Small, *Acta Crystallogr.*, 1973, **B29**, 2585.
49. R. S. George, H. H. Cady, R. N. Rogers and R. K. Rohwer, *Ind. Eng. Chem., Prod. Res. Dev.*, 1965, **4**, 209.
50. R. E. Cobbledick and R. W. H. Small, *Acta Crystallogr.*, 1975, **B31**, 2805.
51. W. S. Rapson, D. H. Saunder and E. T. Stewart, *J. Chem. Soc.*, 1946, 1110.
52. D. H. Saunder, *Proc. R. Soc. London, Ser. A*, 1947, **190**, 508.
53. J. N. van Niekerk and D. H. Saunder, *Acta Crystallogr.*, 1948, **1**, 44.
54. M. J. S. Dewar and B. D. Nahlovsky, *J. Am. Chem. Soc.*, 1974, **96**, 460.
55. F. H. Herbstein and M. Kaftory, *Z. Kristallogr.*, 1981, **157**, 1.
56. R. F. Hunter, R. A. Morton and A. T. Carpenter, *J. Chem. Soc.*, 1950, 441.
57. J. C. Sheehan and J. J. Ryan, *J. Am. Chem. Soc.*, 1951, **73**, 1204.
58. J. Milgrom, *J. Phys. Chem.*, 1959, **63**, 1843.
59. J. H. Chapman, J. E. Page, A. C. Parker, D. Rogers, C. J. Sharp and S. E. Staniforth, *J. Pharm. Pharmacol.*, 1968, **20**, 418.
60. M. Ohmasa, M. Kinoshita and H. Akamatu, *Bull. Chem. Soc. Jpn.*, 1971, **44**, 391.
61. H. Kuroda, I. Ikemoto, K. Yakushi and K. Chikaishi, *Acta Crystallogr.*, 1972, **A28**, 515.
62. K. Yakushi, I. Ikemoto and H. Kuroda, *Acta Crystallogr.*, 1974, **B30**, 1738.
63. T. G. Miller, J. J. Hiller, A. C. Stavitsky and D. A. Wunsch, *J. Chem. Eng. Data*, 1972, **17**, 254.
64. S. K. Arora, R. B. Bates, R. A. Grady and N. E. Delfel, *J. Am. Chem. Soc.*, 1975, **97**, 5752.

65. (a) T. Sakurai, K. Kobayashi, K. Tsuboyama and S. Tsuboyama, *Acta Crystallogr.*, 1978, **B34**, 1144. (b) S. Tsuboyama, K. Tsuboyama, S. Mano, K. Kobayashi and T. Sakurai, *Nippon Kagaku Kaishi*, 1983, 281.
66. S. A. Puckett, I. C. Paul and D. Y. Curtin, *J. Chem. Soc., Perkin Trans. 2*, 1976, 1873; R. A. Booker, D. Y. Curtin and I. C. Paul, *Acta Crystallogr.*, 1978, **B34**, 2882.
67. F. Nakanishi, S. Yamada and H. Nakanishi, *J. Chem. Soc., Chem. Commun.*, 1977, 247.
68. R. L. Cobb, *Chem. Abs.*, 1977, **86**, 139824r; 1977, **87**, 152046d.
69. M. Lahav, L. Leiserowitz, L. Roitman and C. P. Tang, *J. Chem. Soc., Chem. Commun.*, 1977, 928.
70. (a) R. Bishop and I. Dance, *J. Chem. Soc., Chem. Commun.*, 1979, 992. (b) R. Bishop, S. Choudhury and I. Dance, *J. Chem. Soc., Perkin Trans. 2*, 1982, 1159.
71. M. A. Peterson, H. Hope and C. P. Nash, *J. Am. Chem. Soc.*, 1979, **101**, 946.
72. G. Zanotti and A. Del Pra, *Acta Crystallogr.*, 1980, **B36**, 313.
73. F. M. Dean, H. Khan, N. Minhaj, S. Prakash and A. Zaman, *J. Chem. Soc., Chem. Commun.*, 1980, 283.
74. T.-L. Chan, T. C. W. Mak and J. Trotter, *J. Chem. Soc., Perkin Trans. 2*, 1980, 672.
75. E. W. Gowling and R. F. N. Mallinson, *Inorg. Chim. Acta*, 1979, **34**, L259.
76. (a) F. Toda, D. L. Ward and H. Hart, *Tetrahedron Lett.*, 1981, **22**, 3865. (b) F. Toda and K. Akagi, *Tetrahedron Lett.*, 1968, 3695.
77. A. Freer, C. J. Gilmore, D. D. MacNicol and D. R. Wilson, *Tetrahedron Lett.*, 1980, **21**, 1159.
78. D. D. MacNicol and A. Murphy, *Tetrahedron Lett.*, 1981, **22**, 1131.
79. (a) A. Allen, V. Fawcett and D. A. Long, *J. Raman Spectrosc.*, 1976, **4**, 285. (b) J. E. Gustaven, P. Klaeboe and H. Kvila, *Acta Chem. Scand., Ser. A*, 1978, **32**, 25.
80. (a) A. Weiss, A. Weiss and U. Hofmann, *Z. Anorg. Allg. Chem.*, 1953, **273**, 129. (b) J. B. Wilford, N. O. Smith and H. M. Powell, *J. Chem. Soc. (A)*, 1968, 1544.
81. G. W. Beall, W. O. Milligan, J. A. Petrich and B. I. Swanson, *Inorg. Chem.*, 1978, **17**, 2978.
82. P. Cartraud, A. Cointot and A. Renaud, *J. Chem. Soc. Faraday Trans. 1*, 1981, **77**, 1561.
83. J. Lewis, R. Nyholm and P. Smith, *J. Chem. Soc.*, 1961, 4590.
84. J. R. Knox and K. Eriks, *Inorg. Chem.*, 1968, **7**, 84.
85. H. P. Fritz and J. Manchot, *Chem. Ber.*, 1963, **96**, 1891.
86. R. Gronbaek and J. Dunitz, *Helv. Chim. Acta*, 1964, **47**, 1889.
87. J. E. D. Davies and A. M. Maver, unpublished results.
88. T. J. Westcott, A. H. White and A. C. Willis, *Aust. J. Chem.*, 1980, **33**, 1853.
89. A. F. Wells, *Proc. R. Soc. London, Ser. A*. 1938, **167**, 169.
90. J. R. Dacey, J. F. Smelko and D. M. Young, *J. Phys. Chem.*, 1955, **59**, 1058.
91. R. M. Barrer, J. Drake and T. V. Whittam, *Proc. R. Soc. London, Ser. A*, 1953, **219**, 32.
92. (a) A. Bhattacharjee and A. N. Basu, *J. Chromatogr.*, 1972, **71**, 534. (b) A. Bhattacharjee and A. Bhaumik, 1975, **115**, 250.
93. S. E. Livingstone, *J. Chem. Soc.*, 1956, 1994.
94. J. F. Steinbach and J. H. Burns, *J. Am. Chem. Soc.*, 1958, **80**, 1839.
95. C. M. Harris, B. F. Hoskins and R. L. Martin, *J. Chem. Soc.*, 1959, 3728.
96. (a) K. N. Semenenko and G. M. Kurdyumov, *Russ. J. Inorg. Chem.* (*Engl. Transl.*) 1961, **6**, 1298; (b) V. S. Bogdanov and M. F. Khabibullin, *Zh. Prikl. Khim.*, 1968, **41**, 598 (*Chem. Abs.*, 1968, **69**, 13363v).
97. Y. Amano, K. Osaki and T. Watanabé, *Bull. Chem. Soc. Jpn.*, 1964, **37**, 1363.

98. L. F. Lindoy, S. E. Livingstone, T. N. Lockyer and N. C. Stephenson, *Aust. J. Chem.*, 1966, **19**, 1165.
99. A. K. Majumdar and R. G. Bhattacharyya, *J. Inorg. Nucl. Chem.*, 1967, **29**, 2359.
100. R. G. Bhattacharyya, *J. Inorg. Nucl. Chem.*, 1975, **37**, 579.
101. K. Folting, M. M. Cox, J. W. Moore and L. L. Merritt, *J. Chem. Soc., Chem. Commun.*, 1968, 1170.
102. R. Hems, T. J. Cardwell and M. F. Mackay, *Aust. J. Chem.*, 1975, **28**, 443.
103. A. G. Krüger and G. Winter, *Aust. J. Chem.*, 1971, **24**, 1353.
104. A. G. Krüger and G. Winter, *Aust. J. Chem.*, 1972, **25**, 2497.
105. J. P. Beale and N. C. Stephenson, *Acta Crystallogr.*, 1972, **28B**, 557.
106. R. G. Solomon and J. K. Kochi, *J. Chem. Soc., Chem. Commun.*, 1972, 559.
107. M. B. Dines and P. H. Bird, *J. Chem. Soc., Chem. Commun.*, 1973, 12.
108. M. B. Dines, *Sep. Sci. Technol.*, 1973, **8**, 661.
109. M. R. Churchill and A. H. Reis, *Inorg. Chem.*, 1972, **11**, 2299.
110. M. R. Churchill and A. H. Reis, *J. Chem. Soc., Dalton Trans.*, 1973, 1570.
111. I. W. Bassi, R. Scordamaglia, and A. Greco, *Chim. Ind. (Milan)*, 1973, **55**, 241.
112. I. W. Bassi and R. Scordamaglia, *J. Organomet. Chem.*, 1973, **51**, 273.
113. M. Abrouk, *Rev. Chim. Min.*, 1974, **11**, 726.
114. M. F. Pilbrow, *J. Chem. Soc., Dalton Trans.*, 1975, 2432.
115. (a) J. M. Adams and R. G. Pritchard, *Acta Crystallogr.*, 1978, **B34**, 1428. (b) J. M. Adams, V. Ramdas and A. W. Hewat, 1981, **B37**, 915.
116. S. P. Rozman and G. I. Dvoretskov, *Russ. J. Inorg. Chem. (Engl. Transl.)* 1976, **21**, 67.
117. K. D. Buse, H. J. Keller and H. Pritzkow, *Inorg. Chem.*, 1977, **16**, 1072.
118. S. Kitagawa and H. Tanaka, *Chem. Pharm. Bull.*, 1978, **26**, 1026.
119. S. Kitagawa and H. Tanaka, *Chem. Pharm. Bull.*, 1978, **26**, 3223.
120. S. Kitagawa and H. Tanaka, *Chem. Pharm. Bull.*, 1978, **26**, 2793.
121. H. Garcia-Fernandez, *C.R. Hebd. Seances Acad. Sci.*, 1978, **287C**, 269.
122. I. W. Bassi, C. Benedicenti, M. Calcaterra, R. Intrito, G. Rucci and C. Santini, *J. Organomet. Chem.*, 1978, 144, 225.
123. J. C. J. Bart, I. W. Bassi and M. Calcaterra, *J. Organomet. Chem.*, 1980, **193**, 1.
124. G. D. Andreetti, G. Bocelli, P. Sgarabotto, G. P. Chiusoli, M. Costa, G. Terenghi and A. Biavati, *Transition Met. Chem.*, 1980, **5**, 129.
125. D. L. Kepert, B. W. Skelton and A. H. White, *Aust. J. Chem.*, 1980, **33**, 1847.
126. (a) J. Galy, R. Enjalbert and J.-F. Labarre, *Acta Crystallogr.*, 1980, **B36**, 392. (b) F. H. Herbstein and R. E. Marsh, *Acta Crystallogr.*, 1982, **B38**, 1051.
127. T. S. Cameron, J.-F. Labarre and M. Graffeuil, *Acta Crystallogr.*, 1982, **B38**, 168.
128. K. J. Takeuchi, D. H. Busch and N. Alcock, *J. Am. Chem. Soc.*, 1981, **103**, 2421.
129. D. H. Busch *et al.*, *J. Am. Chem. Soc.*, 1981, **103**, 5107.
130. A. J. Finney, M. A. Hitchman, C. L. Raston, G. L. Rowbottom, B. W. Skelton and A. H. White, *Aust. J. Chem.*, 1981, **34**, 2095.
131. A. J. Finney, M. A. Hitchman, C. L. Raston, G. L. Rowbottom and A. H. White, *Aust. J. Chem.*, 1981, **34**, 2113.
132. A. J. Finney, M. A. Hitchman, C. L. Raston, G. L. Rowbottom and A. H. White, *Aust. J. Chem.*, 1981, **34**, 2125.
133. M. Bukowska-Strzyzewska and A. Tosik, *Acta Crystallogr.*, 1982, **B38**, 265 and 950.
134. M. Mikami-Kido and Y. Saito, *Acta Crystallogr.*, 1982, **B38**, 452.
135. R. W. H. Small and I. J. Worrall, *Acta Crystallogr.*, 1982, **B38**, 932.
136. E. Sinn, *Inorg. Chem.*, 1976, **15**, 369.

137. R. D. Bereman, M. R. Churchill and D. Nalewajek, *Inorg. Chem.*, 1979, **18**, 3112.
138. T. C. Woon, M. F. MacKay and M. J. O'Connor, *Inorg. Chim. Acta*, 1982, **58**, 5.
139. C. Brabant, B. Blanck and A. L. Beauchamp, *J. Organomet. Chem.*, 1974, **82**, 231.
140. P. C. Christidis and P. J. Rentzeperis, *Acta Crystallogr.*, 1979, **B35**, 2543.
141. R. B. English, *Acta Crystallogr.*, 1981, **B37**, 939.
142. K. Kobayashi, T. Sakurai, A. Hasegawa, S. Tsuboyama and K. Tsuboyama, *Acta Crystallogr.*, 1982, **B38**, 1154.
143. M. B. Hossain, D. J. Hanlon, D. F. Marten, D. Van Der Helm and E. V. Dehmlow, *Acta Crystallogr.*, 1982, **B38**, 1457.
144. M. Elder, S. F. A. Kettle and T. C. Tso, *J. Crystallogr., Spectrosc. Res.* 1982, **12**, 65.
145. R. Kamara, J. P. Declercq, G. Germain and M. Van Meerssche, *Bull. Soc. Chim. Belg.*, 1982, **91**, 171.
146. H. Gies, F. Leibau and H. Gerke, *Angew. Chem. Int. Ed. Engl.*, 1982, **21**, 206.
147. H. Gies and F. Leibau, *Acta Crystallogr.*, 1981, **A37**, C187.
148. R. Cini, P. Orioli, M. Sabat and H. D. Gillman, *Inorg. Chim. Acta.*, 1982, **59**, 225.
149. J. A. R. Hartman, B. M. Foxman and S. R. Cooper, *J. Chem. Soc., Chem. Commun.*, 1982, 583.
150. C. Gaffney, P. G. Harrison and T. J. King, *J. Chem. Soc., Dalton Trans.*, 1982, 1061.
151. C. P. Kubiak, C. Woodcock and R. Eisenberg, *Inorg. Chem.*, 1982, **21**, 2119.
152. J. Chatt, B. A. L. Chrichton, J. R. Dilworth, P. Dahlstrom, R. Gutkoska and J. Zubieta, *Inorg. Chem.*, 1982, **21**, 2383.
153. J. C. Jansen, M. Verhage and H. van Koningsveld, *Cryst. Struct. Commun.*, 1982, **11**, 305.
154. M. Lanfranchi, A. Tiripicchio, L. Maresca and G. Natile, *Cryst. Struct. Commun.*, 1982, **11**, 343.
155. R. E. Rundle and E. J. Holman, *J. Am. Chem. Soc.*, 1949, **71**, 3264.
156. W. Hinrichs, D. Mandak and G. Klar, *Cryst. Struct. Commun.*, 1982, **11**, 309.
157. N. Herron, J. J. Grzybowski, N. Matsumoto, L. L. Zimmer, G. G. Christoph and D. H. Busch, *J. Am. Chem. Soc.*, 1982, **104**, 1999.
158. B. Capelle, A. L. Beauchamp, M. Dartiguenave, Y. Dartiguenave and H.-F. Klein, *J. Am. Chem. Soc.*, 1982, **104**, 3891.
159. D. A. Buckingham, C. R. Clark, M. G. Weller and G. J. Gainsford, *J. Chem. Soc., Chem. Commun.*, 1982, 779.
160. D. W. Hoffman and J. S. Wood, *Cryst. Struct. Commun.*, 1982, **11**, 685.
161. K. J. Moynihan, P. M. Boorman, J. M. Ball, V. D. Patel and K. A. Kerr, *Acta Crystallogr.*, 1982, **B38**, 2258.
162. M. Di Vaira, M. Peruzzini and P. Stoppioni, *J. Chem. Soc., Chem. Commun.*, 1982, 894.
163. P. L. Bellon, A. Ceriotti, F. Demartin, G. Longoni and B. T. Heaton, *J. Chem. Soc., Dalton Trans.*, 1982, 1671.
164. R. McCrindle, G. Ferguson, A. J. McAlees, M. Parvez and P. J. Roberts, *J. Chem. Soc., Dalton Trans.*, 1982, 1699.
165. D. Braga, K. Henrick, B. F. G. Johnson, J. Lewis, M. McPartlin, W. J. H. Nelson and J. Puga, *J. Chem. Soc., Chem. Commun.*, 1982, 1083.
166. C. Bianchini, A. Meli, F. Nuzzi and P. Dapporto, *J. Organomet. Chem.*, 1982, **236**, 245.
167. P. Dapporto, P. Stoppioni and P. M. Maitlis, *J. Organomet. Chem.*, 1982, **236**, 273.

168. P. L. Dedert, T. Sorrell, T. J. Marks and J. A. Ibers, *Inorg. Chem.*, 1982, 21, 3506.
169. D. Dudis and J. P. Fackler, *Inorg. Chem.*, 1982, 21, 3577.
170. G. Fachinetti, C. Floriani, A. Chiesi-Villa and C. Guastini, *J. Am. Chem. Soc.*, 1979, 101, 1767.
171. R. D. Rogers, R. V. Bynum and J. L. Atwood, *J. Crystallogr. Spectrosc.. Res.*, 1982, 12, 239.
172. H. M. Colquhoun, J. F. Stoddart and D. J. Williams, *J. Am. Chem. Soc.*, 1982, 104, 1426.
173. R. Weber, K. Dehnicke, E. Schweda and J..Strähle, *Z. Anorg. Allg. Chem.*, 1982, 490, 159.
174. L. Banci, A. Bencini, C. Benelli, M. DiVaira and D. Gatteschi, *Inorg. Chem.*, 1982, 21, 3801.
175. R. J. Burt, J. R. Dilworth, G. J. Leigh and J. A. Zubieta, *J. Chem. Soc., Dalton Trans.*, 1982, 2295.
176. B. Demerseman, P. H. Dixneuf, J. Douglade and R. Mercier, *Inorg. Chem.*, 1982, 21, 3942.
177. D. Grdenič, B. Korpar-Colig, M. Sikirica and M. Bruvo, *J. Organomet. Chem.*, 1982, 238, 327.
178. D. A. Tocher and M. D. Walkinshaw, *Acta Crystallogr.*, 1982, B38, 3083.
179. F. Diederich and K. Dick, *Tetrahedron Lett.*, 1982, 23, 3167.
180. M. Armand and F. Jeanne, *Chem. Abs.*, 1981, 95, 9378y.
181. F. Vögtle, H. Puff, E. Friedrichs and W. M. Müller, *J. Chem. Soc., Chem. Commun.*, 1982, 1398.
182. E. G. Rozantsev, *Theor. Exp. Chem. (Engl. Transl.)*, 1966, 2, 218.
183. K. Sasvari and L. Parkanyi, *Cryst. Struct. Commun.*, 1980, 9, 277.
184. R. E. Stenkamp, L. H. Jensen, T. B. Murphy and N. J. Rose, *Acta Crystallogr.*, 1982, B38, 1169.
185. C. J. Suckling, *J. Chem. Soc., Chem. Commun.*, 1982, 661.
186. K. H. Shankara Narayana and G. S. Krishna Rao, *J. Am. Oil Chem. Soc.*, 1982, 59, 240.
187. J. Galy, R. Enjalbert and J.-F. Labarre, *Acta Crystallogr.*, 1980, B36, 392.
188. F. W. Einstein and T. Jones, *Can. J. Chem.*, 1982, 60, 2065.
189. T. Pilati, M. Simonetta and S. Quici, *Cryst. Struct. Commun.*, 1982, 11, 1027.
190. G. D. Andreetti, *J. Mol. Struct.*, 1981, 75, 129.

AUTHOR INDEX

Numbers in parentheses are reference numbers and italic numbers are pages in the reference section at the end of a chapter.

SUBJECT INDEX

Abiotic receptors, approach to, 363
Acetic acid, guest dimers of, 150
Acetylenic diol adducts, 437
π-Acid π-base attractions, host–guest, in crown adducts, 290
Active transport of ADP, by means of synthetic carrier molecule, 384
Alkalides, 388
4-p-Aminophenyl-2,2,4-trimethyl-chroman, 27
4-p-Aminophenyl-2,2,4-trimethylthia-chroman, 29
Ammonium-containing anion receptors, 378–384
Anion complexation, 331, 373, 375–384, 394
  active transport of ADP, 384
  applications of, 394
  by ammonium-containing (macro-polycyclic) anion receptors, synthesis and stability constants for complexes, 379–380
  by diaza-macrobicycles, as halide receptors, 378–379
  by ditopic receptors designed for dicarboxylate anions, 383–384
  by guanidinium-containing anion receptors, synthesis and stability constants for complexes, 377–378
  by lipophilic diammonium salts, dis-criminative extraction of adenosine phosphates and active transport of ADP, 384
  by macrobicyclic anion receptors, 380–381
  by macrocyclic polyaza anion recep-tors, range of anionic guests, and stability constants for complexes, 381–384
  crystal structure of anion complex [azide $\subset$ bis-tren-6H^{+}], 380–381
  crystal structure of Cl^{-} anion complex, 379

discriminative extractions, 384
general features of, 376–377
stability constants for complexes with anionic guests, 377, 379–380, 382–384
Anion radii, 376
Anionic activation, employing cryp-tands, 389–390
Anionic polymerization, application of cryptands in, 391
9-Anthranoates, host properties of, 431, 436
Arenediazonium salts
  crystal structures, crown complexes of, 282–283
  factors affecting stability of crown complexes of, 282–283
2-(2-Arylindan-1,3-dion-2-yl)-1,4-naph-thohydroquinone hosts, 35–36
O$\cdots$C=O Attractive interactions in crown compounds and acyclic analogue, 308–310, 326

Benzene-1,3,5-tricarboxylic acid, *see* Trimesic acid
Benzo-27-crown-9
  crystal structure, with guanidinium perchlorate guest, 307
  crystal structure, with uronium nitrate guest, 307
3,3'-(1,1'-Bi-2-naphthol)-21-crown-5
  ligand, crystal structure, water adduct of, 304
1,11-Bis(2-acetylaminophenoxy)-3,6,9-trioxaundecane,
  crystal structure, complex with KSCN, 321–322
  crystal structure, free ligand, 321
Bis(2'-carboxy-1',3'-xylyl)-24-crown-6
  ligand
  crystal structure of, 309–310
  *intra*molecular carboxy dimer in, 309–310